Water-Soluble Polymers for Petroleum Recovery

Water-Soluble Polymers for Petroleum Recovery

Edited by

G. A. Stahl
Exxon Chemical Company
Baytown, Texas

and

D. N. Schulz
Exxon Chemical Company
Linden, New Jersey

Plenum Press • New York and London

Library of Congress Cataloging in Publication Data

American Chemical Society. Meeting (1986: Anaheim, Calif.)
Water-soluble polymers for petroleum recovery.

"Proceedings of the National Meeting of the ACS entitled Polymers in Enhanced Oil Recovery and the Recovery of Other Natural Resources, held September 7-12, 1986, in Anaheim, California"—T.p. verso.

Includes bibliographies and index.

1. Water-soluble polymers—Congresses. 2. Secondary and Tertiary recovery of oil—Congresses. I. Stahl, G. Allan. II. Schulz, D. N. III. Title.

QD382.W3A43 1987 665.5′3 88-12531

ISBN 0-306-42915-2

Proceedings of a National Meeting of the ACS entitled Polymers in Enhanced Oil Recovery and the Recovery of Other Natural Resources, held September 7-12, 1986, in Anaheim, California

A Division of Plenum Publishing Corporation
233 Spring Street, New York, N.Y. 10013

Printed in the United States of America

PREFACE

We both found ourselves working on water-soluble polymers for oil recovery in the early 1980's. Our previous backgrounds involved the synthesis and characterization of hydrocarbon polymers for everything from elastomers to plastics. As such, we were largely unprepared for the special difficulties associated with water soluble polymers in general, and their use in enhanced oil recovery (EOR), in particular. Oil patch applications have a jargon and technical heritage quite apart from that usually experienced by traditional polymer scientists. At that time, no books were available to help us "get up to speed" in the polymers for oil recovery field. Since then, there have been a number of symposia on this topic, but still few books, especially from the polymer (rather than the field-applications) perspective.

Synthetic water soluble/swellable polymers have commercial importance in such application as water treatment, cosmetics, and foods. Yet, these polymers have not received the scientific/technological attention they deserve. The application of water soluble polymers to oil recovery has, in fact, highlighted the need for new water based materials, and a fundamental understanding of their structure and use. Interest has been spurred not only for the potential economic credits from enhanced oil recovery and an augmented polymers business, but also by the challenge of designing water soluble polymers for harsh environments. Even though the economic driving force for enhanced oil processes is not as high as it was in the early 1980's, academic and industrial interest in water soluble/swellable polymers persists. Furthermore, given the unpredictability of oil supply and pricing, higher oil prices will undoubtedly return again. With higher oil prices will come a renewed (and probably hysterical) interest in polymers for oil recovery.

This book is based primarily upon papers presented at a symposium held by the American Chemical society, Division of Polymeric Materials: Science and Engineering, in Anaheim, California, August 1986. The focus of the symposium (and, hence, this book) was polymers for oil recovery. However, it is obvious that the symposium and book reflect as much or more about recent advances in water soluble polymer synthesis and characterization as application to oil recovery. For example, how does one prepare a water soluble polymer with hydrophobic groups? Or, how does one determine the MWD of polar polymers that exceed the exclusion or polarity limits of conventional GPC?

This book begins with an Overview Section, concerning the design and use of polymers for oil recovery. The next chapters emphasize the preparation and characterization of water-soluble/gelled polymers, especially for the demanding conditions present in the oil field. Included herein are some of the most recent preparative techniques for such polymers, as well as the newest analytical methods for defining their structures. The preparative and characterization sections could well stand alone as valuable assessments of the state of the art in water soluble polymers.

Both practitioners and newcomers to either the oil recovery or water soluble polymer fields could benefit from this book. The volume will also be useful for stimulating the imagination and problem-solving ability of those polymer scientists who have spent their life in "relative ease" working on hydrocarbon elastomers and plastics for above ground applications.

G. A. Stahl
D. N. Schulz

December, 1987

CONTENTS

OVERVIEW

POLYMER CHEMISTRY/PREPARATION

POLYMER CHARACTERIZATION

CROSSLINKED POLYMERS

THE ROLE OF POLYMERS IN ENHANCED OIL RECOVERY

W. M. Kulicke, N. Böse, and M. Bouldin

Institut für Technische und Makromolekulare Chimie Universität
Bundesstr. 45. 2000 Hamburg 13. West Germany

1. Introduction and Objective

Oil is an important energy source and an important chemical raw material. 95% of world-wide oil production is used to provide energy and 5% for chemical purposes (2% of the latter for the synthesis of polymers). However, oil resources are being depleted rapidly. Annual oil consumption was approximately 2.7 billion T in 1985. With proven oil reserves of 105 billion T and non-proven of ca. 90 billion T, it can be assumed that worldwide oil resources will be exhausted in approximately 70 years time.

In addition to reducing the proportion of oil used for energy (1973: 49%; 1984: 41%), other measures must be applied to maintain a steady supply of oil for the long term. It has been shown that for each 1% increase in drilling yield, an increased in supply of 1.7 years results.(1) One of the most important methods for enhancing oil recovery is the use of methods like polymer flooding.

It has been proven in field projects that polymer flooding can improve the oil yield. This is reflected by a 50% increase in the number of projects from 1984-1986 in the United States.(2). The results of these polymer flood projects have confirmed that, if designed properly, polymers can also be applied successfully to highly saline reservoirs, recovering ca. 10-20% OOIP (original oil in place).(3,4)

However, there are still problems which reduce oil recovery effectiveness.(5) In the presence of bivalent cations (e.g. Ca^{2+}), many of the polymers employed tend to flocculate or exhibit a drastic loss of viscosity. Other difficulties are adsorption, long term instabilities, thermal effects, etc.

The purpose of this paper is to describe structure-property correlations for relating polymer solution viscosity (η) to fundamental molecular (e.g. polymer MW, MWD) and rheological (e.g. shear rate, normal stresses) and porous media flow(5,10) parameters. After discussing these topics, a listing of potential EOR polymer candidates is given.

2. <u>Molecular parameters of non-ionic polymers, polyelectrolytes and biopolymers</u>

2.1 <u>Molecular weight, concentration, size and shape</u>

The effectiveness of a polymer for oil displacement in enhanced oil recovery is determined by the mobility ratio (M) of the polymer slug and the oil. The mobility ratio (M) is defined as the ratio $k_w \bullet \eta_p / k_p \bullet \eta_w$, where η = viscosity and k = permeability. Values of M = 1 stabilize the displacement process.

The influence of molecular weight (M_w) and concentration (c) on zero-shear-viscosity η_o of polyacrylamides (PAAm), which are statistical coils in solution, was first investigated by Kulicke and Klein [see ref. 174 in (7)] in 1978. Figure 1 shows that the desired zero shear viscosity of 10 mPa•s is not achieved even with the highest polymer molecular weights investigated.

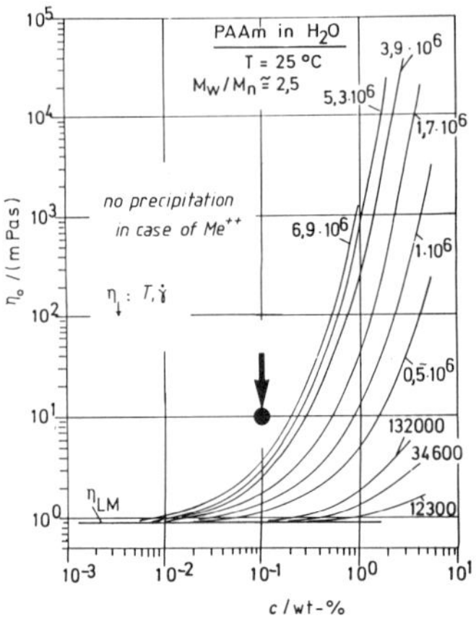

Fig. 1
The dependence of the zero-shear-viscosity of aqueous solutions of non-ionic polyacrylamides on concentration for various polymer molecular weights. The large point indicates the desired efficiency of a polymer used in EOR.

An easily implemented method for ascertaining the molecular weight is to determine the intrinisic viscosity $[\eta]$ and then use a $[\eta]$-M-relationship, which gives the mean value of M_η.(6) Measurement of effective elongational viscosity by monitoring the pressure drop is also a method for determining the mean molecular weights of ultra high molecular weight polymers > 10^7g/mol.(10,26) In this case, one has to observe the onset-point of the increase of the elongational viscosity as a function of the molecular weight. Beyond this onset-point, degradation may occur.(26) In all methods of determination, a critical concentration limit may not be exceeded.(6,7,13)

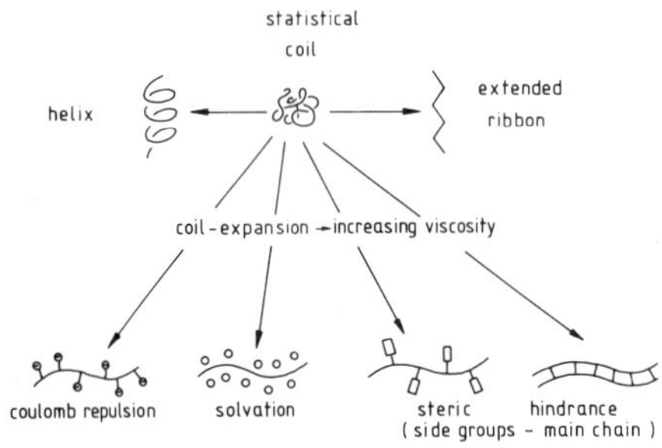

Fig. 2
Structural principles for polymers which lead to an expansion of the coil and thereby to an increase in viscosity [based on a molecular model first introduced by Klein et al. [see ref. 16 in.(29)]

The zero-shear viscosity can be increased by expansion of the polymer coil, in solution. Figure 2 shows the main possibilities for coil expansion, which are further discussed for selected polymers in Section 4.

Primary interest focuses on clarifying the influence of the various molecular parameters on the viscosity level. Figure 3 shows the attained zero-shear viscosity(7,19) (η_0) for polyacrylamnide in water, for varying concentrations and molecular weights. This figure shows that by doubling the concentration, a strong increase in zero-shear viscosity occurs. However, no such effect arises from doubling the molecular weight. Therefore, with c = 0.2% the required value of η_0 = 10mPa•s can be attained with pure PAAm. Moreover, since this polymer is nonionic, it has no sensitivity to brine at all.

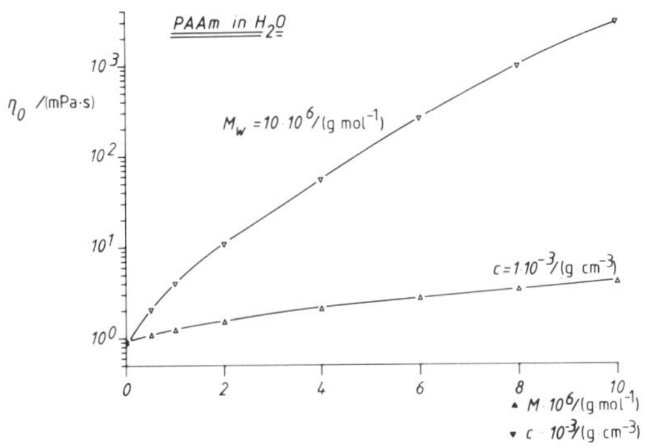

Fig. 3
Influence of the concentration on η_0 with constant molecular weight and influence of the molecular weight on η_0 with constant concentration.

2.2 Molecular weight distribution

Because a polymer sample is very rarely monodisperse, any specification of molecular weight must necessarily be an average value. It is also important to know how the molecular weight was determined. The average molecular weight (e.g. M_w, M_n, M_z, M_i) and, hence, the MWD depends both on the method of determination and the polymer itself. Increasing the breadth of the distribution is also accompanied by an increase in the breadth of the relaxation time spectrum. Consequently, a sample with a broad MWD shows earlier transition from the zero-shear rate range to the shear-rate dependent range of viscosity than for a narrowly distributed sample of the same molecular weight.

The determination of the molecular weight distribution can be performed via size exclusion chromatography (SEC). Here, a small amount of polymer is carried by a stream of solvent with a constant flow rate over a column containing porous beads. Depending on the size of the polymer particles, they can either diffuse into the pores of the beads or not. The polymer molecules with higher M_i and larger hydrodynamic volumes, therefore, exhibit shorter retention times. The result, a separation of polymer particles according to size, is obtained at the end of the column. This analysis is carried out by continuous monitoring of the concentration and subsequent conversion from a previously recorded calibration relationship for molecular weight. Recently, an absolute molecular weight distribution determination was carried out for the first time on ultra-high molecular weight polyacrylamides in aqueous solution.(8) In this case, the solution flowing out of the column was passed continuously through a concentration detector and a light-scattering photometer. Figure 4a shows the signals from both detectors from which the absolute weight can be calculated for any point in time, see Fig. 4b.

It can be shown that even at typical flow rates of 1 ml/min degradation occurs in the high molecular weight range of the distribution ($M_w \geq 5 \cdot 10^5$ $g \cdot mol^{-1}$) because of elongational stress (c.f. Section 3).[8,26]

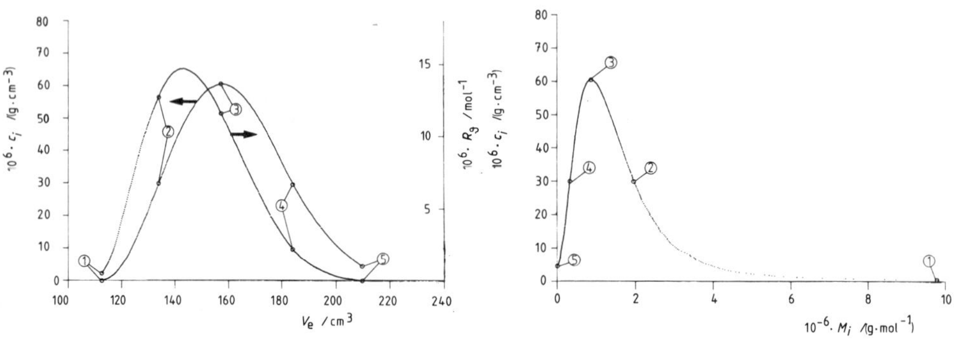

Fig. 4 Readings from the concentration and photometer analysis for a polyacrylamide in aqueous solution and the absolute molecular weight distribution calculated by using the Zimm-Debye equation for low-angle-laser-light-scattering.

2.3 Maxima in Viscosity-Composition Responses

With reference to ionic polymers: the concept of coil expansion (c.f. Section 2.1) depends upon the assumption that the coil volume grows due to intramolecular coulombic repulsion from the ionic groups, causing the viscosity of the solution to increase. In order to be able to investigate a range of polymers with an equal degree of polymerization and distribution, but a different level of ionic groups, a non-ionic polyacrylamide was taken as the base polymer and subjected to alkaline hydrolysis. Levels of hydrolysis ranging from almost zero to 67% were recorded; these were determined by IR-spectroscopy, as well as by elemental analysis. The introduction of ionic groups cannot be increased beyond 67%, even at long alkaline hydrolysis times. With increasing amounts of carboxylic acid groups, the intrinsic viscosity rises strongly because of the increase in the coil dimensions. This was proven using light-scattering measurements (i.e. radii of gyration).[11] This condition also persists in the presence of an electrolyte. The amount of the increase depends upon the type of salt, its ionic power and the polymer concentration. The intrinsic viscosity of the pure polyacrylic acid, which was separately synthesized, lies considerably lower than the value of the poly(acrylamide-*co*-acrylate) with 67% acrylic units. The stages between 67% and 100% were produced by further hydrolysis with acids under mild conditions to avoid mechanical or thermal degradation. The outcome is surprising (c.f. Fig. 5). After reaching a maximum at 67% acidic groups, the intrinsic viscosity, thereafter continually decreases until it reaches the value of the pure poly(acrylic acid).

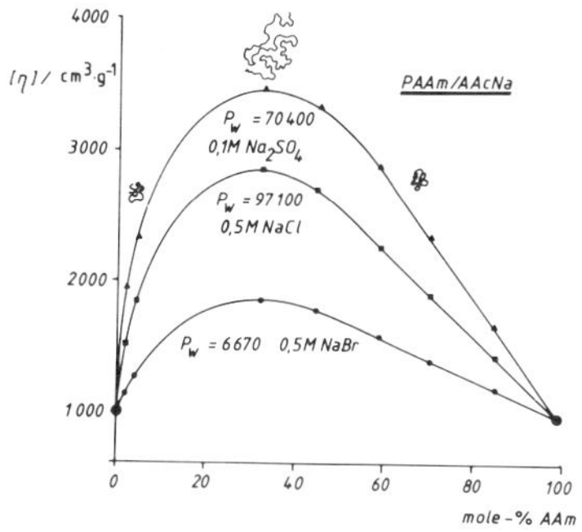

Fig. 5
Responses of the intrinsic viscosity $[\eta]$ (c_p = 0) for a homologous series of poly(acrylamide- <u>co</u>-acrylate) at varying degress of hydrolysis (at three different salt concentrations).

The height of this maximum is strongly influenced by the polymer concentration, the salt concentration, the chain length and the molecular weight distribution.

This result is consistent with the formation of acid-amide-acid triads during alkaline hydrolysis. These triads protect the enclosed amide group against further attack from OH^- and so a maximum degree of 67% cannot be exceeded. These triads are strongly stabilized by H-bonds and result in a maximum coil expansion.(11,13) Further hydrolysis in acid medium leads to the destruction of these triads and to coil shrinkage.(11,33) The existence of these triads has also been shown by NMR-spectroscopy.(39) A viscosity maximum is also found for poly(2-acrylamido-2-methylpropanesulfonate) sodium salt [c. f. ref. 12 in (17)], as well as in polysaccharides.(12)

2.4 <u>Aging of polymers in solution</u>

The phenomenon of aging can be observed for many EOR polymer solutions. In many cases, these aging effects lead to solutions with completely different rheological properties than those of the unaged ones. Therefore, it is important to understand the aging mechanisms, in order to find ways to control them. In Fig. 6, different types of aging are shown.

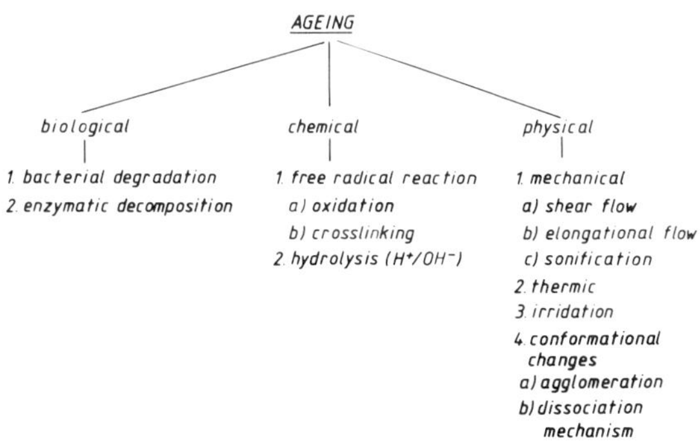

Fig. 6. Overview of chemical, physical and biological aging.

2.4.1 Biological aging

Solutions of polymers (especially, polysaccharides like xanthan or hydroxycellulose) are very sensitive to biological aging . In the case of xanthan, [c.f. ref. 8 in [17]] it has been shown, that many bacteria found in oil deposits are able to decompose this polysaccharide under anaerobic conditions. This degradation is manifested by a drastic loss in the apparent viscosity because of chain degradation. By addition of bactericides, e.g. NaN_3, an infestation can be avoided. Unfortunately, these agents are extremely toxic, so that they cannot be used in every case. In the same paper, the resistance of polyacrylamide to bacterial degradation was reported. In order to combine the stability of the all carbon vinyl-chain with the bulkiness (chain extension) and the intermolecular interactions known from biopolymers, a group of the vinyl-saccharides have been synthesized. Another kind of biological aging is found in in enzymatic decomposition reactions, e.g., the cracking of starch to α-glucose with α-amylose.

2.4.2 Chemical aging

This type of aging leads either to a change in the chemical structure (e.g. hydrolysis) or to free radical reactions (e.g. chain-scission). At the high temperatures, which are to be found in the oil deposits, Rinaudo and co-workers [14] observed a significant loss in polymer solution viscosity over a period of some weeks. This effect was explained as oxidative decomposition of the macromolecules. Diradical oxygen can also trigger cross-linking reactions because oxygen can form highly reactive hydroperoxides. Tanaka et al.[15] found that catalytic amounts of alkali can lead to a slow hydrolysis of polyacrylamide. Because of the maximum in viscosity for partially hydrolysed PAAm, small changes in the reactivity ratio can lead to significant changes in the properties of the solution. In the case of xanthan, Rehage et al.[16] reported acid-catalysed degradation. This reaction is coupled with a reduction in the mean degree of polymerisation and in the viscosity.

2.4.3 Physical aging

In EOR, physical aging is another important aging phenomenon. While dissolving and injecting the polymer solutions, mechanical and thermal aging (degradation) can occur. But even when suitable precautions are taken to suppress biological (NaN_3), chemical (highly purified samples, neutral pH) and the normal causes of physical aging (absence of light, constant temperature of 25° C, no mechanical stress), time dependent effects can be detected. After being dissolved, the polymer is not in thermodynamic equilibrium. A conformational change can lead, as shown in Fig. 7, to either an increase or a decrease in viscosity, depending upon the storage time and the polymer-solvent system.[7,17,18]

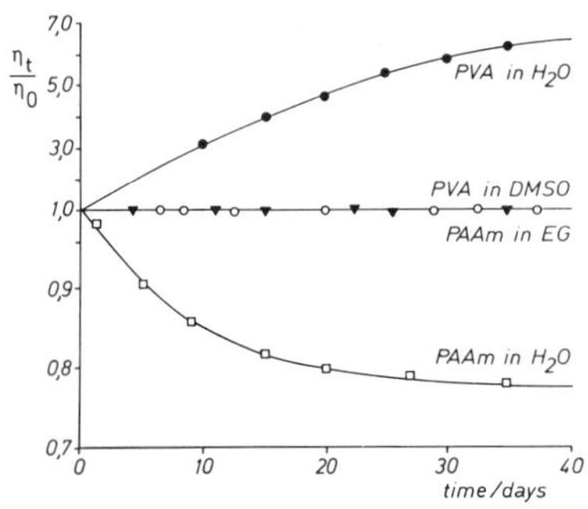

Fig. 7
The solution viscosity of polyvinylalcohol and polyacrylamide in different solvents as a function of the storage time.

Examples of the former have been published by various authors [see ref. 3 and 4 in[17]] for aqueous solutions of polyvinylalcohol. Using dynamic light-scattering techniques, Lechner und Mattern were able to prove that this effect is due to the tendency of PVA to agglomerate [see ref. 5 in [17]]. Therefore, the mean molecular weight of the particles increases drastically with the storage time. On the other hand, a decrease in viscosity has been obtained with no detectable change in the average molecular weight. Such a dissociation process has been observed for many non-ionic[17], cationic[17] and anionic[17,18] water soluble macromolecules.

Table 1. Time-dependent viscosity behavior under the influence of molecular weight.

M_w / g•mole^{-1}	Time Dependent Viscosity
$6.9 \bullet 10^6$	yes
$5.3 \bullet 10^6$	yes
$2.2 \bullet 10^6$	yes
$1.6 \bullet 10^6$	yes
$1.0 \bullet 10^6$	yes
794 000	no
500 000	no
165 000	no
60 000	no
38 000	no

An interpretation of these results based upon molecular level understanding is difficult. A disentanglement mechanism seems improbable because the effect also occurs in dilute solutions. On the other hand, the light-scattering results show that no entangled aggregates are present. This viscosity loss may be explained by an intramolecular conformational transition. Above a critical molecular weight (c.f. Table 1) the solvent (H_2O) needs a long time to penetrate or order structures (e.g. helical structures). This model is also supported by the fact that aging depends upon the polarity of the solvent. Solvents with a lower dielectric constant ε do not stabilize intramolecular H-bonds. In such a case, aging effects are detected [18] (c.f. Fig. 7).

3. Characteristics of Porous Media Flow

Porous media flow of polymer solutions has been described in terms of the mobility ratio, with special emphasis on increasing the viscosity at low values of M. To be able to compare different polymer samples, it is necessary to establish structure-property relationships. Furthermore the pronounced elasticity of high molecular weight polymer solutions cannot be ignored. Moreover, porous media flow cannot be quantitatively described by simple shear flow, because of the elongational flow pattern which occurs within the converging/diverging passages of porous media.

3.1 Structure-property-relationships in shear flow

In order to obtain solutions which are suitable for polymer flooding, the mobility ratio should be close to unity. For this reason, the polymer solutions used should exhibit the highest possible viscosity. The chemical structure is one of the deciding factors in the flow behavior of polymer solutions. It is, therefore, important to know which structural parameters determine the viscosity level.

Early approaches for characterizing the zero-shear-viscosity were accomplished for melts by Bueche. In this case, one can obtain a master curve if the viscosity is plotted against the mean molecular weight. For highly concentrated solutions, a modified Bueche-plot (η_0 vs C•M) was shown to be useful [see ref. 11 in [19]]. In such polymer fluids, the viscosity level is only determined by the number of entanglements. For moderately concentrated solutions, the Bueche concept fails.[19,20] In this concentration regime, the viscosity depends not only on the entanglement density but also on the quality of the solvent and the volume of the polymer coils. As we have shown, a linear relationship is obtained if one plots the specific viscosity ($\eta_{sp} = (\eta - \eta_s)/\eta_s$) against the overlap parameter $c \bullet [\eta]$. The viscosity of dilute and moderately concentrated solutions can then be easily estimated with a modified Huggins-equation:

$$\eta_{sp} = c \bullet [\eta] + k_H \bullet c^2 \bullet [\eta]^2 + K_b \bullet c^b \bullet [\eta]^b \qquad (1)$$

We could prove that the exponent b is a function of the solvent quality ($b = 3.4/a$).[19] Using the Mark-Houwink equation (c.f. Section 2), it is possible to write η_{sp} as a function of the solvent power (expressed by the Mark-Houwink exponent a), the mean molecular weight M_w and the concentration:

$$\eta_{sp} = k \bullet c \bullet M^a + k^1 \bullet c^2 M^{2a} + k^{11} \bullet c^b \bullet M^{b \bullet a} \qquad (2)$$

In plot 8, the shear rates during the injection and field flow stages are shown.

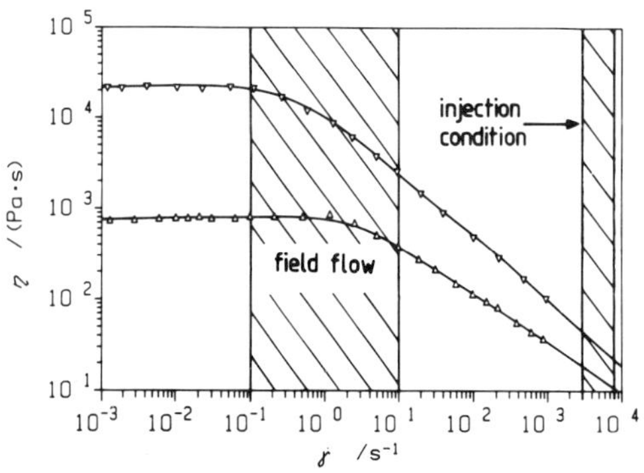

Fig. 8.
The viscosity level of polymer solutions during the different stages of polymer flooding (field flow and injection conditions).

Particularly, during the injection stage, the viscosity of the flooding fluid exhibits a shear-rate dependence. It is, therefore, important to be able to evaluate the viscosity level of such pseudoplastic fluids. A reduced viscosity (η/η_0) is obtained if it is plotted against the reduced shear-rate β ($\dot{\gamma}/\dot{\gamma}_{crit}$) as proposed by Graessley [see ref. 11 in [19]]. Unfortunately all parameters in such a plot have to be derived experimentally. In order to obtain β from molecular parameters, it has been proposed that it should be calculated using the Beuche-equation (see ref. 13 in[19]:

$$\beta = \dot{\gamma} \bullet \tau_R = \frac{12}{\pi^2} \frac{(\eta_0 - \eta_s) \bullet M}{c \bullet R \bullet T} \bullet \dot{\gamma} \qquad (3)$$

For moderately concentrated solutions, this formula does not lead to a satisfactory result. The reason for this is that the characteristic relaxation time in this concentration regime is not only governed by the coil dimensions but is also a function of the entanglement density and the polymer-solvent-interactions.[20] To solve this very interesting problem, we have developed an equation in which the reduced viscosity can be estimated on the basis of molecular parameters:[19]

$$\beta' = \frac{(\eta_0 - \eta_s) \cdot M^a}{c^{1/a} \cdot R \cdot T} \qquad \frac{\eta}{\eta_0} = f(\beta') \tag{4}$$

In order to obtain solutions with the desired flow properties, degradation must be avoided. It is, therefore, important to have a knowledge of the conditions, under which chain-scission occurs. Narrow MWD polystyrene was chosen for mechanical degradation experiments, in order to simplify the experiment.

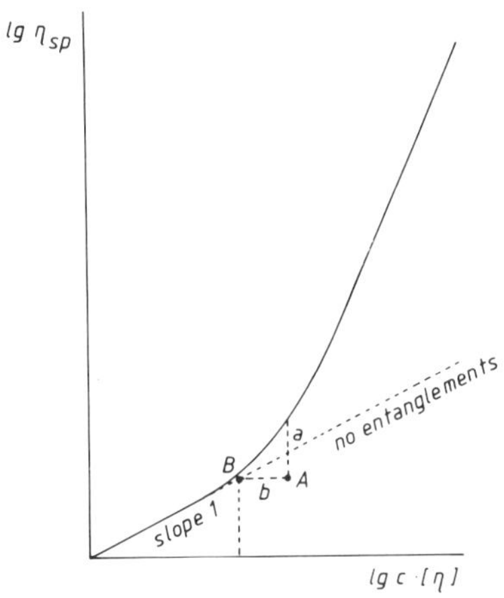

Fig. 9.
Schematical plot of the specific viscosity vs. the overlap parameter c•[η]. The points of intersection A and B indicate the critical values.

The results show that when the shear-stress reaches a critical value all entanglements are detached.[19] By increasing the shear-rate, chain-scission can be detected. These results are shown schematically in plot 9. The critical specific viscosity of a sheared polymer solution is the point of intersection (A) between the vertical (a) and the line with a slope of one (theoretical viscosity of a non-entangled fluid). Higher shear-stress leads to degradation. The mean molecular weight of a degraded compound can easily be estimated by the point of intersection (B) between the horizontal (b) and the line with the slope of one. Knowledge of such conditions can lead to a turning of the injection conditions so as to avoid or minimize shear degradation. The results for different polymers have been published.[7,19]

3.2 Elasticity in shear flow

Elasticity is one of the most characteristic properties of polymer fluids. To date, this parameter has been largely overlooked. Thus, the elastic material function may lead, under high shear-rates (e.g. injection) to a loss of energy and to material defects. A measure of the storable elastic energy is the storage modulus G' (G" = loss modulus). The ratio G"/G' indicates whether a body is mainly elastic or viscous.

For an ideal Newtonian fluid, the ratio is infinite and zero for an ideally elastic body. All polymer solutions exhibit viscoelastic behavior, i.e, $0 \leq G''/G' \leq x$. At high frequencies, the solutions show an increasing elastic portion, i.e. only a small part of the energy is transformed into viscous flow (small G''/G'). Together with this linear-viscoelastic behavior, polymer solutions also exhibit, in contrast to Newtonian fluids, non-linear viscoelastic properties. The two most important functions are the first and second normal stress differences, N_1 and N_2. These normal stresses arise from a deformation of the macromolecules in the velocity field. Accordingly the resulting normal stresses are a function of the distorted coil distribution [see ref. 11 in[19]]. At high shear-rates (c.f. Fig. 10), N_1 can be the determining material function ($N_1 \gg \sigma_{12}$).[7,21]

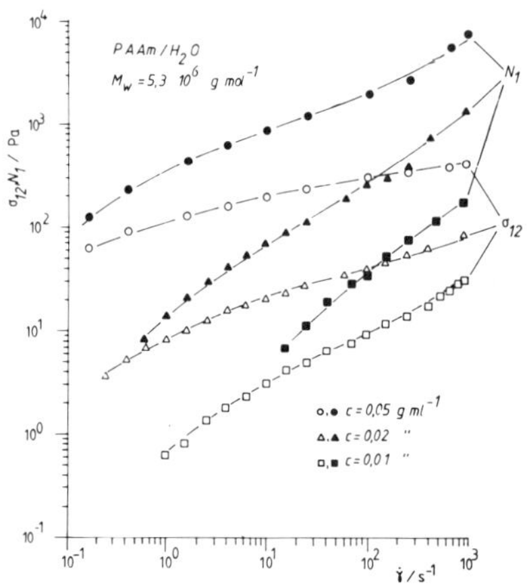

Fig. 10:
Shear stress σ_{12} and first normal stress difference N_1 plotted as a function of the shear-rate for different concentrations of polyacrylamide in water.

For moderately concentrated polymer solutions, we have found that N_1 depends on the average molecular weight, the concentration (c.f. Fig. 10) and the solvent power.[7,21] In order to minimize non-linear viscoelastic effects, it is important to know whether the concentration or M_w has more influence on N_1. With increasing energy costs, the elastic behavior of polymer solutions must be taken more into consideration in the future.

3.3 Indirect determination of elongational flow conditions

The most important limiting condition in EOR is the porous media flow itself. Figure 11 schematically shows the governing processes for such a medium. For polymer solutions, the elongational viscosity η_e cannot be measured directly, as in the case of polymer melts. However, it is easy to detect the pressure drop over a given length of porous media, which is a function the solvent viscosity (the latter is the sum of the shear viscosity η_s and the elongational viscosity η_e in porous media).

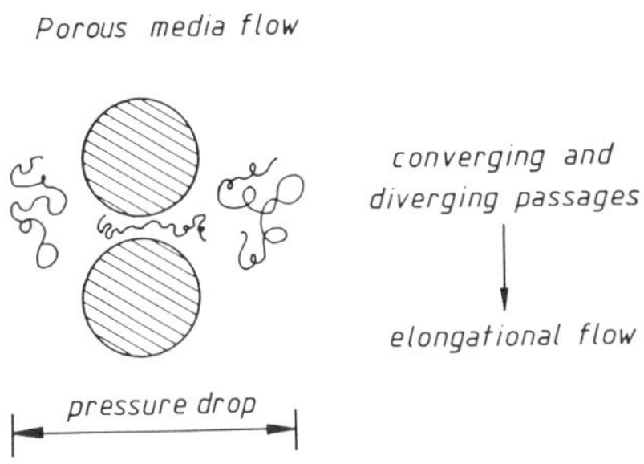

Fig. 11.
Schematic diagram of elongational flow in porous media

The coil is continually extended and relaxed. If the time between two deformations ($\dot{\varepsilon}^{-1}$) is longer than the double relaxation time τ no significant changes of the flow properties can be observed. If, however, the velocity is so high that the molecules do not have the time to relax completely between the elongations, then drastic increases in the pressure drop can be detected. In this case, the polymer is permanently deformed in the velocity field. Consequently, the effective hydrodynamic radius of the molecule is higher than that of the random coil in equilibrium. This is easily understood if one models the polymer as a non-linear viscoelastic dumbbell as proposed by Bird.(22,23) A dimensionless identification number for this mechanism is the Deborah-Number, De. It is defined as the product of the relaxation time τ and elongation rate $\dot{\varepsilon}$ ($De = \dot{\varepsilon}\cdot\tau$). As proposed by Durst and Haas, the onset of elongational viscosity, therefore, occurs when De is equal to 1/2.(22,23) A very useful representation of this effect can be obtained by plotting the resistance coefficient Λ against the Reynolds-number Re:(10)

$$f = \frac{\Delta p\cdot d\cdot n^3}{\Delta l\cdot\bar{v}^2\cdot(1-n)} \qquad Re = \frac{\bar{v}\cdot d\cdot\rho}{\eta(1-n)} \qquad \Lambda = Re\cdot f \tag{5}$$

The critical value Re_0 can be calculated on the basis of molecular parameters as presented in the following ($De = 0.5 = \dot{\varepsilon}\cdot\tau$):

$$\dot{\varepsilon} = \frac{k\cdot(1-n)\,\eta\,Re}{\rho\cdot d^2} \qquad \tau = \frac{\eta_s\cdot[\eta]\cdot M}{R\cdot T} \tag{6}$$

$$Re_0 = \frac{\rho\cdot R\cdot T\cdot d^2}{16\cdot\eta_s^2\cdot(1-n)\cdot(1+[\eta]\cdot c)\cdot k\cdot M^{1+a}} \tag{7}$$

In plot 12, the concentration dependence of Re_0 is shown. One of the essential aspects of this plot is that for Re-numbers greater than Re_0, the elongational viscosity is the governing material parameter. In this flow regime, η_s is completely negligible ($\eta_e \gg \eta_s$). Unfortunately the calculations (5-7) are only valid for statistical coils. They, therefore, cannot be used to estimate the critical data for rod-like systems; e.g. polyelectrolytes, xanthan.

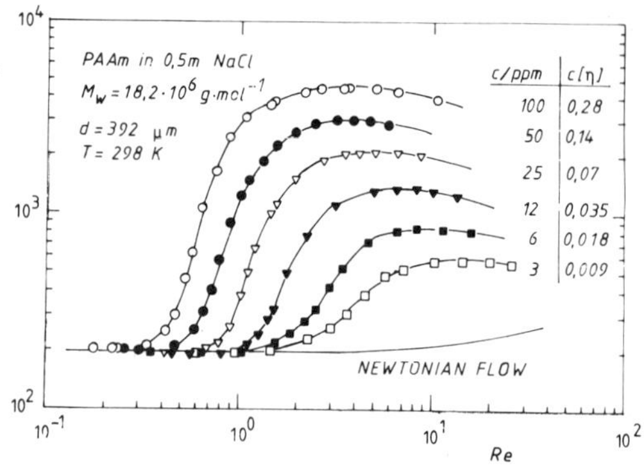

Fig. 12.
Concentration dependence of
Re_0 of polyacrylamide in water

Absent additional experimental data, it is impossible to make more precise statements about the behaviour of rod-like macromolecules in elongational flow.

As we have recently shown, degradation must be taken into consideration for De-numbers greater than 1/2.(28) Under these conditions, the chain is permanently extended in the velocity field. If such a polymer is elongated again, then the energy cannot be used to extend the chain, and chain-scission occurs. For polyacrylamide with $M_w > 10^6$ g•mol^{-1}, we have demonstrated this effect using size-exclusion-chromatography (SEC).(8) The same results were obtained using narrowly distributed polystyrene with toluene as solvent.(26) Fortunately, the Re-numbers which are common in EOR are mostly much smaller than Re_0, so that degradation is normally negligible. Only in cracks and holes in the rock does the fluid velocity exceed Re_0. This leads to a drastic increase in pressure resistance and, therefore, the flooding fluid is able to displace the oil from such inhomogenities.

3.4 Other parameters which influence the characteristics of flow behavior of flooding fluids

3.4.1 Influence of salts

The influence of salts on the viscosity level is important because the oil deposits are often extremely saline. Most water-soluble polymers, e.g. polyacrylamide(7), cationic(37,38) and anionic poly-(acrylamide-co-acrylate)s(11,37,38), show an extreme loss of both zero-shear and intrinsic viscosity upon addition of salts. This effect is caused by a decrease of the dielectric constant ε with increasing salt concentration. The result is that the energy of the H-bonds and, consequently, also the solvent-power is reduced. Bivalent cations (Ca^{2+}, Mg^{2+}) have an especially strong influence. It has been reported by McCormick [see pp 366- in (5)] that small amounts of Ca^{2+} can lead to flocculation. The polysaccharides tend to be resistant to electrolytes. Below a critical concentration (c) the helical structure is stable and no change in the apparent viscosity is observed. If, however, the salt concentration is higher than c, a rod-coil transition can be detected.(27) This phenomenon is accompanied by a dramatic loss of viscosity.

Polyampholytes have recently been introduced by Salomone [see pp 269- in[5]] as electroyte resistant polymers. Early reports seem to indicate that their solution structure does not change with salt concentrations of up to 4 mol•l^{-1}.

Bi- and trivalent cations can also lead to the formation of ionotropic gels [see McCormick p 336 in [5]]. In these gels, the metal ions function as cross-linking agents. This can also be used to form gels in the core which function as blockers [some examples are given in ref.[5]).

3.4.2 Influence of temperature

In oil deposits with an average temperature of ca. 80°C, the temperature dependence of the rheological material functions cannot be neglected. In all cases, a loss of the viscosity with increasing temperature is observed. Polysaccharides (e.g. xanthan) exhibit only a minor temperature dependence on η, in the quotient ($d\log\eta/dT$).[28] This stems from the fact that the flow-activation-energy in rod-like molecules is much smaller than that in coiled ones.

The influence of temperature on elongational viscosity is very complex because the viscosity, solvent-power and the characteristic relaxation time are all functions of T. In most cases, an increase of Re_0 is to be expected. The temperature dependence of the other material functions have rarely been investigated. Thus, final conclusions cannot be made.

3.4.3 Influence of adsorption

The adsorption of the dissolved polymer on rock surfaces is one of the main problems in EOR. This situation can lead to a drastic reduction of the polymer concentration of the flooding fluid. The mechanism of polymer adsorption in rock has, therefore, been thoroughly investigated. Synthetic [29] as well as biopolymers [29-31] have been examined. In both cases, it has been found that with increasing salinity, the specific adsorption increases. Klein et al. found for xanthan that above a critical concentration of 10 g•l^{-1}, the saturating adsorption rises rapidly.[29] According to a patent of German Texaco AG[30], the adsorption of polysaccharides can be effectively suppressed [see also VOLZ on p 855 in [5]] by preflooding the core with a solution of polyethyleneglycol (PEG). By such action, the active centers of the rock are coated irreversibly with a monomolecular layer of PEG, thereby, minimizing adsorption during an actual flood.

4. Comparison of selected polymers

This section describes some of the advantages and disadvantages of potential EOR polymers. Moreover, the desired properties of an ideal EOR polymer are described. Some of these water soluble polymers approach this ideal. Table 2 shows the various polymers and polymer groups which have recently been investigated.

Of these, polyacrylamide (PAAm) has been the most thoroughly studied[7,8,10,17,29,37,38]. Its main property debit is its relatively low viscosification efficiency; its primary advantage is its immunity to multivalent cations (brine). A negligibly small viscosity loss is caused by decreasing the thermodynamic quality of the solvent with increasing salt concentration.

Polyacrylamide is available in linear or branched form [see ref. 48 in (7)]. Commercial samples, for economic and ecological reasons (i.e. content of residual monomer), are usually polymerized to 100% conversion, resulting in hydrolysis and branching. Moreover, PAAm is subject to conformational changes with increasing storage time in aqueous solution (aging). The N-substituted PAAm, which were synthesized in our laboratory, are not subjected to aging , but their viscosity level is considerably poorer than PAAm.

Table 2. Compilation of polymers and polymer groups thus far investigated for EOR application, together with their advantages and disadvantages.

	ADVANTAGES	DISADVANTAGES
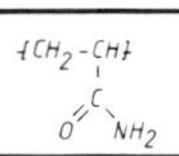 linear and branched	- good injectability - no sensitivity to brine	- ageing - sensitive to oxygen - poor viscosity level
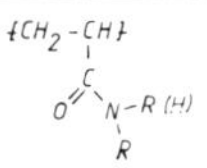	- no aging - no sensitivity to brine	- very poor viscosity level
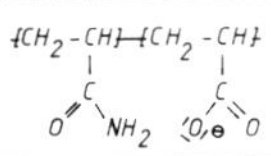	- good injectability - high productivity in fresh water	- ageing - no resistance to brine - sensitive to oxygen
Polyacrylamid-co-sodium-2-sulfoethylmethacrylate	- no sensitivity to Me^{++}	- ageing
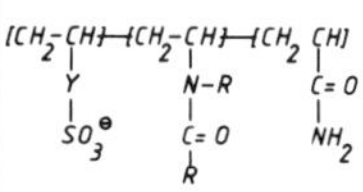 Y = carbonamide derivatives R = organic groups	- stability at high temperatures - no sensitivity to Me^{++}	- viscosity level still improvable
BIOPOLYMERS Xanthan Scleroglucan	- high productivity in brine - resistant at high shear rates and temperature - adsorption (salt)	- problematic injectability - sensitive to microbial degradation - sensitive to oxygen
Cellulose Derivatives	- good solubility - resistant to brine	- sensitive to pH - " " Fe^{+++} - low temperature stability - no shear thinning
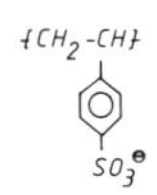	- high viscosity level - no sensitivity to Me^{++}	- ageing - possibly side reactions
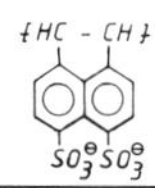	- stability even in presence of high concencentration of Me^{++}	- possibly side reactions - possibly degradation
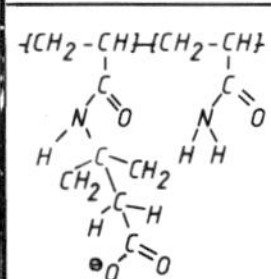	- low viscosity loss in brine - chain stiffening - no precipitation with Ca^{++}	- possibly cross-linking and phase-separation
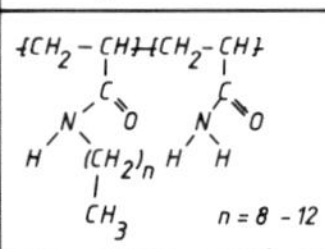	- viscosity increase due to association - low influence of salt and high temperature	- low viscosity level
Polyvinylsaccharides 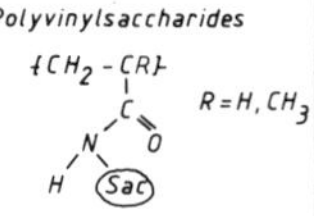	- stability against microorganisms - stability against brine	- low viscosity level
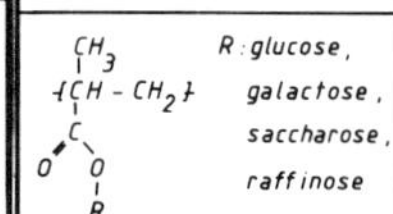	- mechanical stability - microbiological " - stability against brine, including Ca^{++}	- low viscosity level

Partially hydrolysed polyacrylamides, however, exhibit a high viscosity level due to coil expansion, but succumb to aging and are not stable to multivalent cations (precipitation).

Nevertheless, if the oilfield can be preflushed by fresh water, satisfactory results can be obtained with hydrolyzed polyacrylamides. On the other hand, poly(acrylamide-co-sodium-2-sulfoethyl-methacrylate) shows no sensitivity to bivalent cations although it has anionic groups. Yet, it is subject to aging [see Neidlinger et al. ref. 12 in (17)]. Terpolymers which contain vinylamine groups, sulfonate and acrylamide side-groups exhibit good stability at high temperatures, but the viscosity level could still be improved [p 263in(5)].

Biopolymers such as xanthan and scleroglucan exhibit many desirable properties(28,34,36) (i.e. temperature behavior and viscosity level),but they are not stable to microbial degradation [see ref. 8 in(17)]. Cellulose in its water-soluble form, hydroxyethylcellulose (HEC), has good resistance to brine but the required concentration for a solution viscosity of 10 mPa•s is too high. Poly(styrene-sulfonate) also has been investigated(33), [see ref. 14 in(17)], but its sensitivity to brine, its viscosity loss by aging , and the possibility of side-reactions limits its use in EOR. Polymers such as polyacenaphthylene sulfonate (PAcSAc) attain a high viscosity level because of steric hindrance and restricted rotation about the backbone, as well as the polyelectrolyte effect. Moreover, despite its acidic character, PAcSAc is resistant to bivalent cations33). More recent developments by McCormick et al. indicate that sodium-3-acrylamido-3-methyl-butanoate- acrylamide copolymer also exhibits a high viscosity due to stiffening of the chain and absence of sensitivity to brine [see p 366 in(5)]. Associative polymers also investigated by Bock and McCormick show a viscosity increase due to an increase in the apparent macromolecular weight which, in turn, is due to cooperative intermolecular aggregation. This aggregation is reversibly destroyed, if shear-stress is applied to the solution [see pp. 355, 366 in(5)].

Another interesting polymer type is the class of polyvinylsaccharides. These polymers display a high resistance to microbial attack because of an all carbon backbone. In addition, they exhibit a stable solution structure in the presence of salts.(32,33) Yet their viscosification efficiency could still be improved.

In principle, cationic polymers could be employed equally well. However, they exhibit strong adsorption on the surfaces of silicate materials and rocks and thereby easily give rise to pore blockage in the oil reservoirs.

SUMMARY

Obviously, no optimal EOR polymer currently exists. It is difficult for one single polymer to meet all of the requirements. This situation is caused by the various physical conditions (e.g. salinity, temperature, porosity, clay, rock formation, etc.) which the polymer is subjected to in the underground formations. Therefore, it is necessary to choose a synthetic polymer which exhibits the desired behavior for the specific oil bearing formation. Section 2 deals with the characteristic molecular parameters of a polymer sample; e.g. M_w, M_w/M_n, size and shape, as well as with the phenomenon of aging. The viscosity maxima behavior of partially hydrolysed PAAm (c.f. Section 2) has been noted. Samples with 67 mole % acrylic acid attain maximal viscosity at a minimal M_w.

Section 3 describes how the flow characteristics of polymer solutions can be estimated from fundamental rheological properties. Using established structure-property relationships, it is possible to calculate the rheological behavior of polymer solutions in the Newtonian and non-Newtonian regions of the flow curve. Moreover, it is desirable to take the pronounced elasticity of these fluids into consideration.

Porous media flow has been the subject of a large number of investigations. The influence of the elongational viscosity, i.e. the resistance coefficient, can be estimated for solutions of coiled polymers. It is still difficult to treat biopolymers (e.g. polysaccharies), which have a rod-like conformation in aqueous solutions.

One of the essential requirements of an EOR application is that the solution structure of the dissolved polymers be stable under the conditions to which they are subjected. Polymer flooding should, therefore, exhibit the highest possible mechanical, chemical and biological stability. Besides this, the polymer should show no tendency toward rock adsorption. The macromolecules should also be stable to long term instabilities, i.e. aging , as discussed in Section. 2.4. Another factor of importance is the need to lower the interfacial tension between the polymer slug and the oil. In order to achieve this, recently, a two stage flooding process has been applied. First, a microemulsion slug was used to mobilize the oil in the micropores (microscopical displacement), followed by a polymer slug for macroscopic displacement of the oil.

References

1. Sfligiotti, G.M., 3rd European Meeting on Improved Oil Recovery, Rom, Italy, April 16th 1985
2. Leonard, J., Oil & Gas Journal, April 14th 1986, page 71
3. Kesselm D., Ceremonial Test prepared for the tenth anniversary of German Texavo AG, Hamburg, Feb. 1986
4. Maitinm B.K., Volz, H., SPE/DOE 9794, 1981
5. Proceedings of the ACS Division of Polymeric Materials, Science and Engineering, 55, Anaheim, California, Sept. 1986
6. Vollmert, B., Grundriss der Makromolekularen Chimie, E. Vollmert-Verlag, Karlsruhe, 1980
7. Kulicke, W.-M., Kniewske, R., Klein, J., Prog. Polym. Sci., 8:373, 1982
8. Kulicke, W.-M., Böse, N., Coll. & Polym. Sci., 262:197, 1984
9. Lecourtier, J., Chauveteau, G., Xanthan, Macromolecules, 17:1340, 1984
10. Kulicke, W.-M., Haas, R., Ind. & Eng. Chem., Fundam., 23/3:308, 1984
11. Kulicke, W.-M., Hörl, H.H., Coll. & Polym. Sci., 263(7):530, 1985
12. Pfannenmüller, B., Neue Polymere aus Polysacchariden, page 280 in "Polysaccharide", W. Burchard (Ed.), Fließverhalten von Stoffen und Stoffgemischen, Huethig und Wepf, Basel, 1986
13. Kulicke, W.-M., (Ed.) "Fließverhalten von Stoffen und Stoffgemischen", Huethig und Wepf, Basel, 1986
14. Lampert, F., Rinaudo, M., Polymer, 26:1549, 1985
15. Tanaka, T., et al., Phy. Rev. Lett., 45:1636, 1980
16. Rehage, G., Block, H., Lange, W., BMFT-Vorhaben 03E-3067-A
17. Kulicke, W.-M., Macromol. Chem., Macromol. Symp., 2:137, 1986
18. Kulicke, W.-M., Kniewske, R., Makromol. Chem., 182:2277, 1981
19. Kulicke, W.-M., Kniewske, R., Müller, R.J., Prescher, M., Kehler, H., Angew. Makromol. Chem., 142:29, 1986
20. Kulicke, W.-M., Kehler, H., Bouldin, M., in preparation
21. Kulicke, W.-M., Wallbaum, U., Chem. Eng. Sci., 40:961, 1985
22. Bird, R.B., Armstrong, R.C., Hassager, O., "Dynamics of Polymer Liquids I", Wiley & Sons, New York, 1982
23. Haas, R., Durst, F., Rheol. Acta, 21:566, 1982
24. Yamakawa, H., Macromol., 8:339, 1975

25. Mathsen, R.R., Macromol., 13:643, 1980
26. Hashemzadeh, A., Kulicke, W.-M., Chem. Ing. Tech., 58:325, 1986
27. Morris, V.J., Franklin, D., L'Anson, K., Carbohyd. Res., 121:13, 1983
28. Kulicke, W.-M., Lehmann, J., Chem. Ing. Tech., 12:67, 1986
29. Klein, J., et. al., a) Chem. Tech. Lab., 27:310, 1979. b) 21.-24.10.1980. Adsorptions-und Retensionsverhalten von Polymeren an Quarzsand. ÖGEW/DGMK-Gemein-schaftstagung. München
30. GERMAN Patent. DE 32111 68cl. Deutsche Texaco AG
31. Sorbie, K.S., Parker, A., Chifford, P.I., SPE14231, 1985
32. Klein, J., Herzog, D., Hajibegli, A., Poly(vinylsaccharide)s. Part I Makromol. Chem., RC6:675, 1985 Part II Makromol. Chem. in Press
33. Kulicke, W.-M., Klein, J., ACS Polymer Preprints, 22(2):88, 1981
34. Chauveteau, G., Zaitonn. A., European Symp. on EOR, Bournemouth, UK. 1981
35. Münstedt, H., Erdöl-Erdgas 99/12:403. 1983
36. Davison, P., Mentzger, E., Polymer Flooding in North Sea Oil Reservoirs, SPE 9300. 55th Technical Conference, Dallas, Texas, 1980
37. Molyneux, P., "Water-Soluble Synthetic Polymers: Properties and Behavior," Vol. 1+2. CRC Press. Boca Raton. 1983/1984
38. Bekturov, E.A., Bakanova. Z.Kh., Huethig & Wepf. Basel-Heidelberg-New York, 1986
39. Halverson, F., Lancaster, I.E., O'Connor, M.N., Macromolecules 18:1139, 1985

The support of this work by the Bundesminister für Forschung und Technologie (BMFI) and by the Fonds der Chemischen Industrie (FCI) is gratefully acknowledged.

A REVIEW OF SYNTHESIS, CHARACTERIZATION, AND PROPERTIES OF COMPLEX POLYMERS FOR USE IN THE RECOVERY OF PETROLEUM AND OTHER NATURAL RESOURCES

John J. Meister

Department of Chemistry
University of Detroit
Detroit, MI 48221-9987

INTRODUCTION

Many processes that are basic to the extraction of natural resources are facilitated by addition of polymers. To be useful, the polymers must meet an interrelated list of chemical and physical properties as well as economic criteria. The chemical and physical properties demanded of the polymers are:

1. Solubility. Solubility in the primary process solvent is a mandatory criteria which is easily met in most extraction processes conducted in water. But, new applications such as polymeric viscosifiers for supercritical carbon dioxide flooding in enhanced oil recovery present enormous solubility problems for candidate polymers. If the polymer must remain soluble in secondary fluids encountered during the extraction process, then solubility requirements on the polymer will span large ranges of solvent pH, solvent-and-other-solute activity, and solvent polarity.

2. Rheology. Polymers are often added to change solvent or process flow properties. The addition of polymers almost always causes nonnewtonian flow behavior in the resulting fluid. In these fluids, fluid viscosity becomes a function of shear rate of measurement and flow normal to the direction of shear becomes common. Further, alignment of polymer molecules during flow can produce changes in fluid rheology over times ranging from several seconds to several hours. Flow properties therefore become a function of time as well shear rate.

3. Stability. Polymers must be chemically stable, mechanically stable, and thermally stable for long-term applications. Chemical stability to polymer chain scission can be built into the molecule by 1. increasing backbone bond strength, 2. attaching sterically hindering groups to the backbone to protect it, and 3. making a "ladder-backbone" polymer such as is shown in Figure 1. A ladder backbone polymer has 2 chains which must be broken between the same interconnections before the molecule is cleaved. Typical reactions with the polymer which must be inhibited are illustrated by the redox degradation of a polymer by an iron/oxygen couple (1).

$$\text{Poly} + Fe^{3+} \rightleftharpoons \text{Poly}_{ox} + \text{Poly}\cdot + H^{+} + Fe^{2+} \qquad (1)$$

The molecular weight of either Poly_{ox} or Poly· is less than that of Poly since chain scission has occurred.

$$2.\ \mathrm{Poly}\cdot + \mathrm{R}\cdot \rightleftharpoons \mathrm{Poly{-}R} \tag{2}$$

$$3.\ Fe^{2+} + O_2 \rightleftharpoons Fe^{3+} + O_2^- \tag{3}$$

Stability of the polymer to mechanical forces is required because extension and shear forces exerted on the polymer tend to be concentrated, by slippage of the polymer chain, in the center of the molecule (2). For carbon backbone polymers with molecular weights below 10^5, it is difficult for chain slippage to do sufficient work in the center of a molecule to produce energies of the order of the .42 megaJ/mole need to break backbone bonds. In carbon backbone polymers of 10^6 or more molecular weight, however, extension and shear forces commonly encountered in process flow can and do funnel sufficient energy to the center of the polymer to break bonds. Methods to avoid this mechanical polymer decay are 1. use micellizing polymers, 2. use ladder-backbone polymers, 3. use polymers with stronger backbone bonds, and 4. use lightly crosslinked polymers.

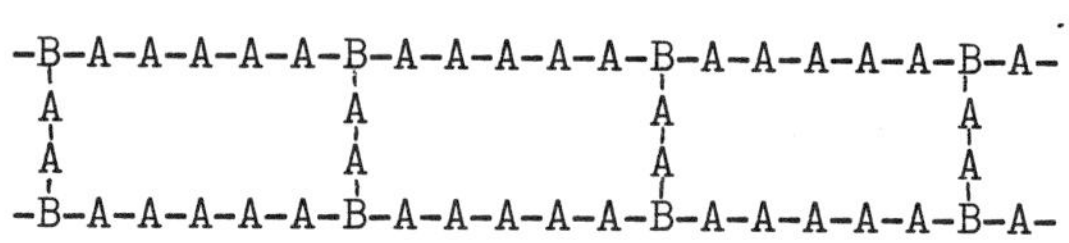

Figure 1. Ladder backbone copolymer.

A requirement that the polymer be thermally stable over a given range of temperature is actually just a demand for chemical stability as a function of temperature. As temperature increases, there is an exponential increase in the rate of any reaction. The reaction-specific temperature (T) is given by

$$T = \Delta E_a/R \tag{4}$$

where ΔE_a is the energy of activation of the reaction and R is the gas constant (3). Thus, reactions which are never seen in a cold process because the "cold" temperature is below the energy of activation of the reaction divided by R, occur quite rapidly when the system is heated. Thermal stability can, therefore, be imparted in the same ways as chemical stability.

4. Surface Behavior. Most extraction processes deal with several phases. At the boundaries between these phases, an interface exists which can be populated with or depopulated of polymer. Situations in which the polymer should accumulate at the surface of one phase are 1. the flocculation of paper pulp or 2. the formation of foams, while situations in which the polymer should depopulate the surface of the phase boundary are 3. minimizing adsorption in mineral acid leaching or 4. minimizing surface tension with surfactants in oil recovery by miscible flooding. In paper production, the pulp is flocculated onto the screens as a dense sheet by cationic polymers which adsorb onto the pulp surface to minimize particle-to-particle repulsion and maximize particle-particle adhesion and entanglement.

In foam formation, polymer adsorbed at the liquid-gas interface increases interfacial viscosity and thereby promotes the formation of thicker, more durable foams. In acid leaching, thickening agents added to increase flow conformance are useless if adsorbed out on the surface of the formation while in enhanced oil recovery, minimum boundary tensions are obtained by maximizing the molar surface excess of surfactant. Polymer molecules which adsorb at the surface reduce this surfactant excess and increase boundary tension.

Methods of controlling surface behavior are to: 1. create polar and nonpolar regions in the molecule thus producing a hydrophilic-lipophilic balance in the molecule, 2. charge the molecule by introducing ionic sites with the same or opposite charge as the boundary, or 3. introduce or remove functional groups in the molecule which produce binding reactions, such as a. hydrogen bond creation or b. nitrogen lone-pair donation, with the surface.

This multitude of properties the polymer must possess dictate that better polymer performance will be obtained from materials with complicated structures. Such polymers are complex polymers: 1) random copolymers, 2) block copolymers, 3) graft copolymers, 4) micellizing copolymers, and 5) network copolymers. There has been a dramatic increase in the past decade in the number and complexity of these copolymers and it is these newly discovered polymers and their chemistry which will be described here. The synthesis, analysis, and testing of these polymers, with particular emphasis on those polymers designed for enhanced oil recovery, will be presented.

COMPLEX POLYMERS FOR RESOURCE RECOVERY

Early applications of polymers to extraction processes were based on the ability of polymers to alter solution rheology. Pye (4) and Sandiford (5) showed that fluid mobility could be effectively reduced by the addition of small amounts of polymer to the solvent. Similarly, Thompson (6) showed that the addition of trace quantities of high molecular weight polymer to solvent reduced friction loss during turbulent flow of the solution. As knowledge of the characteristics needed to do more than change solution rheology grew, complex polymers were prepared and applied.

Poly(1-amidoethylene-co-sodium 1-carboxylatoethylene)

One of the first benefits obtained by the use of complex polymers was lowered adsorption with a simultaneous increase in solution viscosity. Keith R. McKennon of Dow Chemical Company showed (7) that poly((1-amido-ethylene)-co-sodium(1-carboxylatoethylene)) adsorbed less on silica (sandstone) surfaces than did poly(1-amidoethylene).*

Data from (7) showing a minimum in adsorption loss to silica and increasing viscosity as a function of the percent of negatively charged repeat units in the molecule (degree of hydrolysis) is given in Figure 2. The complex, anionic copolymer was formed from poly(1-amidoethylene) by base hydrolysis.

These two benefits occur because 1. in aqueous solution with a pH above 2.0, silica surfaces have a negative charge (8) which would repell a negatively charge molecule but not a neutral one and 2. the high negative charge density in the polyanionic copolymer allows charge repulsion to expand the copolymer molecule in any polar solvent which produces dissociation of the sodium carboxylate unit, $-CO_2^-Na^+$. In processes such as mineral-fines flocculation where surface adsorption is desired, adsorption can be maximized by using a positively charged copolymer such as poly((1-amidoethylene)-co(1-(N-(1,1,N'4,N'4-Tetramethyl-4-ammoniumpentyl chloride))amidoethylene)) as a flocculating agent. For this molecule, the multiple positive charges on the copolymer will aid in binding it to negatively charged surfaces. As these examples show, optimum application of a complex polymer to an extraction process only occurs when the underlying chemical and physical features of the process are understood.

*A list of trivial names corresponding to the correct chemical names used here is given in Table I.

TABLE I

A List of Trivial Names for the Complex Polymers Discussed.

Chemical Name	Trivial Name
Poly(1-amidoethylene)	polyacrylamide
Poly(1-amidoethylene)-co-sodium-(1-carboxylatoethylene))	partially hydrolyzed acrylamide acrylamide-acrylic acid copolymer anionic polyacrylamide
Poly((1-amidoethylene)-co-(1-N-(1,1,N4,N4-Tetramethyl-4-ammonium-pentyl chloride))amidoethlene))	acrylamide-3-acrylamidomethyl butyl-trimethyl ammonium chloride copolymer
Poly((1-hydrogencarboxylato)-ethylene)-b-(amidoethylene)-b-(1-(hydrogen carboxylato)ethylene))	acrylamide/acrylic acid block copolymer
Poly(1-(N-alkylamido)ethylene -co-(1-(sodium carboxylato)-ethylene))	N-alkylacrylamide/sodium acrylate copolymer
Poly(1-(N-methylamido)ethylene -co-sodium 1-carboxylatoethylene)	N-methylacrylamide/sodium acrylate copolymer
Poly(1-(N-(methylethyl)amido) ethylene-co-sodium 1-carboxylato-ethylene)	N-isopropylacrylamide/sodium acrylate copolymer
Poly(1-(N-butylamido)ethylene-co-sodium 1-carboxylatoethylene)	N-butylacrylamide/sodium acrylate copolymer
Poly(1-(N,N-dimethylamido)ethylene -co-1-amidoethylene)	N,N-dimethylacrylamide/acrylamide copolymer
Poly(1-(N,N-diethylamido)ethylene -co-1-amidoethylene)	N,N-diethylacrylamide/acrylamide copolymer
Poly(1-(N,N-diethylamido)-1-ethylethylene-co-1-amidoethylene)	N,N-diethyl-2-ethyl acrylamide/ acrylamide copolymer
Poly((1-amido-1-alkylethylene)-co-(1-(sodium carboxylato)ethylene))	methacrylamide/sodium acrylate copolymer
Poly(1-azo-2,6-oioxo-3,5-dimethyl-cyclohex-3,5-diyl)-co-(1-sodium carboxylato)-1-methylethylene)-co-(1-amido-1-methylethylene)))	imidized and hydrolyzed polymethacryl-amide
Poly(1-amidoethylene-co-(1-(sodium 2-methylprop-2N-yl-1-sulfonate)-amidoethylene))	acrylamide/AMPS copolymer
Poly(1-amidoethylene-co-(1-methyl-1-(sodium 1-oxo-2-oxybutyl-4-sulfonate))ethylene))	acrylamide/ethylsulfonate methacrylate ester copolymer
Poly(starch-g-(1-amidoethylene))	starch/acrylamide graft copolymer

(continued)

Poly(starch-g-((1-amidoethylene)-co-(1-sodium carboxylato)ethylene)))	starch/(acrylamide-sodium acrylate) graft copolymer
Poly(starch-g-((1-amidoethylene)-co-(1-(sodium 2-methylprop-2N-yl-1-sulfonate)amidoethylene)))	starch/(acrylamide-AMPS)graft copolymer
Poly(dextran-g-(1-amidoethylene)))	dextran-acrylamide graft copolymer
Poly(2,3-dihydroxypropylcellulose-g-(1-amidoethylene))	hydroxpropylated cellulose-acrylamide graft copolymer
Poly(lignin-g-(1-amidoethylene))	lignin-acrylamide graft copolymer
Poly(lignin-g-((1-amidoethylene)-co-(1-(sodium 2-methyl-2N-yl-1-sulfonate)amidoethylene)))	lignin-(acrylamide-AMPS) graft copolymer

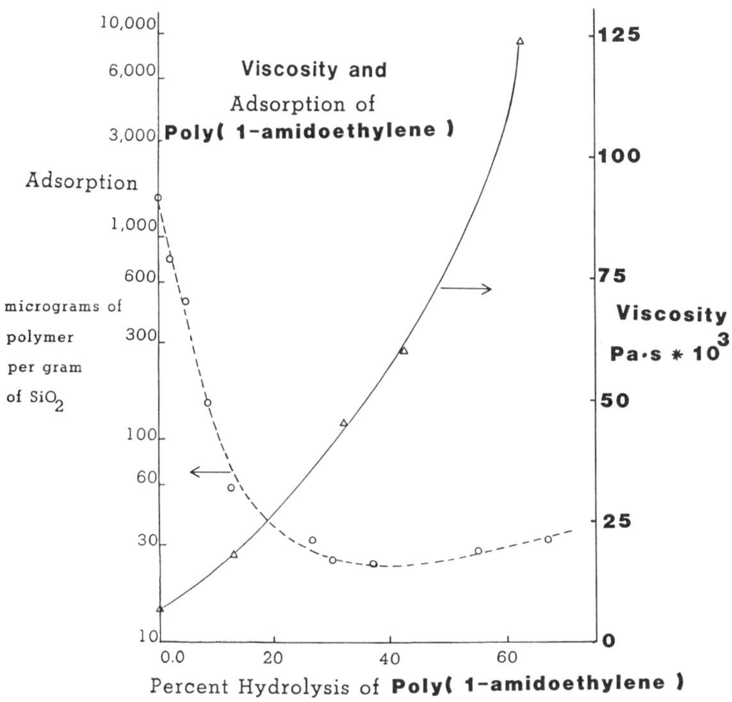

Figure 2.

The conversion of poly(1-amidoethylene) homopolymer to a complex, random copolymer by base hydrolysis* alters several properties of the polymer. Martin et al (9) investigated 1. the change in solution viscosity as a function of sodium chloride concentration and 2. copolymer susceptibility to shear degradation for poly((1-amidoethylene)-co-sodium(1-carboxylato-ethylene)). Table II gives the results from this investigation and shows that as carboxylate content of copolymer increases, 1. viscosity of a fixed concentration solution of the copolymer in sodium chloride brine decreases and 2. fraction of molecules in aqueous solution degraded by shear decreases. Both changes in copolymer properties appear to plateau at 30% conversion of amide groups to carboxylate groups.

TABLE II

[a]Screening Data for Poly(1-amidoethylene-co-sodium 1-carboxylatoethylene), $\overline{M}_w$ = 5 to 6 x 10^6.

Wt. of Acid: Sodium Salt (%)	Viscosity Number 0.01% NaCl (dL/g)	Viscosity Number 2.0% NaCl (dL/g)	Viscosity Retention[b] Shear (%)	Viscosity Retention[b] Brine (%)
<1[c]	10.0	8.3	67.8	83.7
5.6	23.6	9.2	82.8	38.8
8.0	63.6	16.9	78.9	26.6
13.5	85.7	12.7	84.6	14.9
16.1	106.2	14.8	91.4	13.9
28.8	122.7	12.7	94.1	10.4
38.6	118.0	13.0	93.2	11.0
40.5	120.7	13.7	95.5	11.4

a) Starting material was poly(1-amidoethylene) -- Polyscience Lot 3-1590.

b) Viscosity retention is the percent of viscosity number retained under new conditions when compared to capillary flow viscosity number at 25°C of a copolymer solution in 1.71×10^{-3} molar NaCl with a pH of 9.5. Viscosity number is defined as

$$\text{viscosity number} = \frac{\eta - \eta_o}{\eta_o * C_{poly}}$$

where η is solution viscosity, η_o is solvent viscosity, and c_{poly} is copolymer concentration in g/dL. For viscosity retained in brine, the new solution had a NaCl concentration of 0.342 molar. For viscosity retained after shear, the "new" solution had been passed through a screen-shear apparatus (37).

c) The carboxyl content observed was within the experimental error of the determination.

*Base hydrolysis describes the process of heating a polymer solution with a pH above 10 to temperatures of 40 to 70°C to convert amide groups to carboxylic acid groups.

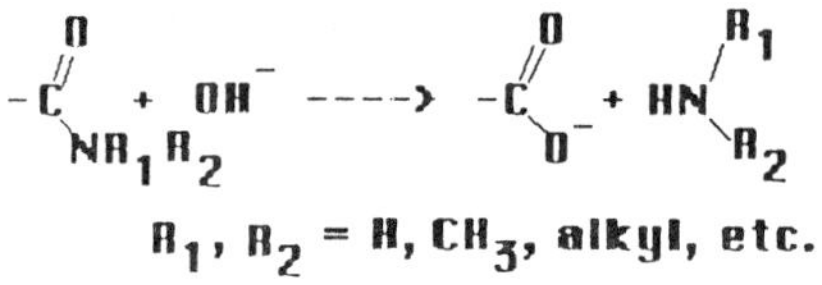

This reaction occurs randomly on the homopolymer chain to create a random amide(carboxylic acid) copolymer. Hydrolyzed amide polymer is prepared by adding the amount of sodium hydroxide needed to hyrolyze the desired fraction of amide units to a 1 weight percent solution of amide polymer and maintaining that solution at 60°C for 2 hr. The reaction mixture is precipitated into 5 times its volume of 2-propanone, filtered, the precipitate washed with 2-propanone, ground in a blender for 2 minutes, and dried under vacuum to constant weight. This procedure is that of Meister, et al. (31).

Poly((1-hydrogen carboxylatoethylene)-b-(1-amidoethylene)-b-(1-hydrogen carboxylatoethylene))

There are some indications that altering the anionic polymer structure from random copolymer to A-B-A acid-amide-acid, block copolymer produces a polyelectrolyte which does not undergo coil collapse in electrolyte solutions (brine) as readily as the random copolymer. Although there is significant scatter in the data (9), shown in Figure 3, the block copolymer retains 20% more of original viscosity than does a randomly copolymerized copolymer, and 25% more of original viscosity than does a hydrolyzed copolymer. These comparisons must be taken as very tentative, however, because of several uncontrolled variables in the experiment. The compared copolymers are not proven to be of equal molecular weight nor is it established that the copolymers have equal degrees of hydrolysis. The change in limiting viscosity number, $[\eta]$, (which reflects molecular size in solution and thus, solution viscosity) as a function of number of charged sites, CS, in the molecule (10) and number of backbone carbon atoms, B, in the molecule (10), is

$$[\eta]^{5/3} - [\eta] = CS^2 * \sqrt[2]{B} \qquad (5)$$

Solution viscosity is therefore a power function of degree of hydrolysis, D_H

$$CS = D_H * B/2 \qquad (6)$$

and polymer number-average molecular weight, $\overline{M}_\eta$.

$$\overline{M}_\eta = \left[\begin{pmatrix}\text{g mole weight of}\\ \text{of amide repeat unit}\end{pmatrix} * (1-D_H) + \begin{pmatrix}\text{g mole weight of}\\ \text{carboxylate repeat unit}\end{pmatrix} * DH\right] * B/2 \qquad (7)$$

In the three copolymers compared, the hydrolyzed copolymer will, on the basis of the behavior of the monomers in the synthesis, have the highest molecular weight, the block copolymer will have the next highest molecular weight, and the randomly synthesized copolymer should have the lowest molecular weight. This prediction is based on the fact that the ethene carboxylate monomer tends to give lower molecular weights to homo- or copolymers than does the ethene amide monomer. Since solution viscosity depends in a complex way on molecular weight and molecular composition, data obtained by comparing polymers of different sequence structure without control of these two variables is of doubtful validity.

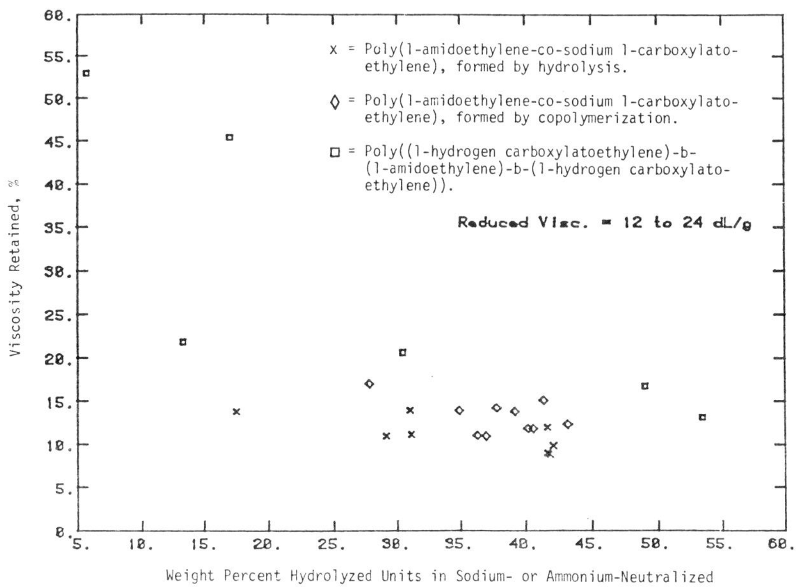

Figure 3. Viscosity Retained when Salinity is Increased from 0.01% NaCl to 2% NaCl.

Poly(1-(N-alkylamido)ethylene-co-[(sodium 1-carboxylato) or (1-amido)] ethylene]

A series of these copolymers were made (9) by placing weighed amounts of monomer in an appropriate amount of distilled, deaerated water in a 50 ml serum bottle. The bottle was flushed and sealed under a nitrogen stream and the initiator solution was introduced into the bottle by use of a syringe. As each reaction progressed, the solution would become more viscous. Copolymerizations were terminated when degree of conversion in the reaction mixture was 30 wt % or less. The reaction was terminated by pouring the reaction mixture into a stirred, nonsolvent for the polymer. Optimum temperature for these reactions is 15 to 25°C. To generate copolymers with a larger limiting viscosity number in water and, most likely, a higher molecular weight, mass ratios of monomer to potassium persulfate initiator of 100 to 0.5 were used. Larger amounts of initiator (smaller monomer/initiator mass ratio) decreased limiting viscosity number of the copolymers in water while smaller amounts of initiator (larger mass ratio) altered $[\eta]$ differently for the different N-alkylsubstituted copolymers. For N-methyl copolymers, smaller amounts of initiator decreased copolymer capacity to thicken water while for N-(1-methylethyl) copolymers, smaller initiator amounts produce copolymers with increased capacity to thicken water. The copolymers synthesized and tested are:

poly(1-(N-methylamido)ethylene-co-sodium 1-carboxylatoethylene) = AA (9),
poly(1-(N-(methylethyl)amido)ethylene-co-sodium 1-carboxylatoethylene) = BB (9),
poly(1-(N-butylamido)ethylene-co-sodium 1-carboxylatoethylene) = CC (9),
poly(1-(N,N-dimethylamido)ethylene-co-1-amidoethylene) = DD (11),
poly(1-(N,N-diethylamido)ethylene-co-1-amidoethylene) = EE (11), and
poly(1-(N,N-diethylamido)-1-ethylethylene-co-1-amidoethylene) = FF (11).

The major effects of adding alkyl groups to the amide unit of a polyamide chain are to 1. make the amide more resistant to hydrolysis (9), and 2. make the resulting polymer less hydrophilic, and thus less soluble, than its unsubstituted precursor. The effect on solubility is clearly seen in the behavior of polymer CC when the fraction of N-butyl-substituted, repeat units is high. Such products are water insoluble.

Properties: Most of the testing done on the copolymers determined rheological properties of solutions of the copolymers in water. Since similar data exists for copolymers without N-substitution, general comparisons of the effect of N-substitution on solution flow can be made. These comparisons cannot be taken very far, however, because the molecular weights and repeat-unit-composition of the copolymers were not determined.

Data for poly(1-amidoethylene-co-sodium 1-carboxylatoethylene) = AC are given in Table III. Data for copolymers AA, BB, and CC are given in Table IV while data for copolymers DD, EE, and FF are given in Table V and Figures 4 and 5. These data show that N-substitution generally lowers the viscosity number, $(\eta-\eta_o)/(\eta_o{*}c_{poly})$, of the copolymer, making it a poorer viscosifier. This generalization of the data may merely be a reflection of lower molecular weights in the N-substituted copolymers since solution concentrations are based on weight instead of molar units. Adding electrolyte (salt) to aqueous solutions of N-substituted copolymers does cause reduction in solution viscosity, but the reduction in viscosity caused by adding an equal amount of salt to a simple amide/carboxylate copolymer is significantly greater. Thus, the N-substituted copolymers tend to maintain the same viscosity in water irrespective of salt content while the simple amide/carboxylate copolymer undergoes more coil collapse and viscosity loss. The addition of an alkyl group to the amide unit of the copolymer produces small, if any, increases in the maintenance of solution viscosity in the face of extensive shearing of the solution. Replacing a hydrogen on the amide of a copolymer with an alkyl group, R, does little, if anything, to increase the mechanical stability of the copolymer to shear. Generally, the viscosity of a solution of an N-substituted copolymer is less than that of an equal concentration, unsubstituted copolymer with the viscosity of the N-substituted copolymer solution decreasing as the number of methylene units (CH_2's) added to the amide nitrogen increases.

TABLE III
Screening Data for Poly(1-amidoethylene-co-(sodium 1-carboxylatoethylene)).

Polymer Conc. (g/dL)	Wt. of Acid: Sodium Salt (%)	Viscosity Number (dL/g) 0.01% NaCl	Viscosity Number (dL/g) 2.0% NaCl	Viscosity Retention (%) Shear	Viscosity Retention (%) Brine
0.0170	16.1	106.19	14.76	91.4	13.9
0.0187	18.8	11.72	0.86	94.4	7.7
0.0152	14.2	14.43	1.34	98.6	9.3
0.0608	19.6	103.92	9.07	94.5	8.7
0.0179	28.8	122.71	12.72	94.1	10.4
0.0207	31.5	15.16	1.25	98.7	8.2
0.0173	41.7	115.57	10.46	92.4	9.0
0.0146	41.7	103.71	11.11	95.9	10.7
0.0069	43.0	151.33	16.41	96.0	10.8
0.0086	50.8	143.86	13.66	90.3	9.5
0.0126	52.9	64.02	4.93	94.3	7.7

TABLE IV
A. Comparative Screening Data: (1-(N-methylamido)ethylene-co-(sodium 1-carboxylatoethylene)).

Polymer Conc. (g/dL)	Wt. of Acid: Sodium Salt (%)	Viscosity Number (dL/g) 0.01% NaCl	Viscosity Number (dL/g) 2.0% NaCl	Viscosity Retention (%) Shear	Viscosity Retention (%) Brine
Polymers made by Copolymerization					
0.0151	9.5	35.69	12.97	80.7	36.3
0.0151	16.1	74.94	13.17	75.1	17.6
0.0160	24.3	63.46	10.00	89.8	15.8
0.0147	28.4	76.66	10.11	89.1	13.2
0.0169	31.8	24.96	4.30	92.6	17.2
0.0141	36.2	81.20	9.89	91.0	12.2
0.0149	36.4	102.73	13.00	80.4	12.7
0.0157	36.8	36.49	5.38	83.4	14.8
0.0151	42.7	103.85	13.04	80.0	12.6
0.0152	46.6	113.12	14.79	73.3	13.1
Polymers made by Hydrolysis					
0.0117	20.5	85.90	10.36	96.8	12.1
0.0129	22.3	40.58	3.66	99.5	9.0
0.0132	24.7	85.65	10.02	97.2	11.7
0.0127	30.6	94.81	10.72	98.4	11.3
0.0125	41.5	87.05	9.10	97.2	10.4

(continued)

B. Comparative Screening Data: (1-(N-(1-methylethyl)amido)ethylene-co-(sodium 1-carboxylatoethylene)).

Polymer Conc. (g/dL)	Wt. of Acid: Sodium Salt (%)	Reduced Viscosity (dL/g) 0.01% NaCl	Reduced Viscosity (dL/g) 2.0% NaCl	Viscosity Retention (%) Shear	Viscosity Retention (%) Brine
0.0151	17.6	16.13	2.27	97.6	14.1
0.0169	28.9	71.64	12.25	89.0	17.10
0.0125	31.5	40.58	4.81	89.5	11.9
0.0153	33.0	48.83	6.19	95.5	12.7
0.0152	45.3	85.96	11.64	92.0	13.5
0.0154	46.8	88.83	11.27	88.1	12.7
0.0081	51.3	85.54	12.38	85.5	14.5

TABLE V-
Reaction Parameters for the Synthesis of Poly(1-N,N-dimethylamido) ethylene-co-(1-amidoethylene)) in 30% Methanol in Water.

Sample #	2-propenamide Concentration	N conc. (M)	$K_2S_2O_8$ Conc. (M) $Mx10^3$	Mole Fraction of II in Feed	Reaction Temp. (°C)	Reaction Time (hr)	Convers. (%)
1	2.111	-	4.245	0.00	40	3	98.8
2	1.688	0.2982	3.971	0.15	40	6	98.4
					60	3	
3	1.322	0.5664	3.774	0.30	40	8	97.7
					60	2	
4	0.8886	0.8870	3.578	0.50	40	8	99.8
					60	2	
5	--	1.514	3.089	1.00	40	8	96.3
					60	2	

N = N-methyl-2-imino-3-oxopent-4-ene(II).

Limited studies of copolymer retention (9) in a consolidated silica, berea sandstone, were done with copolymers AC, AA, and BB. Firm conclusions about adsorption of copolymer from solution can not be gained from this data since the fluid was flowed through the porous rock. Occlusion and filtration phenomena may therefore produce a major portion of the retention seen in the tests. With the distinction established that retention is not equivalent to adsorption, the data show that adding a methyl group to the amide unit decreases copolymer retention while adding a 1-methylethyl unit (isopropyl) to the amide increases copolymer retention. These data are shown in Figure 6.

Poly((1-amido-1-alkylethylene)-co-(sodium 1-carboxylatoethylene)

The addition of an alkyl group to every other carbon atom of one repeat unit of a copolymer should stiffen the copolymer chain. The group limits the number of positions the chain can adopt without crushing the added alkyl group into some other part of the copolymer. This is stiffening the polymer chain by a "steric" or "excluded volume" method. The steric (part-to-part, repulsive) forces in the molecule then keep the copolymer from occupying some of the volume (excluded volume) inside the polymer coil with the result that the alkyl-substituted molecule is larger in volume than an unsubstituted polymer with the same number of repeat units.

A series of 1-methyl-substituted poly(1-amidoethylenes) were synthesized (9) and found to have limiting viscosity numbers in hydrogen-bond-blocking solvents that were equal to those of poly(1-amidoethylenes) of 10 or more times higher molecular weight. These data show that the concept of

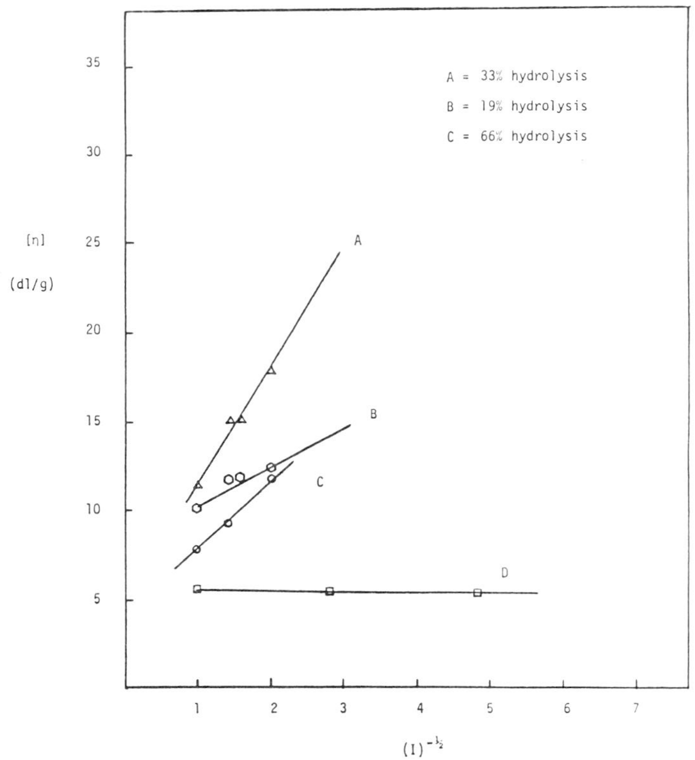

Figure 4. Intrinsic Viscosity as a Function of (solution ionic strength)$^{-\frac{1}{2}}$ for sodium chloride solutions of various hydrolyzed poly(1-amidoethylenes) [A,B,C,] and poly (1-(N,N-dimethylamido) ethylene-co-1-amidoethylene), (11). (T=30°C). Poly(1-(N,N-dimethylamido)-ethylene-co-1-amidoethylene) = (D).

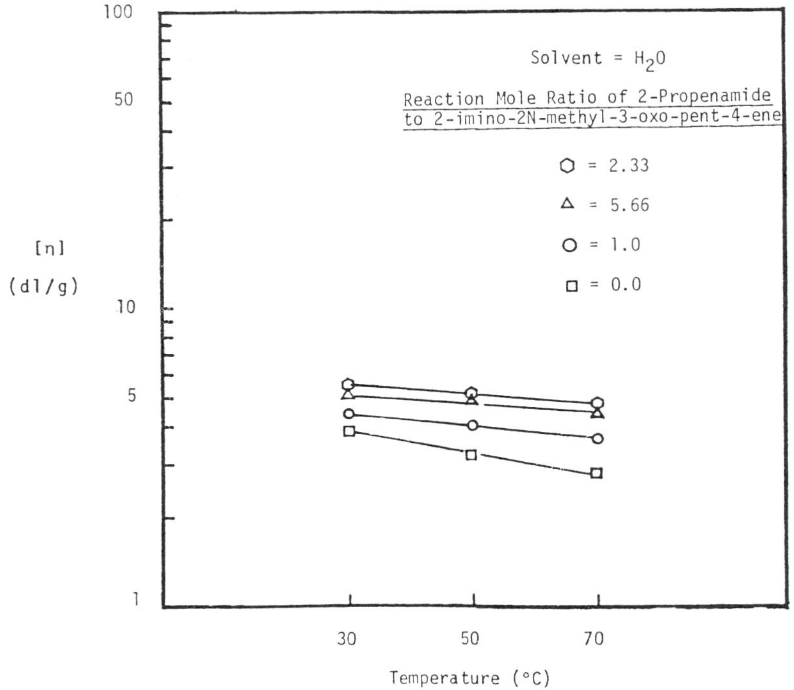

Fig. 5. Intrinsic Viscosity of Poly (1-N,N-dimethylamido) ethylene)-co-1-amidoethylene). Samples in Water as a Function of Temperature.

stiffening a polymer chain to increase ability to viscosify a solvent, worked. When 2-methyl-2-propenamide was copolymerized with 2-propenoic acid, copolymers of significantly lower molecular weight were obtained (9) than would be obtained in a similar 2-propenamide polymerization. These molecular weight comparisons were based on comparison of limiting viscosity numbers in water. Since the 1-amido-1-methylethylene repeat unit is more hydrophobic than a 1-amidoethylene repeat unit, differences in limiting viscosity numbers of these two types of polymers do not correspond to exact molecular weight differences but are approximate measures of molecular weight difference.

Table VI contains synthesis and rheological data for poly((1-amido-1-methylethylene)-co-(sodium 1-carboxylatoethylene)) = GG. These data show that the lower-molecular-weight, methyl-substituted GG are somewhat 1. poorer viscosifiers, 2. more prone to shear degradation in high shear applications, and 3. more prone to lose thickening capacity when electrolyte is added to the copolymer solution than is the unsubstituted copolymer.

TABLE VI

A: Copolymerization Conditions: Poly((1-amido-1-methylene)-co-(sodium 1-carboxylatoethylene)).

2-methyl 2-propenamide	2-methyl 2-propenoic acid	NaOH	$K_2S_2O_8$
10.09 g	10.62 g	5.00 g	0.1326 g
15.33	5.31	2.50	0.1321
19.26	1.06	0.50	0.1331
4.81	16.57	10.50	0.1364
16.93	3.67	2.00	0.1330
8.15	12.25	7.75	0.1318

Conditions: 9 to 10% solids, 25°C, 24 hours.

B. Comparative Screening Data.

Carboxyl Content as Na salt (wt. %)	Viscosity Number (dL/g)		Viscosity Retained (%)	
	0.01% NaCl	2.00% NaCl	Shear	Brine
Poly((1-amido-1-methylethylene)-co-(sodium 1-carboxylatoethylene))				
5.7	15.4	3.4	89.8	22.4
17.9	24.4	4.1	90.2	16.6
22.6	25.2	3.7	92.9	14.8
51.9	45.7	4.0	94.5	8.8
61.0	40.8	4.1	83.8	10.1
76.1	41.9	4.9	89.1	11.7
Poly(1-amidoethylene-co-(sodium 1-carboxylatoethylene))				
6.0	14.8	4.9	96.3	32.9
19.2	14.4	1.3	98.6	9.3
23.6	83.9	9.8	94.5	11.7
52.9	64.0	4.9	94.3	7.7
54.4	102.0	11.5	86.4	11.3

Poly((1-aza-2,6-dioxo-3,5-dimethylcyclohex-3,5-diyl)-co-(sodium 1-carboxylato-1-methylethylene)-co-(1-amido-1-methylethylene))

When poly(1-amido-1-R-ethylene) is dissolved in solutions with a pH less than 2, it starts to link adjacent amide units into 6-member rings. The reaction, imidization, for head-to-tail linked, repeat units is

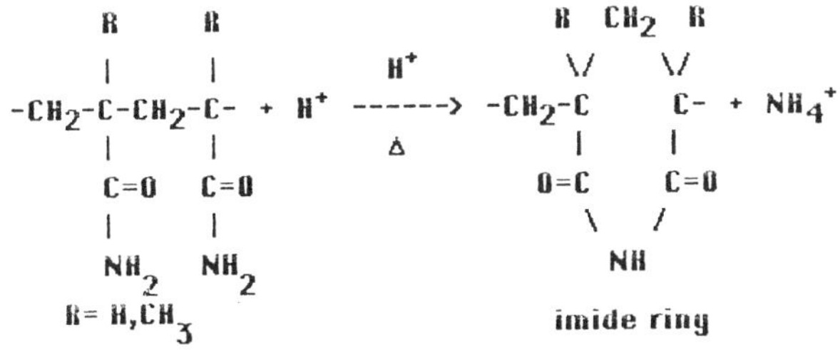

Head-to-head-linked, repeat units would form 5-membered rings by this reaction. As the fraction of repeat units converted to ring structures increases, the polymer becomes less soluble in water. Insolubility in water occurs at a lower degree of imidization for the 1-methyl ($R=CH_3$) polymer than for the 1-hydro (R=H) polymer. Polymers containing imide rings which precipitate from water can be made water soluble by treatment of a solution of the polymer in 1,2-ethandiol with sodium hydroxide. This hydrolysis treatment breaks most rings in 1-hydropolymers but converts amide unis to carboxylic acid units with little imide group loss in the 1-methyl polymers.

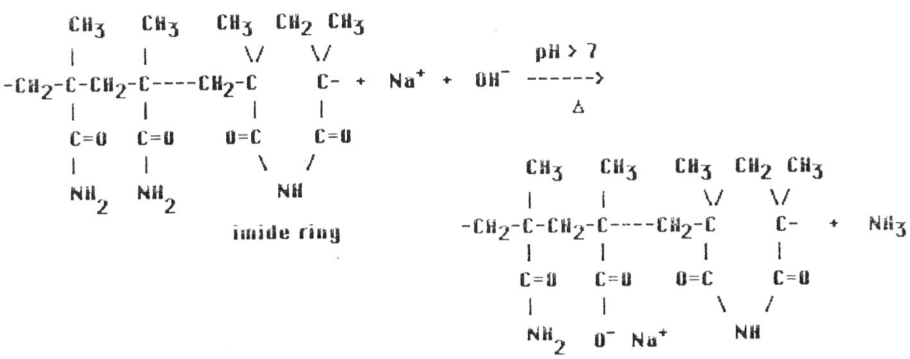

Since the presence of an imide ring in a linear polymer chain 1. stiffens the chain to produce an expanded molecule in solution and 2. introduces a 2-chain, ladder-like unit to withstand shear stress, structures containing this ring system have been synthesized and tested (9).

Acid-catalyzed imidization was produced in 5 to 6 million molecular weight (1-amidoethylene) by dissolving over a 2.5 to 5 day period 2 to 4 weight percent polymer in 6 M hydrochloric acid. Polymer was recovered after 4 to 10 days in the solution. The recovered polymers were dissolved in 1,2-ethanediol at 90°C and treated with a stoichiometric amount of sodium hydroxide. The solution was allowed to sit at room temperature for 5 to 10 days to complete the hydrolysis. Infrared studies of the produced copolymer show that 6-member rings predominate over 5-member rings in the product. This is to be expected since head-to-tail addition of repeat units in amidoethene polymerizations is more common than head-to-head addition (12). Analytical data for carboxyl and imide content for a series of poly(1-amidoethylene)-derived copolymers is contained in Table VII along with rheological test data for aqueous solutions of these samples. The same types of data are given in Table VIII for a series of copolymers derived from poly(1-amido-1-methylethylene).

TABLE VII
Screening Data: Partially Imidized Poly((1-amidoethylene)-co-(sodium 1-carboxylatoethylene)).

Carboxyl Repeat Unit Content (wt. %)	Imide Content (wt. %)	Viscosity Number 0.01% NaCl (dL/g)	Viscosity Number 2.00% NaCl (dL/g)	Viscosity Retained Shear (%)	Viscosity Retained Brine (%)
14.2	-	88.5	17.3	88.1	19.5
19.2	18	109.2	29.9	91.5	27.4
35.7	12	64.4	10.3	80.5	15.9
37.7	10	155.1	19.6	77.4	12.7
42.8	20	132.1	26.6	94.5	20.2
51.0	-	115.6	22.3	97.2	19.2
58.4	5	80.8	9.2	91.4	11.4
81.2	3	109.9	7.3	95.4	6.7

TABLE VIII
Screening Data: Partially Imidized Poly((1-amido-1-methylethylene)-co-(sodium 1-carboxylato-1-methylethylene)).

Carboxyl Repeat Unit Content (wt. %)	Imide Content (wt. %)	Viscosity Number 0.01% NaCl (dL/g)	Viscosity Number 2.00% NaCl (dL/g)	Viscosity Retained Shear (%)	Viscosity Retained Brine (%)
28.0	-	196.7	43.2	74.2	22.0
29.2	16	159.6	23.8	90.5	14.9
30.3	22	183.9	38.2	75.3	20.7
30.4	24	136.8	31.7	88.4	23.2
31.2	21	155.9	34.8	85.4	22.3
31.3	27	140.9	39.8	93.0	28.3
34.8	25	170.1	41.4	79.5	24.3
45.9	15	123.7	16.8	89.0	13.6
58.4	6	130.2	13.5	88.4	10.4
64.8	-	129.6	14.4	86.7	11.1
75.1	9	142.7	19.1	76.4	13.4

These data show that imide rings are much more stable when formed in 1-methyl-group-containing polymers instead of the 1-hydro group containing polymer. Imide rings made in 1-hydro group containing polymers are readily hydrolyzed to carboxylic acid units during treatment of the polymer with base. Because imide rings are readily destroyed in subsequent hydrolysis of 1-hydro-group containing polymers, the production of rings in these molecules has less effect on rheological properties than in 1-methyl-group containing polymers. The rheological properties of the polymers containing imide rings are equivalent, if not better, than those of equal carboxylic-acid-content, poly(1-amidoethylene-co-(sodium 1-carboxylatoethylene) copolymers in 1.74×10^{-3} M (0.01 weight percent) sodium chloride brine. Further, the viscosity of solutions of polymer processing imide rings does not decrease as extensively as does that of solutions of imide-free copolymer (9) when the sodium chloride content is increased to 0.3448 M (2.0 weight percent). Thus, the goal of adding ring structures to the polymer, inhibiting the collapse of the polymer coil by stiffening the backbone, was achieved. Unfortunately, this success is quite limited. When this comparison is repeated in brines containing calcium or magnesium ions in place of sodium ions, the polymers containing imide rings collapse more than do the imide-free copolymers in identical brines (9).

Copolymers Containing 1-Amidoethylene Repeat Units and Sulfonate Groups

The sulfonate groups, $-SO_3^-$, is a strong, polar ionic group which has a much greater ability to dissociate in polar solvents than does the carboxylate group, $-CO_2^-$. This chemical behavior would be a highly desired property in polymers prepared as thickening agents for concentrated salt solutions or in polymers prepared as surface active agents in high-electrolyte-concentration solutions. With a strong ionic group like the $-SO_3^-$ unit spotted repeatedly on the backbone of a polymer molecule, the molecule will develop a large number of negative charges in a polar solvent. These charges, produced by dissociation of the positive cation from the negative sulfonate group,

$$R\text{-}SO_3^{\ominus}Na^{\oplus} + H_2O \xrightleftharpoons{\text{in } H_2O} R\text{-}SO_3^{\ominus}(aq) + Na^{\oplus}(aq) + H_2O,$$

will repell one another and cause the solvated polymer to occupy a large volume in solution. The statement that the sulfonate group is a stronger ionic group than the carboxylic group is based on the fact that most dissociation constants, K_dSO_3, for the sulfonate group are larger than those, K_dCO_2, for an identically neutralized carboxylic group,

$$K_d^{SO_3} = \frac{[RSO3^-]^n[X^{+n}]}{[(RSO_3)_nX]} \gg K_d^{CO_2} = \frac{[RCO_2]^n[X^{+n}]}{[(RCO_2)_nX]}$$

Since molecular expansion, α_e, depends (10) on the number of charged sites on the molecule, CS, and the number of backbone atoms, B

$$\alpha_e^5 - \alpha_e^3 = CS^2 * \sqrt{B},$$

sulfonate-containing polymers will be larger molecules in virtually all brines than identical carboxylate-containing polymers because the more highly dissociated, highly charged sulfonate molecule will expand more than will the less charged carboxylate.

To gain this behavior in polymers for resource recovery, complex polymers containing sulfonate groups have been prepared. Calgon has a commercially available, sulfonate-containing copolymer called Polymer 340. This product is poly(1-amidoethylene-Co-(1-(sodium 2-methylprop-2-N-yl-1-sulfonate)amidoethylene)). This product was tested in comparison with a number of other commercially-available polymers by Szabo (13). The data gathered is viscosity in water and 3 different sodium chloride solutions, limited data on viscosity change as a function of time for polymers dissolved in sodium hydroxide solution, and solution viscosity loss as a function of time during which the solution underwent high mechanical shear. The copolymer performed well in comparison to the other 11 polymers tested but was ranked in the middle of the group when rated on a cost-effective performance basis.

Researchers at the University of Southern Mississippi have also investigated this copolymer (14). The copolymer was prepared in free radical reactions containing sulfonate-amide monomer to amide monomer ratios of from 1 to 9 to 7 to 3 and an initiator to monomer ratio of 1 to 1000. The reaction was run in a 3-neck flask at 30°C by dissolving the monomers in water and saturating with nitrogen for 15 minutes. A solution of potassium persulfate in 10 mL of water was introduced and the reaction was allowed to proceed for up to 4 hr. The reaction was terminated and the polymer recovered by precipitation in 2-propanone. Representative synthesis data are given in Table IX.

Repeat unit composition data from a series of these copolymers was used to calculate reactivity ratios for 2-propenamide, M_1, and sodium 2,2-dimethyl-3-imino-4-oxohex-5-ene-1-sulfonate, M_2, of $r_1 = 1.00 \pm 0.08$ and $r_2 = 0.50 \pm 0.04$. These results may have a determinate error, however, because 1.

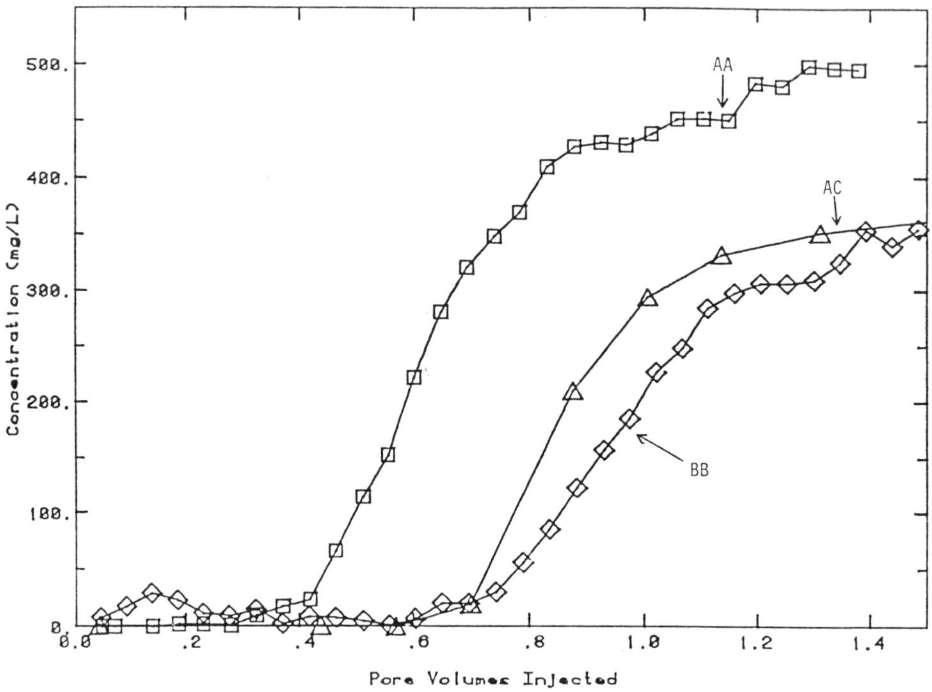

Figure 6. Polymer Breakthrough Curves for Copolymers.

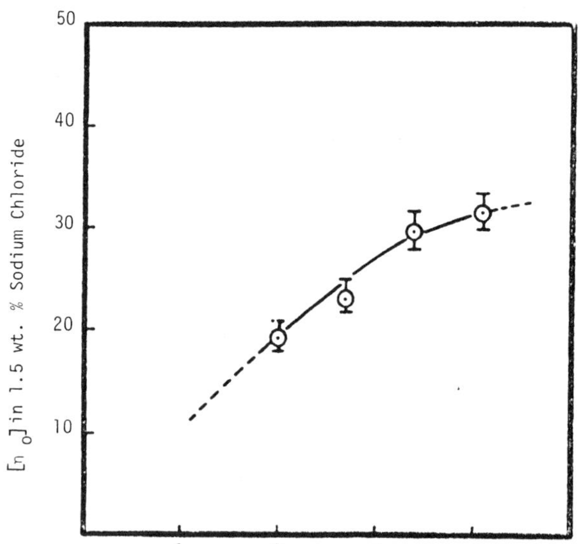

Figure 7. Zero-shear intrinsic viscosity as a function of $\ln[M][I]^{-\frac{1}{2}}$ at 30°C for the copolymerization of 2-propenamide with sodium 2,2-dimethyl-3-imino-4-oxohex-5-ene-1 sulfonate at a 9/1 feed ratio.

the reactivity ratios were calculated using a linear approximate method and 2. most of the reactions were run at a pH of 9.0 and detectable hydrolysis of the primary amide may have occurred during synthesis. Studies were also done on the dependence of zero shear limiting viscosity number in 0.265 M sodium chloride solution on the natural log of the monomer concentration divided by the initiator concentration to the one half power. Results of these tests are shown in Figure 7. The data prove that hydrodynamic radius of the copolymer in solution (and thus 1. limiting viscosity number and 2. capacity to produce a given viscosity in solution at a fixed concentration) can be controlled by fixing $\ln[M][I]^{-1/2}$ for a synthesis.

TABLE IX
Reaction Parameters for the Copolymerization of M_1 with M_2 of M_2.

Feed radio M_1/M_2	Reaction time (min)	Conversion (%)	Weight (% N)	Weight (% N)	M_1 in copolymer (mol %)
9:1	20	6.2	14.4	3.11	90.3
9:1	120	28.3	14.73	3.14	90.7
9:1	240	67.9	14.03	3.05	90.5
8:2	15	5.5	13.00	5.66	81.0
8:2	120	49.2	11.93	5.16	81.0
8:2	240	70.2	12.15	5.28	81.0
7:3	15	5.6	11.45	6.78	74.0
7:3	120	49.5	10.65	6.74	72.3
7:3	240	74.1	10.65	6.78	72.1
6:4	20	14.4	9.68	7.53	66.0
6:4	60	16.3	9.45	7.41	65.7
6:4	240	95.2	9.65	8.30	62.4
3:7	20	9.1	7.14	10.20	37.5
3:7	60	50.3	6.94	10.15	36.0
3:7	240	98.0	6.94	10.43	34.3
0:10	60	22.5	6.76[a]	15.46[a]	0
0:10	120	62.4	6.76[a]	15.46[a]	0
0:10	180	92.4	6.76[a]	15.46[a]	0
0.10	240	99.0	6.76[a]	15.46[a]	0

[a]Theoretical Value.

Limiting viscosity number can also be controlled by adjusting mole ratios of the two monomers in the reaction vessel. Figure 8 shows zero shear, limiting viscosity number in 0.256 M sodium chloride solution plotted versus mole percent M_2 in the reaction which produced the polymer. Maximum $[\eta]$ is obtained at approximately 12 mole percent M_2. Number average molecular weight, $\overline{M}_n$, is also altered by changes in mole ratios of the monomers in the synthesis reaction. Figurc 9 shows that $\overline{M}_n$ drops to 69 percent of its 100 mole percent M_1 value when the reaction contains 100 mole percent M_2. Further data on characterizing these copolymers by size exclusion chromatography and calculating repeat unit sequence distributions are given in reference (14).

These data show that 2-propenamide and sodium 2,2-dimethyl-3-imino-4-oxohex-5-ene-1-sulfonate form a random copolymer which 1. gives a larger hydrodynamic radius in aqueous solution with increasing value of $\ln[M][I]^{-1/2}$, 2. has a maximum hydrodynamic radius in aqueous solution when the sulfonated monomer is approximately 12 mole percent of the monomer charge for synthesis, and 3. loses 31 percent of number average molecular weight as mole percent sulfonated monomer in the reaction changes from 0 to 100 mole percent.

The research group at the University of Southern Mississippi has also, with the aid of a Department of Energy grant, studied the synthesis and properties of poly((1-amidoethylene)-co-(1-methyl-1-(sodium

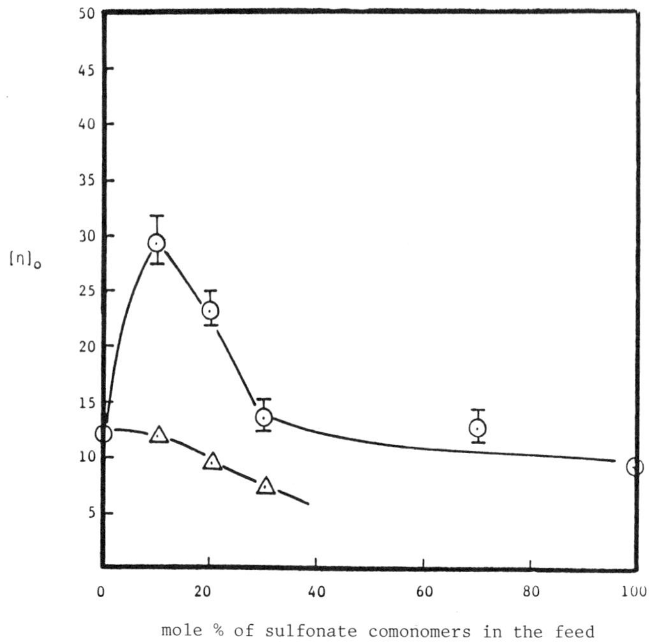

Figure 8. Effect of feed composition on zero-shear intrinsic viscosity in 1.5% NaCl of Poly(1-amidoethylene-co-(1-(sodium 2-methylprop-2N-yl-1-sulfonate)amidoethylene)) (Φ) and poly-((1-amidoethylene)-co-(1-methyl-1-(sodium 1-oxo-2-oxybutyl-4-sulfonate)) ethylene)) (Δ) copolymers at high conversion.

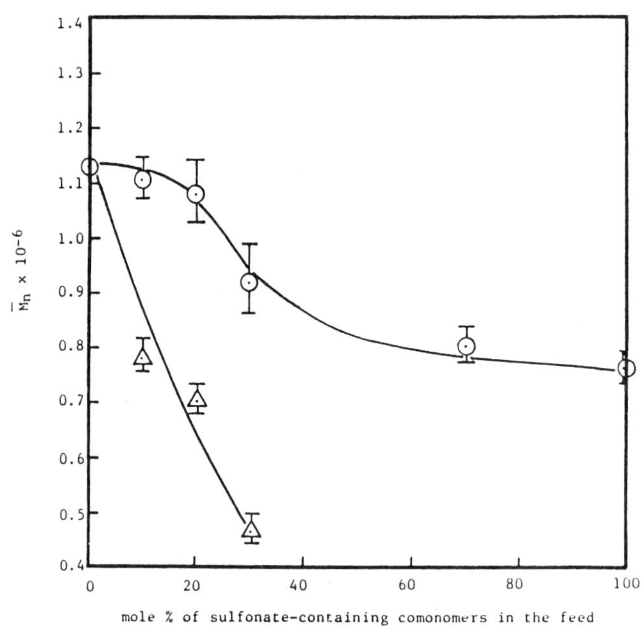

Figure 9. Effect of feed composition on $\bar{M}_n$ of Poly(1-amidoethylene-co-(1-(sodium 2-methylprop-2N-yl-1-sulfonate)amidoethylene)) (Φ) and poly((1-amidoethylene)-co-(1-methyl-1-(sodium-1-oxo-2-oxybutyl-4-sulfonate))ethylene)) (Δ) copolymers at high conversion.

1-oxo-2-oxybutyl-4-sulfonate))ethylene)) (14). These materials were synthesized as just described for the amide/(propylsulfonate-amide) copolymer with reaction pH at 7.4, temperature controlled at 30°C, and potassium persulfate as initiator. The reactivity ratios between 2-propenamide, M_1, and sodium 5-methyl-4-oxo-3-oxyhex-5-ene-1-sulfonate M_2,, are $r_1 = 0.38 \pm 0.05$ and $r_2 = 3.64 \pm 0.55$. No study was done on control of hydrodynamic radius of these copolymers by varying $\ln[M][I]^{-1/2}$ of the synthesis mixture. The effect of increasing mole percent M_2 in the reaction mixture on zero shear, limiting viscosity number in 0.256 M sodium chloride solution was examined and the results are shown in Figure 8. Similarly, the effect on $\overline{M}_n$ of increasing mole percent M_2 in the synthesis monomer charge was checked and the results are given in Figure 9. These data show that the copolymer is random with additions of the sulfonated monomer to the polymer steadily reducing $[\eta]_0$ and sharply reducing $\overline{M}_n$. In all applications where higher molecular weight polymers are needed, adding this sulfonated monomer to the reaction will reduce product performance.

GRAFT COPOLYMERS

As a technical data base was developed to aid in explaining the role of polymers in resource recovery, the many chemical and physical properties needed in a polymer for resource recovery were documented by reviewers (1). Goggarty was one of the first applications specialists to propose, in 1978, that graft copolymers be used for resource recovery (15). A number of such graft copolymers have now been made and tested and the knowledged gained about these products will be summarized below.

Most of the work done on graft copolymers has produced graft copolymers with a natural polymer as backbone and a synthetic polymer as sidechain. The impitus for making this type of polymer is that a sizeable fraction of the mass of the finished product is a cheap, renewable, natural product. If product performance can be enhanced while a major portion of the product is obtained from agricultural or forest waste, a very great cost savings is possible with natural backbone graft copolymers. Most research in this area has concentrated on starch, dextrin, and lignin with a small amount of work done on cellulose.

Starch or Polysaccharide Graft Copolymers

Several groups have conducted graft copolymerizations on starch or its purified components, amylose and amylopectin. The syntheses are based on attack of the anhydroglucose unit of starch by cerium (+4) ion. The free radical produced from this attack is then immersed in a monomer solution polymerizable by free radical, chain polymerization and a graft copolymer is formed. A typical synthesis procedure (10,16) is as follows.

Starch was dispersed thoroughly in two-thirds of the water used in the reaction. The mixture was heated with stirring to 100°C and held there for 3 min. This gelatinizes the starch and produced a clear, visually homogeneous sample that will cloud and begin to retrograde within an hour after it cools. The reaction must be initiated within 1 hr of boiling. The starch solution was bubbled with N_2 and stirred while it is allowed to cool to 25°C. The total amount of monomer to be added to the reaction mixture was calculated from the restrictions that 1) the final, complete reaction mixture should contain 1-2 m monomer with 1.5 m the preferred concentration, 2) with 100% grafting efficiency and 100% reaction, product will have a specified design molecular weight, and 3) the wt % side chain desired in the product should be maintained in the reaction mixture.

An appropriate amount of monomer(s) was dissolved in the remaining 1/3 of the water. Both solutions were bubbled with N_2 for 20 min. An appropriate volume of ceric ion solution was added to the starch solution and the flask is sealed with a septum stopper and stirred. The volume of cerium (+4) solution added was determined by the mole ratios of starch to Ce(+4) or monomer(s) to Ce(+4) desired in the reaction mixture. The monomer solution was bubbled with N_2 for 15 more min and added to the starch sample under a N_2 blanket. The reaction mixture was capped, stirred for 1 min, and placed in a 30°C bath. The mixture was then stirred every hour until it became too viscous to stir.

After 48 hours, the reaction was terminated by injection of 0.5 mL of aqueous, 1.0 wt % hydroquinone solution. The thick or gelatinous reaction mixture was scraped into a 1-L beaker using ceramic or plastic tools. The flask was washed with water, and the wash is placed in the beaker. This mixture was stirred until a thick, uniform solution has formed. Product was precipitated by dropwise addition of the product solution to five times its volume of vigorously stirred nonsolvent.

The precipitated copolymer was filtered from nonsolvent, slurried in a blender for 30 s in 4 times the reaction mixture volume of nonsolvent, filtered, and dried under vacuum to constant weight.

Poly(starch-g-(1-amidoethylene))

This graft copolymer is readily made by the above synthesis procedure. Yields of product are 90 wt % or above (16) and high-molecular-weight samples can be made by using molal concentrations of 1.5 or above to insure gelling of the sample during polymerization. Limiting viscosity numbers for a series of graft copolymers and graft-free poly(1-amidoethylene) are given in Figure 10. This comparison shows that the graft copolymers with few sidechains per backbone molecule outperform the homopolymer. However, the graft copolymers are plotted based on a "calculated" molecular weight which has been shown by subsequent investigations to be low. Therefore, the performance superiority demonstrated by the graft copolymer may not be as great as implied by the Figure. The copolymers are also good drag reducing agents for promoting pipeline flow (17). Data demonstrating drag reduction by parts-per-million concentrations of these copolymers is given in Table X. Some unpublished work has been performed in industrial laboratories to determine if these copolymers are effective beneficiating agents for bauxite ore.

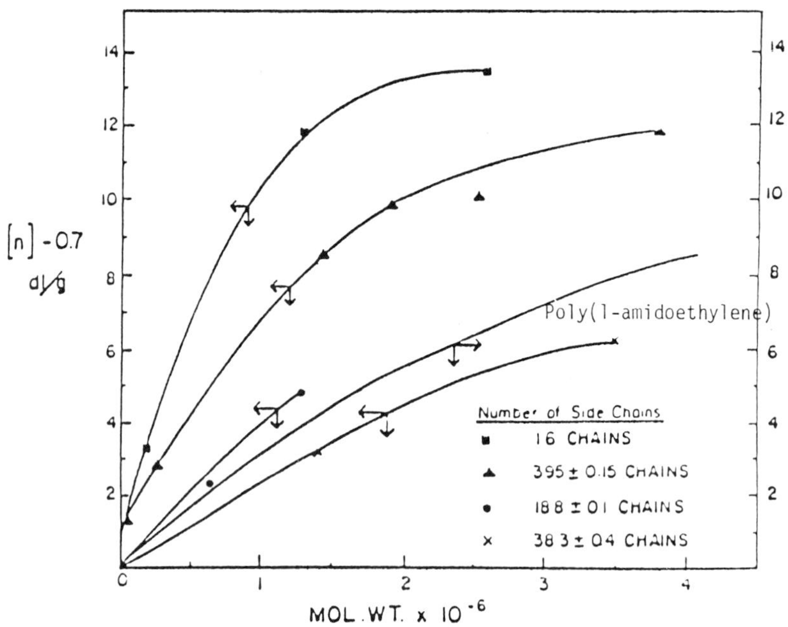

Figure 10. Limiting viscosity number versus calculated molecular weight for samples of poly(starch-g-1-amidoethylene) with different numbers of sidechains in the molecule.

TABLE X
Reduction of Drag in the Flow of Water Caused by the Addition of Starch-1-amidoethylene Copolymer.

Example Number	Fluid	Graft Copolymer Concentration (wppm)	Passage Time[1] (seconds)	Percent of Drag Reduction (%)
33	Copolymer Solution	3.72	29.3 ± 0.49	17.75
34	Copolymer Solution	6.10	33.48 ± 3.35	6.02

[1]Passage times given are the means of 4 determinations.

Poly(starch-g-((1-amidoethylene)-co-(sodium 1-carboxylatoethylene)))

The graft copolymer just discussed can be hydrolyzed to a polyanionic, graft terpolymer by solution in nitrogen-saturated, aqueous sodium hydroxide. Limiting viscosity number data in Table XI shows that, in the absence of oxygen, this hydrolysis process sharply increases the size of the polymer in solution without degrading the molecule (18). These graft copolymers possess all the same benefits and blemishes as simple carboxylic-acid containing polymers when applied to resource recovery. The copolymers will 1. adsorb less on negatively charged surfaces, 2. adsorb more on positively charged surfaces, 3. produce higher viscosity at a given concentration in aqueous solution, 4. lose solution viscosity when electrolyte concentration increases, 5. precipitate in the presence of di- or tri-valent cations, and 6. gel in the presence of low solubility, di- or tri-valent cations, more than the nonionic precursor.

TABLE XI
Limiting Viscosity Number of Poly(starch-g-(1-amidoethylene)) Copolymers with One Graft per Starch Molecule.

Number[a]	Unhydrolyzed (Dist. H_2O)	Hydrolyzed (1 M $NaNO_3$)
3	3.99	5.0
6	5.84	6.08
9	12.2	*3.69

*O_2 present during hydrolysis.

[a]Synthesis described in reference 36.

Poly(starch-g-((1-amidoethylene)-co-(sodium 1-(2-methylprop-2N-yl-1-sulfonate)amidoethylene)))

Strongly anionic, highly water-soluble, graft copolymers of starch can be made by adding 2-propenamide and sodium 2,2-dimethyl-3-imino-4-oxohex-5-ene-1-sulfonate (Na DMIH) to the polymerization reaction. The resulting graft terpolymers are water-soluble, thickening agents with a solution rheology controlled by the ratio of unsulfonated to sulfonated repeat units in the synthetic chains of the molecule. Aqueous solutions of these products are shear thinning and have power law exponents which decrease 1) with increasing product concentration or 2) as the unsulfonated to sulfonated repeat unit ratio moves toward 3/1. Limiting viscosity number of product in water decreases with increasing shear rate of measurement or with increasing

TABLE XII

Analysis of Terpolymer Reaction Products.

Sample	Elemental Analysis (wt. %)				Weight Percent Composition of Reaction Product				Mole Fraction Composition			Limiting Viscosity Number, Water (30°C)
Number	C	H	N	S	Starch	WP_N	WP_S	$R_{N/S}^{(a)}$	MF_N	MF_S	$MF_\tau x10^4$	$\dot{\gamma}=225\ s^{-1}$ (dL/g)
1	37.11	6.12	3.72	8.04	22.4	1.05	57.4	0.059	0.056	0.943	6.7	26.6
2	39.77	6.07	3.48	4.40	51.8	7.90	31.4	0.810	0.447	0.551	16.5	58.9
3	40.59	6.18	6.53	3.44	33.1	25.5	24.5	3.34	0.769	0.230	5.6	95.8
4	42.13	6.41	7.16	1.81	36.4	32.3	12.9	8.05	0.927	0.072	5.6	74.9
5	35.97	6.23	4.45	8.99	14.2	2.65	64.2	0.133	0.117	0.882	3.5	95.8
6	37.64	5.99	7.12	7.35	11.6	19.81	52.5	1.22	0.549	0.451	1.8	114
7	40.66	6.22	9.62	4.86	18.9	38.0	34.7	3.52	0.779	0.221	2.2	230
8	41.92	6.66	10.62	2.57	25.3	48.1	18.4	8.46	0.894	0.106	2.6	201
9	35.72	5.91	4.77	9.41	10.2	3.34	67.2	0.160	0.138	0.862	2.4	69
10	38.56	6.07	7.74	7.61	9.2	22.4	54.4	1.33	0.570	0.429	1.3	157
11	40.62	6.23	10.41	5.08	11.7	41.5	36.3	3.69	0.787	0.213	1.2	282
12	42.49	6.63	11.11	2.53	12.5	50.7	18.1	9.05	0.900	0.099	1.3	225

a. Repeat unit mole ratio is that of the reaction product.

WP_N = 1-amidoethylene.
WP_S = 1-sodium 2-methylprop-2N-yl-1-sulfonate)amidoethylene.
MF_N,MF_S = mole fractions of WP_N and WP_S in sample.

salt concentration in the solvent. Limiting viscosity number for products with the same design molecular weight reaches a maximum at an unsulfonated to sulfonated repeat unit ratio of 3/1 (10), as shown by Table XII. These products are biodegradable and are degraded in both acidic or basic solution.

TABLE XIII
Reaction Parameters for the Preparation of Poly(Dextran-g-1-amidoethylene).

Rxn. time (hr)	2-propenamide conc. (mmol)	Dextran* conc. (mmol)	Ce(IV) conc. (mmol)	Rxn. temp. (4°C)	HNO_3 conc. (mmol)	Conc. (%)
0.5	115	7.7	0.050	25	0.50	69.8
1.0	115	7.7	0.050	25	0.50	76.4
1.5	115	7.7	0.050	25	0.50	78.8
2.0	115	7.7	0.050	25	0.50	86.7
2.5	115	7.7	0.050	25	0.50	88.3
3.0	115	7.7	0.050	25	0.50	93.5
3.5	115	7.7	0.050	25	0.50	97.4
3.0	230	7.7	0.050	25	0.50	49.2
3.0	173	7.7	0.050	25	0.50	68.5
3.0	115	7.7	0.050	25	0.50	93.5
3.0	86	7.7	0.050	25	0.50	72.7
3.0	58	7.7	0.050	25	0.50	45.9
3.0	29	7.7	0.050	25	0.50	33.0
3.0	115	23.1	0.050	25	0.50	97.2
3.0	115	15.4	0.050	25	0.50	95.4
3.0	115	11.6	0.050	25	0.50	93.8
3.0	115	7.7	0.050	25	0.50	93.5

*Dextran ($\bar{M}_w$ = 340,000) concentration is based on millimole of anhydroglucose repeating units.

Poly(dextran-g-(1-amidoethylene))

Dextran is an extracellular polysaccharide synthesized by certain bacteria when fed on sucrose (19). The polymer can be extracted from bacterial cultures and purified for use in technical and biomedical applications. It has also been grafted using the cerium ion reaction method to produce a water-soluble graft copolymer (20). Synthesis and product data for a series of graft copolymers are given in Table XIII. Characterization data for some of these polymers are given in Table XIV. These results are very useful in establishing synthesis-property correlations for graft copolymers and have confirmed trends previously seen in other copolymers (16). The compounds will probably only be applied in situations where the purity and biocompatability of the backbone are important performance characteristics which justify the high cost of the backbone.

TABLE XIV

Effects of Variation Ce(IV) Concentration on the Resulting Poly(Dextran-g-1-amidoethylene) Structures.

(mmol)	% Nitrogen	$\overline{M}_w$ of PAM Graft	$\overline{M}_w$ of Graft Polymer	Number of Graft Sites	[η] at 30°C
0.005	14.03	463,000	1,580,000	2.29	3.21
0.010	14.72	468,000	1,610,000	2.19	3.64
0.025	14.72	476,000	1,620,000	2.31	4.31
0.035	14.98	482,000	1,630,000	2.31	4.72
0.050	15.08	601,000	1,670,000	1.91	4.86
0.065	15.13	630,000	1,690,000	1.85	5.12
0.075	15.15	487,000	1,650,000	2.37	4.57
0.100	15.18	471,000	1,630,000	2.48	4.08
0.500	15.68	340,000	1,520,000	3.24	2.28

Poly(2,3-dihydroxypropylcellulose-g-(1-amidoethylene))

One of the major barriers to forming effective graft copolymers of cellulose is the difficulty of overcoming the strong hydrogen bonds of cellulose which inhibit solution of the polymer. Recent studies have shown that cellulose can be made water soluble by treatment of sodium cellulose in 1,4-dioxacylohexane with 3-chloro-1,2-propanediol at 60°C (21). This cellulose derivative is then grafted with 2-propenamide using the cerium (+4) synthesis procedure previously described. Data on these copolymers are limited, but the data available to date show that molecules with limiting viscosity numbers between 2.6 and 6.3 can be formed.

Lignin Graft Copolymers

Lignin [8068-00-6] is a natural product produced by all woody plants. It is second only to cellulose in mass of polymer formed per annum (22). Lignin constitutes between 15 and 40 percent of the dry weight of wood with variation in lignin content being caused by growing conditions, species type, the parts of the plant tested, and numerous other factors (23). Plants use lignin to 1. control fluid flow, 2. add strength, and 3. protect against attack by microorganisms (24). Each cell of the plant grows its own lignin.

Recent work by Atalla (25) supports the idea that lignin is at least a semi-ordered substance in wood with the plane of the aromatic ring parallel to the cell wall surface. Woody plants synthesize lignin from trans-confieryl alcohol 1 (pines), Trans-sinapyl alcohol 2 (deciduous), and Trans-4-coumaryl alcohol 3 by free radical crosslinking initiated by enzymatic dehydrogenation (26). Structures of these alcohols are given in Figure 11.

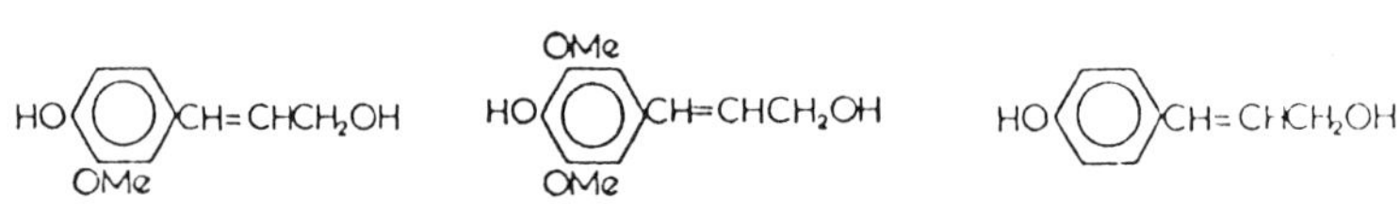

1 2 3

Figure 11.

Methods for recovering lignin are the alkali process, the sulfite process, ball milling, enzymatic release, hydrochloric acid digestion, and organic solvent extraction. Alkali lignins are produced by the kraft and soda methods for wood pulping. They have low sulfur content (1.6 wt. %), have sulfur contamination present as thioether linkages, and are water-insoluble, nonionic polymers of low (2,000 to 15,000) molecular weight. Approximately 20 million tons of kraft lignin are produced in the United States each year. This enormous production of cheap biomass has induced a significant effort, stretching over 40 years, to alter lignin into industrial or commercial products. A series of lignin graft (27,28) copolymers have been made which function effectively as drilling mud additives, flocculating agents, and thickening agents. The synthesis procedure for these copolymers is as follows (29,30).

The grafting reaction is run by adding lignin and calcium chloride to a nitrogen-flushed 125-mL flask containing 20 mL of solvent. The mixture is bubbled with nitrogen for 3 min, a hydroperoxide is added, and the flask contents are bubbled with nitrogen for an additional 2 min. Flask contents are stirred for 20 min and bubbled with nitrogen for 4 min. The nitrogen-saturated monomer or monomer solution is then added, flask contents are bubbled with nitrogen for 10 min while being stirred, 0.15 mL of 0.05 M cerium(IV) sulfate is added, the sample is stirred and bubbled for an additional 10 min, and it is then capped with a septum stopper.

The reaction flask is placed in a 30°C bath and allowed to sit for 2 days. The reaction is then terminated with 0.5 mL of 1% by weight hydroquinone in water. The reaction mixture is diluted with 100 mL of water and stirred until uniform. Reaction product is precipitated by adding the dilute reaction mixture dropwise to 1 L of 2-Propanone.

The properties of these graft copolymers which meet needs in resource recovery will now be presented.

Poly(lignin-g-(1-amidoethylene))

Data for a series of graft copolymers made using 2-propenamide are given in Table XV for copolymers synthesized in dimethylsulfoxide. These data show that maximum yield is obtained when chloride ion to lignin* mole ratio is 492. Data for a series of compounds made in different solvents is given in Table XVI. These data show that the reaction can be run in a series of solvents with the highest product limiting viscosity number obtained from a sample prepared in a 50:50 by volume mixture of dimethylsulfoxide and water. These nonionic molecules are small in size, readily adsorbed on silica surfaces, and prone to complex di- and tri-valent metal ions from aqueous solution (31,32).

Poly(lignin-g-((1-amidoethylene)-co-(sodium-1-carboxylatoethylene)))

The copolymer just described can be converted to an anionic polyelectrolyte by hydrolysis of the amide units with strong base. Hydrolyzed samples are prepared from aqueous solutions containing between 1.0 and 2.5 g/dL of reaction product. A stoichiometric amount of sodium hydroxide that will produce the desired percent hydrolysis is added to the solution, the solution is heated to 50°C with stirring, and is maintained there for 1 hr. Product is recovered by precipitation and drying or dialysis and freeze drying (28). Data from 2 samples of copolymer hydrolyzed to five different degrees of hydrolysis data are given in Table XVII and separation/hydrolysis are given for a purified graft polymer in Table XVIII.

*A number average molecular weight of 9,600 is used in calculating moles of lignin. Proof of copolymerization is obtained by real time spectra acquisition of absorbance for size exclusion chromatography effluent (27).

TABLE XV
Lignin-Co-(1-amidoethylene) samples. DMSO Data[a]

Sample Number	Reaction Contents: 2-propen-amide (g)	Lignin (g)	$CaCl_2$ (g)	RO_2H[b] (g)	Ce(+IV) (ml, .05N)	Yield[e] g	Yield[e] wt.%	[n] (dL/g)	Product Composition (wt.%) N	1-amido[f]	Lignin	Ca
1	3.20	0.50	0.50	0.40/.3532	0.15	4.71/ 2.799	75.68	0.322	15.21	76.73	6.50	2.789
2	3.20	0.50	0.10	0.40/ 3532	0.15	3.30*/ 2.565	69.32	0.56	13.34	67.38	5.38	0.73
3	3.20	0.50	0.0503	0.40/.3532	0.15	2.72*/ 2.14	57.83	0.69	13.45	67.91	4.94	0.415
4	3.20	0.50	0.0102	0.40/.3532	0.15	3.16/ 2.58	70.07	0.77	15.047	76.045	4.601	0.137
5	3.20	0.50	0.50	0.25/.2208	0.15	4.31/ 3.306	89.36	0.35	13.284	66.946	7.00	2.89
6	3.20	0.50	0.50	0.152/.1342	0.15	4.40/ 3.34	90.35	0.44	13.034	65.79	6.03	2.86
7@	3.20	0.50	0.50	0.80/ .7064	0.15	4.73/ 3.283	88.73	0.306	12.37	62.31	7.21	2.48
@	3.20	0.50	0.50	0.416	0.15							
@	3.20	0.50	0.50	0.416	0.15							
@	3.20	0.50	0.50	0.416	0.15							
8	3.20	0.50	0.1	0.15/.1325	0.15	4.07/ 3.307	89.37	0.615	13.55	68.41	5.48	0.757
9	3.20	0.50	0.0515	0.15/.1325	0.15	3.90/ 3.128	84.53	0.666	13.641	68.89	4.94	0.395
10	3.20	0.50	0.010	0.15/ .1325	0.15	3.46/ 2.863	77.38	0.801	13.734	69.42	4.16	0.0949
11	2.00	0.50	0.50	0.15/ .1325	0.15	3.10/ 2.119	84.77	0.372	11.74	59.12	6.88	3.54
12	2.00	0.50	0.10	0.15/ .1325	0.15	3.04/ 2.091	83.64	0.395	11.67	58.81	6.26	1.007
13	2.00	0.50	0.0516	0.15/ .1325	0.15	2.68/ 2.109	84.36	0.478	13.197	66.43	8.13	0.558
14	2.00	0.50	0.0107	0.15/.1325	0.15	1.84/ 1.44	57.58	0.565	14.162	71.43	6.54	0.33
15	1.00	0.50	0.50	0.15/.1325	0.15	2.14/ 1.105	73.64	0.192	7.950	39.53	12.24	5.45
16	1.00	0.50	0.10	0.15/.1325	0.15	1.12/ 1.063	70.87	0.288	11.515	57.52	13.74	1.65
17	1.00	0.50	0.0512	0.15/.1325	0.15	1.49/.857	57.14	0.275	10.63	52.95	14.95	0.88
18	1.00	0.50	0.0113	0.15/.1325	0.15	1.25/.919	61.29	0.276	11.516	57.41	15.52	0.325
19	3.20	0.50	0.5	0.4/.3532	0.15	3.27/	88.38	0.523	15.802	79.43	11.41	0.2
20	3.20	0.50	0.1	0.15	0.15	4.08/ 2.896	78.26	0.56	14.062	70.98	5.62	0.679

a. All reactions run in 20.0 mL of dimethylsulfoxide save for the 4 marked samples (@) run in 1,4-dioxacyclohexane. b. The results given as #1/#2 are #1 = g of crude 2-hydroperoxy-1,4-dioxacyclohexane added to the reaction and #2 = g of pure peroxide added to the reaction.

e. The yields listed, #1/#2, are: #1 = crude product recovered and #2 = pure product recovered. Weight percent yield is based on pure product recovered.

f. 1-amido = 1-amidoethylene repeat units in the polymer. * = Some product lost during recovery.

TABLE XVI

Results of Syntheses[a] Run in Different Solvents

sample	solvent	yield, g	reaction product composition, wt%: lignin	1-amido-ethylene	Ca	Cl	limiting viscosity number dL/g	precipitation test: wt of reaction prod used g	wt of solids recd, g	wt % insol in water	phases in reacted mixture
1	1-methyl-2-pyrrolidinone	4.4	4.74	75.35	3.11	2.85	14.2	0.20075	0.002	1.0	precipitate
2	dimethyl sulfoxide	4.71	6.5	76.73	2.79	2.07	32.2	0.20065	0.001	0.5	gelled soln
3	dimethylacetamide	4.13	5.61	57.43	2.77	1.85	27.6	0.20065	0.0018	0.9	precipitate
4	dimethylformamide	4.25	7.18	73.75	2.82	1.90	32	0.20015	0.0023	1.1	precipitate
5	50:50 (v/v) dimethyl sulfoxide-dioxane	4.19	5.13	72.14			53	0.2005	0.00175	0.9	precipitate
6	50:50 (v/v) dimethyl sulfoxide-water	4.28	6.27	71.52			66	0.2005	0.00585	2.9	two fluid phase sample, gelled bottom layer
7	pyridine	4.00	5.42	71.67			47	0.2001	0.001	0.5	precipitate

[a]Each reaction contained 0.50 g of lignin, 3.2 g of 2-propenamide, 0.4 g of 2-hydroperoxy-1,4-dioxacyclohexane, 0.15 mL of ceric sulfate solution, and 20.0 mL of solvent.

TABLE XVII
Elemental Assay, Percent Hydrolysis, and Limiting Viscosity Number for Partially Hydrolyzed Copolymer Used in Mud Tests.

sample	elemental assay, wt % C	H	N	S	O	Na	degree of hydrolysis from assay[a]	limiting viscosity number (dL/g)
				Batch 1				
a	50.19	7.30	16.19	0.24	29.02	0.095	0	0.32
b	43.73	5.97	10.45	0.14	34.13	7.40	24.8	15.2
c	41.52	5.03	6.37	0.18	38.74	12.49	51.4	15.1
				Batch 2				
e	48.49	7.51	12.76	0.43	28.40	0.02	0	0.13
f	45.69	6.07	8.63	0.57	33.04	7.85	26.6	1.31
g	43.17	5.60	7.14	0.67	33.33	9.72	38.0	4.09
h	46.05	5.94	5.69	0.28	31.47	10.13	50.4	3.84

a) Based on nitrogen loss from samples a or e.

TABLE XVIII
Separated Products from Lignin-(2-propenamide) Reactions.

	Crude Reaction Product
weight, g	18.0
content, wt %	
lignin	7.27
1-amidoethylene units	73.8
[η], water, 30°C	0.51
After Separation; 3.h Graft Copolymer Fraction	
weight, g	9.09
wt % of original sample	50.5
content, wt%	
lignin	9.52
1-amidoethylene units	36.43
degree of hydrolysis	49.4
[η], in water, 30°C	6.77
3.j Homopolymer Fraction	
weight, g	3.20
wt % of original sample	17.8
content, wt %	
lignin	1.53
1-amidoethylene units	36.94
degree of hydrolysis	50.1
[η], in 0.50 M NaOH, 30°C	0.83

Hydrolysis increases the terpolymer limiting viscosity number in water by a factor of up to 45. The tendency of the terpolymer to act as a suspending agent is shown by the test data for terpolymer-containing drilling muds given in Table XIX. The API filtrate volume data shows that the terpolymer produces a negatively-charged clay suspension in aqueous solution which occuludes any porous surface. Other previously published data (28,29,31) show that these terpolymers are nonnewtonian viscosifiers, metal-ion complexing agents, and effective flocculators. These materials are still "small" molecules in aqueous solution, however, and do not function as effectively when used 1. as nonnewtonian viscosifiers or 2. drag reducing agents as do poly(1-amidoethylene-co-(sodium 1-carboxylatoethylene)) copolymers.

TABLE XIX

Properties of Test Muds Before and After Hot-Rolling

	base mud		graft copolymer fraction		homopolymer fraction		chrome lignosulfonate	
	before	after	before	after	before	after	before	after
viscosity in cP at a shear rate of								
1020 s^{-1}	52	69	74	42	74	43	29	50
510 s^{-1}	36	47	49	24	49	24	16	33
340 s^{-1}	30	39	40	17	40	17	12	26
170 s^{-1}	21	28	27	10	27	10	7	18
gel strength in lb/100 ft after mud has set for								
10 s	6	12	5	3	5	2	2	11
10 min	25	35	20	3	20	3	9	24
apparent viscosity, cP	26	35	37	21	37	22	15	25
plastic viscosity, cP	16	22	25	18	25	19	13	17
yield point, lb/100 ft	20	25	24	6	24	5	3	16
API filtrate, volume, mL	12.0	13.8	7.8	8.0	7.8	8.4	11.6	14
pH	9.1	8.6	9.0	8.0	9.0	8.0	9.5	8.2
high-pressure, high-temperature filtrate, mL		62.8		52.4		46.0		64.0

Table XVIII, 3.h. Table XVIII, 3.j.

TABLE XX

Synthesis Data and Physical Characteristics of Graft Terpolymer

	Reactants												Repeat Units (wt %)	
Sample Number	2-propen amide (g)	A (g)	Ce (4+) solution (mL)	Dimethyl sulfoxide (mL)	CaCl2 (g)	B (g)	Yield (wt %)	[η] (dL/g)	Assays (wt. %) C	H	N	S	1-amido ethylene	D
1	1.60	4.66	0.15	20	0.50	0.15	70.12	10.52	35.41	5.94	7.03	9.78	14.1	62.6
2	1.60	4.66	0,15	50	0.50	0.15	86.98	11.40	36.77	6.17	8.39	9.54	21.51	61.1
3	1.60	4.66	0.15	50	0.50	0.25	78.40	7.40	36.88	6.20	8.49	9.19	22.79	58.8
4	1.60	4.66	0.15	40	0.50	0.40	69.82	9.30	37.74	6.39	8.47	9.23	22.60	59.1
5	1.60	5.16	0.15	30	0.50	0.15	78.79	12.59	39.85	6.79	9.39	8.61	28.63	55.1
6	1.60	5.16	0.15	30	0.50	0.15	77.27	6.81	38.03	6.40	8.73	9.95	22.33	63.7
7	1.60	4.66	0.15	30	0.50	0.15	87.28	10.46	36.51	6.33	7.74	10.29	16.56	65.9
8	2.58	1.87	0.15	30	0.50	0.39	67.89	.953	43.31	6.65	10.92	5.65	42.89	40.4
9	2.56	1.86	0.15	30	0.53	0.39	79.49	2.46	42.47	6.29	10.76	5.66	42.1	36.2
10	21.98	15.99	1.28	219	4.35	3.35	91.02	1.97	42.25	6.49	11.34	6.21	43.78	44.4
11	2.56	1.86	0.15	20	0.50	0.36	80.69	-	44.38	6.83	10.84	4.79	44.39	34.2
12	2.56	1.86	0.15	20	0.50	0.343	15.64	-	44.73	6.46	9.62	4.70	38.40	33.6
13	2.57	1.87	0.15	20	0.50	0.34	69.89	-	43.88	6.58	10.82	4.32	45.33	30.9
14	2.56	1.87	0.15	20	0.50	0.34	66.79	-	44.90	6.82	10.73	4.69	44.05	33.5
15	2.57	1.76	0.20	15	0.50	0.34	18.50	-	42.80	6.74	11.41	4.75	47.37	34.0
16	2.57	1.86	0.15	20	0.50	0.34	45.06	-						
												$FeCl_2 \cdot 4H_2O$ g	moles x 10^5	
17	2.57	1.86	0.14	30	0.50	0.34	70.28		42.72	6.79	10.60	0	0	
18	2.56	1.87	0.15	20	0.50	0.335	51.63#		42.55	7.40	11.32	1.5×10^{-3}	.754	
19	2.56	1.86	0.15	20	0.501	0.338	72.72		44.36	6.89	9.91	1.87×10^{-2}	9.41	
20	2.56	1.87	0.15	20	0.51	0.34	18.26		42.26	6.50	9.25	.184	9.26	
21	2.56	1.87	0.15	20	0.50	0.34	68.29		41.77	6.67	10.78	0	0	
22	2.57	1.86	0.26	20	0.50	0.34	71.27		42.84	6.50	10.76	1.55×10^{-3}	.780	

= some product lost during purification.

a) All reactions save #10 contained 0.50 g of lignin. Reaction #10 contained 4.39 g of lignin.

A) = 2,2-dimethyl-3-imino-4-oxohex-5-ene-1-sulfonic acid.

B) = Hydroperoxide. Samples 1 to 7: Values are weight of 1,4-dioxa-2-hydroperoxycycloexane in g. Samples 8 to 2: Values are amount of aqueous solution of 1,2-dioxy-3,3-dimethylbutane in mL. Equivalent/mL = 7.23x10 .

D) = N-substituted 1-amidoethylene.

Poly(lignin-g-((1-amidoethylene)-co-(sodium 1-(2-methylprop-2N-yl-1-sulfonate)amidoethylene)))

A strongly anionic, polyelectrolyte can be made from lignin by conducting a graft polymerization in the presence of 2-propenamide and 2,2-dimethyl-3-imino-4-oxohex-5-ene-1-sulfonic acid or its salts.

A series of reactions were run to determine 1. the dependence of yield on reactant concentrations, 2. a preferred duration for the reaction, and 3. the effect of iron contamination on reaction yield. Data from 23 reactions is given in Table XX. All reactions, with the exception of number 10, contained 0.50 g of Kraft, pine lignin.

Reactions 1 to 7 were run with different mole ratios of lignin to hydroperoxide and chloride ion to hydroperoxide. A maximum yield of 87 wt. % of polymerizable mass in the reaction is obtained when the lignin to hydroperoxide mole ratio is 4.17×10^{-2} and the chloride ion to hydroperoxide mole ratio is 7.21. These ratios occur in sample 2. Previous studies (30) have shown that the mole ratios between lignin, chloride ion, and hydroperoxide control yield in the graft copolymerization of poly(lignin-g-(1-amidoethlene)). These same ratios control yield in the formation of poly(lignin-g-(1-(2-methylprop-2N-yl-1-sulfonic acid)amidoethylene)-co-(1-amidoethylene))). These results imply that the grafting mechanism involves lignin, chloride ion, and hydroperoxide. An initiation reaction which is compatable with all findings to date is the attack on lignin by a hydroperoxide-(chloride ion) complex to create a site for free radical propagation. Such complexes have been seen in ESR studies of hydroperoxides (34).

Samples 8 and 9 of Table XX show that this grafting reaction can also be run with an alternative hydroperoxide, 1,2-dioxy-3,3-dimethylbutane. This commercially available hydroperoxide can be used in place of 1,4-dioxacyclohexane-2-hydroperoxide. Sample 10 shows that amounts of copolymer as large as 40 g can be made in single pot reactions.

Reactions 11 to 16 of Table XX were identical composition tests run for different amounts of time. Sample 11 was terminated after 31 min., sample 12 after 1 hr, samples 15 and 16 after 3 hr., sample 14 after 24 hr., and sample 13 after 48 hr. These data were gathered to determine the minimum duration of the reaction. The results show that high yields (samples 11 and 16) can be obtained in reaction times as short as 30 minutes. Several samples (12 and 15) show low yields after reaction times as long as 3 hr., but these results were obtained from a contaminated reaction and a reaction containing less than the appropriate amount of solvent, respectively. Previously reactions run in 1,4-dioxacyclohexane give high yields in short reaction times (35). Since solvent does not change the rate of reaction, these data support a free radical polymerization mechanism for ethene monomers adding to lignin (27,30). Free radical reactions have rates which are insensititive to change of solvent.

Samples 17 to 22 were identical, 48 hour reactions run with different amounts of iron(2+) in the reaction mixture. Samples 17 and 21 contain only the iron added as a 104 ppm contaminant of lignin. The reaction mixture, in these cases, contains 9.28×10^{-7} moles of Fe(2+). The mole ratio between hydryperoxide and iron is 1,340. Since iron is not reduced in the presence of hydroperoxide, if the hydroperoxide-Fe(2+) were acting as Fenton's initiator for free radical polymerization, only 2% of the lignin in the reaction mixture could be grafted and lignin content of the reaction product would be 0.2 wt. %. Reaction products usually contain between 6 and 10 wt. % lignin, a lignin concentration which could not occur unless extremely large amounts of chain transfer took place. Since lignin forms less reactive, quinoid-type structure when proton abstracted, high propagation rates could not be achieved by chain transfer mechanism mechanisms. To confirm that Fe(2+) was not a reagent active in the initiation of this reaction, several more reactions were run with larger concentrations of Fe(2+). Product was lost during the recovery steps for reaction 18, but reaction 22 was completely successful.

Reaction 22 has an $RO_2H/Fe(2+)$ mole ratio of 160 but has about the same yield as an uncontaminated reaction, #17 and 21. Reaction 19 has a $RO_2H/Fe(2+)$ mole ratio of 13.3 but again shows no change in yield from that of an uncontaminated reaction. Sample 20 has a 1.35 mole ratio of RO_2H to Fe(2+) and shows a sharp decrease in reaction yield and grafting. Here, the approximately 1 to 1 mole of peroxide to iron should produce a high concentration of hydroxide radicals and extensive polymerization if these radicals are part of the polymerization process. Instead of high yield, however, the reaction yield is less than one third of that obtained in the absence of iron. Therefore, a Fenton's initiation mechanism for this reaction is inconsistent with the data and probably does not occur.

The limiting viscosity number values of these graft copolymers show that the molecules are sharply expanded by the addition of an ionic monomer to the chain. Reactions (28) run with the same number of moles of nonionic monomer (0.045) and producing apprroximately the same yield of product gave limiting viscosity numbers of 0.50 dL/g. In reactions 1 to 7 where 50 mole % of the monomer is now sulfonated, ionic monomer, the limiting viscosity number is 12 to 24 times higher. For products 8 to 10, the reaction mixture contained 20 mole % sulfonated monomer. Product limiting viscosity numbers for samples 8 to 10 are 2 to 5 times higher than those of the nonionic copolymer, poly(lignin-g-(1-amidoethylene)) (28). These anionic graft copolymers are 5 to 24 times more effective thickening agents than the nonionic lignin graft copolymers and are active flocculating/deflocculating agents.

SUMMATION

The performance requirements which must be met by polymers used in resource recovery are job-specific and complex. These requirements have been partially met by creating complex polymers with physical properties which can provide the demanded performance.

While performance of resource recovery processes have been significantly improved in the past two decades, further efforts must be directed to 1. minimize shear degradation, 2. control adsorption, 3. control process rheology, and 4. provide appropriate chemical stability. These are the properties which fundamental and applied research must target during the next decade.

References

1. H.L. Chang, Polymer Flooding Technology - Yesterday, Today, and Tomorrow, Proceedings of the Soc. Petrol. Eng. Symposium on Improved Oil Recovery, 4/16-19/78, Tulsa, OK. SPE Paper #7043.

2. F. Bueche, Mechanical Degradation of High Polymers, J. Appl. Polym. Sci., 4, (#10), 101-106, (1960).

3. F.W. Billmeyer, Jr., "Textbook of Polymer Science", 3rd Ed., p.72, John Wiley and Sons, N.Y., (1984), ISBN 0-471-03196-8.

4. D.J. Pye, J. Petrol. Tech, 16, (#8), 911, (1964).

5. B.B. Sandiford, J. Petrol. Tech., 16, (#8), 917, (1964).

6. B.A. Toms, Proc. Internat. Rheological Congr., Holland, (1948), Vol. II, pp 135-141, North-Holland Pub. Co., Amsterdam, (1949).

7. Keith R. McKennon, Secondary Recovery of Petroleum, U.S. 3,039,529, June 19, 1962.

8. John J. Meister, Huey Pledger, Jr., Thieo E. Hogen-Esch, George B. Butler, Retention of Polyacrylamide by Berea Sandstone, Baker Dolomite, and Sodium Kaolinite During Polymer Flooding, Proceedings of the Symposium on Oilfield and Geothermal Chemistry, Soc. Petrol. Eng., SPE Paper #8981, Stanford, CA 5/28-30/80.

9. F. David Martin, L. Guy Donaruma, Melvin J. Hatch, "Development of Improved Mobility Control Agents for Surfactant/Polymer Flooding", U.S. Government Department of Energy Report, DOE/BC/00047-19.

10. John J. Meister, Damodar R. Patil, Margaret C. Jewell, Kyle Krohn, Synthesis, Characterization, and Properties in Aqueous Solution of Poly(starch-g-(1-amidoethylene)-co-(Sodium 1-(2-methylprop-2N-yl-1-sulfonate)amidoethylene))), Accepted by the J. Appl. Polym. Sci., Publication expected 1/87.

11. C.L. McCormick, R.D. Hester, H.H. Neidlinger, G.C. Wildman, U.S. Dept. of Energy Report DOE/BETC/5603-10, (June, 1980).

12. "The Chemistry of Acrylamide", Published by The American Cyanamide Company, Wayne, N.J. (1969).

13. Miklos T. Szabo, J. Petrol. Tech., 31, 553-560, (1979).

14. C.L. McCormick, G.S. Chen, J. Polym. Sci., Poly. Chem. Ed., 20, 817-838, (1982).

15. W. Barney Gogarty, Micellar/Polymer Flooding - An Overview, Proceedings of The Soc. Petrol. Eng. Symposium on Improved Oil Recovery, SPE Paper #7041, Tulsa, OK, 4/16-19/78.

16. H. Pledger, Jr., John J. Meister, T.E. Hogen-Esch, G.B. Butler, Proceedings of the 54th Annual Fall Tech. Confer., Soc. Petrol. Eng., SPE Paper #8422, Las Vegas, NV, 9/23-26/79.

17. G.B. Butler, T.E. Hogen-Esch, J.J. Meister, H. Pledger, Jr., U.S. 4,400,496, 8/23/1983.

18. J.J. Meister, G. Merriman, K. Anderle, Unpublished results.

19. E.A. MacGregor, C.T. Greenwood, Polymers in Nature, p.288, J. Wiley and Sons, Chickester, Eng., (1980) ISBN 0-471-27762-2.

20. C.L. McCormick, R.D. Hester, H.H. Neidlinger, G.C. Wildman, U.S. Energy Dept. Report DOE/BETC/10321-5.

21. D.R. Patil, H. Pledger, Jr., G.B. Butler, T.E. Hogen-Esch, Proceed. Am. Chem. Soc. Div. Polym. Mater., 55, 376-379, (1986).

22. H.I. Bolker, "Natural and Synthetic Polymers, An Introduction", p.580, Marcel Dekker, New York, (1974) ISBN 0-8247-1060-6.

23. Eero Sjostrom, "Wood Chemistry, Fundamentals and Applications", p.69, Academic Press, (1981) ISBN 0-12-647480-X.

24. K.V. Sarkanen, C.H. Ludwig, "Lignins; Occurrence, Formation, Structure, and Reactions", p.1, J. Wiley, (1971) ISBN 0-471-75422-6.

25. J. Haggen, Chem. Eng. News, 63 (#18), p.33-34, (May 6, 1985).

26. T.K. Kirk, T. Higuchi, H. Chang, "Lignin Biodegradation: Microbiology,

Chemistry, and Potential Applications", Vol. 1, p.5, CRC Press, (1980) ISBN 0-8493-5459-5.

27. J.J. Meister, D.R. Patil, L.R. Field, J.C. Nicholson, J. Polym. Sci., Poly. Chem. Ed., 22, 1963-1980, (1984).

28. J.J. Meister, D.R. Patil, H. Channell, J. Appl. Polym. Sci., 29, 3457-3477, (1984).

29. J.J. Meister, J.C. Nicholson, D.R. Patil, L.R. Field, Macromolecules, 19, 803-809, (1986).

30. J.J. Meister, D.R. Patil, Macromolecules, 18, 1559-1564, (1985).

31. J.J. Meister, D.R. Patil, Ind. Eng. Chem. Prod. Res. Devel., 24, 306-313, (1985).

32. J.J. Meister, D.R. Patil, H. Channell, Proceed. Intern. Symp. on Oilfield and Geotherm. Chem., Soc. Petrol. Eng. Paper #13559, Phoenix, AR, 4/9-11/85.

33. J.C. Nicholson, J.J. Meister, D.R. Patil, L.R. Field, Anal. Chem., 56, 2447-2451, (1984).

34. W.T. Dixon, R.O.C. Norman, J. Chem. Soc. (London), (#5), 3119-3124, (1963).

35. J.J. Meister, Review of the Synthesis, Characterization, and Testing of Graft Copolymers of Lignin, p.305-322 of "Renewable-Resource Materials: New Polymer Sources, C.E. Carraher, Jr., L.H. Sperling, Ed., Plenum Press, N.Y. (1985) ISBN 0-306-42271-9.

36. J.J. Meister, Linda (Mu Lan) Sha, Synthesis, Characterization, Properties, and Derivatives of Poly(Starch-g-(1-amidoethylene))-Copolymers, to be published by J. Appl. Poly. Sci., 1/87.

37. N.H. Kilmer, L.G. Donarama, M.J. Hatch, G.D. Khune, F.D. Martin, J.S. Shepitka, K.V. Wilson, Polymer Preprints, 22(#2), 69-71, (1981).

PROPAGATION OF POLYMER SLUGS

THROUGH ADSORBENT POROUS MEDIA

G. Chauveteau and J. Lecourtier

Institut Francais du Pétrole
BP 311 - Rueil - Malmaison 92506 - France

ABSTRACT

The effects of adsorption on the propagation of polymer slugs through oil reservoirs determine the efficiency of polymer flooding for increasing oil recovery. However, up to now the adsorption laws introduced in polymer flood simulation models were oversimplified due to a lack of knowledge. The adsorption properties relevant for polymer propagation, such as instantaneous adsorption, reorganization in adsorbed layer, exchanges of macromolecules between free solution and adsorbed layer, desorption before thermodynamic equilibrium, are analyzed on the basis of both a very careful experimental work and recent polymer adsorption theories taking into account the effects of polymer polydispersity. The influence of surface and adsorbed layer steric exclusion chromatography effects which are in competition with adsorption-desorption chromatography are discussed. The analysis of the respective influence of macromolecular diffusion and hydrodynamic convection shows that this later mechanism governs the overall polymer dispersion. Moreover a determining effect of viscous fingering on the spreading of trailing edge concentration profile is observed. In addition, a theoretical approach of hydrodynamic retention mechanism is proposed. As a practical application, a new methodology to avoid kinetic effect artefacts on the measurement of instantaneous, reversible and irreversible adsorption is described.

INTRODUCTION

High molecular weight and polydisperse water soluble polyelectrolytes, such as xanthans and hydrolyzed polyacrylamides are currently used to enhance oil recovery. They are dissolved in injection water to decrease its mobility and thus to improve its efficiency for sweeping the oil from oil bearing reservoirs.

The optimization of such a process, usually called polymer flooding, requires not only a knowledge of the rheological properties of polymer solutions in porous media, which has been reviewed recently (1), but also an ability to predict both the average propagation velocity and spreading of polymer slugs in reservoirs, taking into account the presence of adsorbent minerals under reservoir conditions (2, 3).

The main phenomena governing polymer propagation properties have been identified, i.e. molecular diffusion and hydrodynamic dispersion, viscous fingering (4 - 7), inaccessible pore volume (8 - 11), surface exclusion chromatography (12 - 13), adsorption (11, 14 - 17) and hydrodynamic retention (9, 14, 18, 19). However, no comprehensive study including the respective influence of all these phenomena has been reported. An investigation aiming at modelling xanthan slug propagation through sandstones has been published recently (20). In this paper, the effects of hydrodynamic dispersion of both polymer and solvent are well taken into account and the fit of viscous fingering is fairly good. In addition, a multi-component model is proposed to predict surface exclusion chromatography effects. Up to now however, the adsorption laws introduced in existing polymer flooding models are somewhat naive and oversimplified, leading to erroneous predictions of oil recovery.

This paper proposes a phenomenological analysis, based on laboratory experimental work, of the effects of adsorption properties on polymer slug propagation. The adsorption properties studied include kinetic aspects, i.e. instantaneous adsorption, reorganization of macromolecules inside adsorbed layer, exchanges between free and adsorbed polymer, desorption as well as properties at thermodynamic equilibrium which can be described by a partially reversible adsorption isotherm. The conditions for hydrodynamic retention are also discussed. In addition, an analysis of the effects of polymer polydispersity on each of these adsorption phenomena shows that these effects cannot be neglected in a predictive simulator.

EXPERIMENTAL

Materials

The properties of the two polymer samples used in this study, a partially hydrolyzed (τ = 30%) polyacrylamide (HPAM) and a fully pyruvated single stranded xanthan (XCPS) have been reported in previous papers (1, 2, 3, 21, 22, 23). The main characteristics of the dissolved macromolecules (average values of viscometric radius of gyration $\overline{R}_{G\eta}$ and zero shear rate intrinsic viscosity $[\eta]$) under the experimental conditions of this work (30°C, 20 g/l NaCl for HPAM, 5 to 50 g/l NaCl for XCPS) are summarized in Table I. Xanthan polydispersity has been determined by surface exclusion chromatography (12) and that of polyacrylamide by a new method of flow rate dependent filtration through non-adsorbent well-calibrated membranes (24). The polymer was gently dissolved using a magnetic stirrer in neutral aqueous solutions having salinities varying from 5 g/l to 50 g/l NaCl. The polymer solutions were slowly filtered according to a method previously described (25) to remove any remaining microgels in order to prevent any partial plugging and microgel-induced hydrodynamic retention. Furthermore, all undesirable low molecular weight impureties such as multivalent cations which could induce aggregation or affect polymer adsorption have been eliminated by extensive ultrafiltration.

Table I. Polymer Solution Characteristics

Polymer	HPAM	XCPS
$\overline{M}_w$	7.5×10^6	1.8×10^6
I_p	2.2	1.4
$\overline{R}_{G\eta}$ (nm)	200	170
$[\eta]$ (cm^3/g)	3800	4300
D (cm^2/s)	1.6×10^{-8}	2.4×10^{-8}

Pure quartzitic sand or a Na-kaolinite-sand mixture containing 2% W/W of clay were packed in glass columns giving porous media with pore volumes varying between 70 and 90 cm^3 and lengths between 25 and 30 cm. The characteristics of these porous media are given in Table II. The BET specific areas of pure sand and Na-kaolinite are respectively 0.1 and 20 m^2/g.

Table II. Porous Media Characteristics

Porous Medium Mineralogy	I Sand + clay	II Sand	III Sand	IV Sand	V Sand
Permeability (μm^2)	2.8	4.7	4.1	4.9	4.9
Porosity (%)	39.8	39.2	38.5	38.9	38.9

Procedure

Successive polymer solution slugs (~ 1.5 VP) separated by water injections at the same salinity were injected into porous media immersed in a thermostatic bath (30°C) at constant flow rates (v = 0.3 m/d and 0.6 m/d). At these flow rates, the wall shear rates in the pore throats $\dot{\gamma}_{PT}$ is low enough ($\dot{\gamma}_{PT}$ = 4.2 s^{-1} and 8.4 s^{-1} calculated as suggested in Ref. 26) to ensure that the flow regime is Newtonian for the polymer solutions used. For every solution sample (2 cm^3) collected at the outlet, polymer concentration was measured by a DC 80 Dorhman carbon analyser and Newtonian intrinsic viscosity was determined by using a Contraves Low Shear viscometer.

RESULTS AND DISCUSSION

The variations of reduced polymer concentration C/Co and intrinsic viscosity $[\eta]_r = [\eta]/\overline{[\eta]}$, where C_o and $\overline{[\eta]}$ are the values at the inlet, are plotted in Fig. 1 to 5 as a function of the volume of polymer solution injected V_p, and of water volume injected V_w, both expressed in terms of pore volume PV.

The curves are analyzed below on the basis of the effects of the different phenomena possibly governing propagation: molecular diffusion and hydrodynamic dispersion, viscous fingering,exclusion chromatography, adsorption and hydrodynamic retention.

Molecular Diffusion and Hydrodynamic Convection

Molecular diffusion and hydrodynamic convection are the only mechanisms responsible for the dispersion of a solute concentration front during its propagation through a porous medium if i) the flow is laminar and not inertial, ii) the solute does not increase solvent viscosity, iii) the solute is not subjected to chromatography. The first condition is actually verified in our experiments. The second is not for the data presented below, but could hold true at lower polymer concentrations. The third condition is not satisfied because EOR polymers are polydispersed. Molecular diffusion is expected to be the dominant factor at very low flow rates as long as the grain Peclet number P_{eg} is lower than 1 for homogeneous porous media. P_{eg} is equal to vd_g/D, where v is frontal velocity, d_g is the grain diameter and D is the molecular diffusion coefficient. Under our experimental conditions (v $\simeq$ 0.3 m/d ; $d_g \simeq$ 100 µm, D = 1.7 x 10^{-5} cm^2/s), P_{eg} for water is close to 0.2. The ratio K/D of dispersion to diffusion coefficient has been found to be 0.8 using potassium iodide as a tracer for water, thus confirming the low dispersivity of our porous media and that molecular diffusion is the dominant dispersive factor for water.

However, the grain Peclet numbers for polymers of high molecular weight such as those used in our experiments are much higher, due to the very low value of bulk molecular diffusion coefficients (Table I). In fact, these molecular diffusion coefficients are expected to be still smaller in confined conditions prevailing in porous media so that the grain Peclet number for the average molecular weights of our samples should be higher than 150 and 230 for XCPS and HPAM respectively. These high values of grain Peclet numbers imply that the spreading of polymer fronts is governed by hydrodynamic dispersion. However, the absolute value of polymer dispersion is expected to be small and close to that of water under our experimental conditions.

Viscous Fingering

Semi-dilute polymer solutions such as those used in polymer flooding have viscosities significantly higher than that of water (around 2.2 times in this study). This higher viscosity of polymer slug is expected to slightly reduce the dispersion at the leading edge but will strongly increase the spreading of the trailing edge due to viscous fingering. Indeed, because of the very slow molecular diffusion of polymer, the growth of viscous fingers is very slightly reduced by transversal dispersion. The effects of viscous fingering, namely the spreading of concentration profiles at the trailing edge of the slugs are clearly shown in Figures 1 to 5. However, since our polymers are polydisperse, chromatographic effects due to solid surface or adsorbed layer exclusion and to adsorption-desorption lead to additional spreading out or modifications of the concentration profiles.

Inaccessible Pore Volume and Surface Exclusion Chromatography

The effects of inaccessible pore volume (IPV) strictly defined as the pore volume where molecules cannot penetrate due to exclusion on the basis of size is very small (IPV << 1%) for microgel-free solutions ($\overline{R}_{G\eta} \sim 0.2$ μm) flowing through unconsolidated packs of large permeability such as those used in this work (k > 3 μm², pore throat radius ~ 20 μm).

However, surface exclusion chromatography which is due to the entropic exclusion of macromolecule centers of mass from pore wall induces a noticeable spreading out of concentration profiles for polydisperse polymers when polymer-to-pore size ratio is not negligible (13). This phenomenon has been used for fractionating polydisperse xanthan samples in low permeability non-adsorbent silicium carbide particle packs, and thus for determining the molecular weight distribution of EOR xanthans (12). This surface exclusion mechanism, the efficiency of which is maximum in the Newtonian regime when macromolecules are neither deformed nor oriented by hydrodynamic forces, explains to a large extent the spreading of leading edge concentration profiles of the last slugs in Figures 1 to 4. For the last slugs, indeed, this spreading due to surface exclusion is less disturbed by the adsorption phenomena. Surface exclusion also spreads the trailing edges but its effects are more difficult to detect due to the dominant effect of viscous fingering.

The molecular weight fractionation due to surface exclusion effects is clearly shown in Fig. 1, 3 and 4 by the variations of intrinsic viscosity versus pore volume injected both at the leading and trailing edges for the last slugs injected. The intrinsic viscosity is directly related to molecular weight through a Mark-Houwink type equation ($[\eta] = K M^{\alpha}$) where $\alpha = 0.8$ and 1 for HPAM and XCPS respectively (see ref. 12 and 21). Thus, this surface exclusion mechanism is effective even in presence of an adsorbed layer, both for xanthan (Fig. 1 and 3) and for polyacrylamide (Fig. 4). This exclusion means an overall repulsion between adsorbed and free molecules but does not exclude exchanges between free and adsorbed polymer. Such an exclusion is expected only for neutral polymers in good solvent or with charged polymers in relatively low salinity waters, but not for neutral polymers in Θ conditions nor for charged polymers in pseudo-Θ conditions which may be realized in some brines containing divalent ions. Due to the relatively low macromolecule-to-pore size ratio, surface exclusion increases average polymer velocity very slightly in our experiments (less than 5%). It should be noticed that adsorbed layer exclusion is effective even before reaching a complete thermodynamic equilibrium for adsorption, but its effects are partly concealed by the effects of preferential adsorption of high molecular species during the reorganization of adsorbed layer, which is analyzed below.

Non-Equilibrium Adsorption

Instantaneous adsorption. During the flow of polymer solutions through porous media, the hydrodynamic convection of macromolecules brings them into contact with solid surface in a mean time t_c which is of the order of a few seconds under our experimental conditions. Since the probability for a macromolecule to be adsorbed during the first contact with the pore surface free of adsorbed polymer is very high, the polymer adsorption can be considered as instantaneous. Indeed, this mean time is negligible compared to the residence time of macromolecules t_r inside the column. ($t_r \sim 24$ h at 0.3 m/d). Since the polymer solutions used do not contain significant low molecular weight fractions which may not be adsorbed, we can assume that every molecule in contact with solid surface will be adsorbed. This molecular weight independent instantaneous adsorption explains the sharper leading edges of the first slugs than those of the following (Fig. 1).

The amount of polymer which is instantaneously adsorbed A_{ins} can be lower or higher than that at thermodynamic equilibrium depending on free polymer concentration. The value of A_{ins} is expected to increase with polymer concentration which determines the number of macromolecules per unit area which can be in contact with the surface at the same time.

The value of instantaneous adsorption A_{ins} is determined by the difference between average velocity of the slug leading edge and that of a slug of the same polymer in strictly non-adsorbing conditions (Tables III and IV). The first slug is the slowest because instantaneous adsorption is maximum for a polymer-free surface (see Fig. 1 to 4 and Tables III and IV). The successive slugs move increasingly faster to reach a maximum value when complete reorganization of the adsorbed layer has been achieved. However, even under these conditions, the leading edge velocity may be lower than that expected for a strictly non-adsorbing polymer, since some adsorption sites of low energy may be accessible at the solid surface. Under these conditions, an instantaneous adsorption may still occur, thus decreasing leading edge velocity, but this adsorption is expected to be entirely reversible when the free polymer concentration on top of the adsorbed

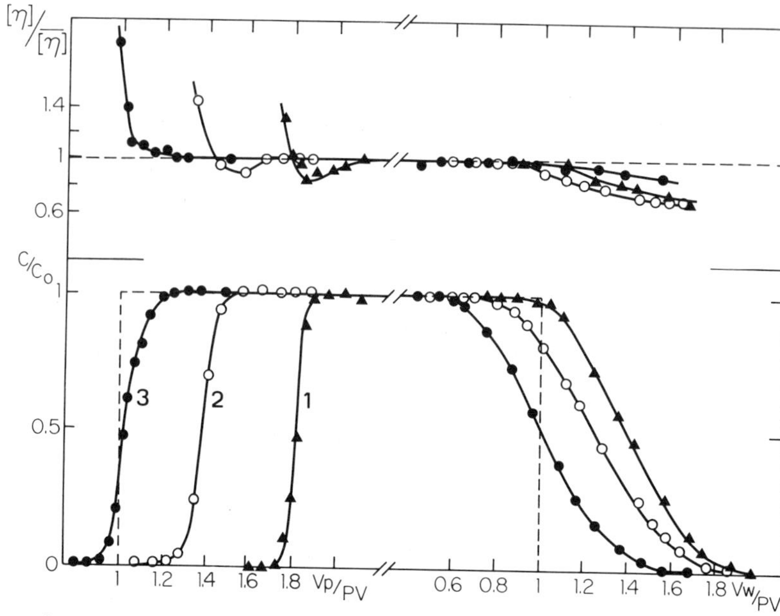

Fig. 1. Injection of xanthan slugs (237 ppm, 20 g/l NaCl) in clayey sand pack I at 0.3 m/d

Table III. Xanthan adsorption during the injection of successive slugs

Porous Medium Mineralogy Salinity	I Sand + kaolinite 20 g/l NaCl			III Pure Sand 50 g/l NaCl		
Adsorption μg/g	A_{ins}	A_{irr}	A_{rev}	A_{ins}	A_{irr}	A_{rev}
Slug n° 1	59	23	36	6.6	2.4	4.2
Slug n° 2	28	5	23	4.8	1.7	3.1
Slug n° 3	9	0	9	3.4	1.3	2.1
Slug n° 4	0	0	0	1	0	1
Total	-	28	-	-	5.4	-

layer decreases at the trailing edge of the slug. The irreversible part of adsorption A_{irr} which is determined by mass balance for each slug decreases from one slug to another up to the end of adsorption process (Tables III and IV). The total irreversible adsorption is obtained by adding the values of A_{irr} for each slug.

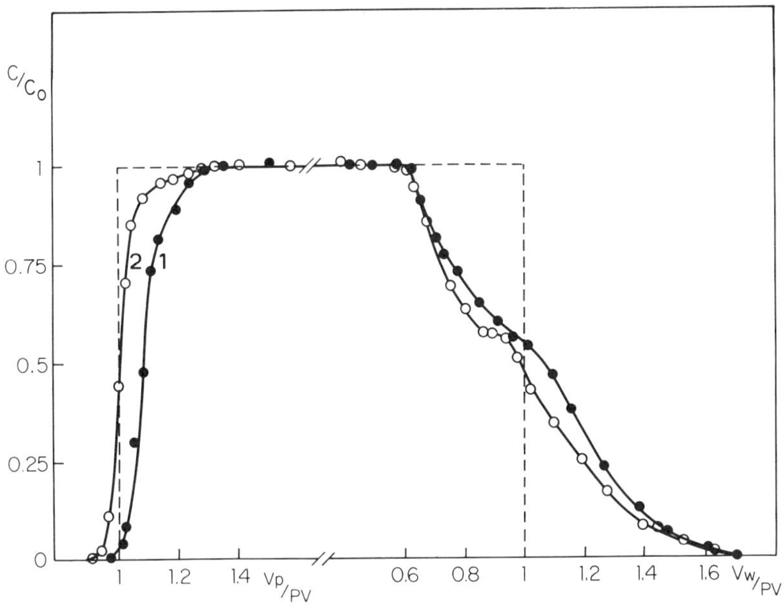

Fig. 2. Injection of xanthan slugs (243 ppm, NaCl 20 g/l) through sand pack II at 0.3 m/d

Table IV. Polyacrylamide adsorption during the injections of successive slugs

Porous Medium Mineralogy Salinity	IV Sand 20 g/NaCl		
Adsorption μg/g	A_{ins}	A_{irr}	A_{rev}
Slug n° 1	11	9	2
Slug n° 2	2	0	2
Slug n° 3	1	0	1
Total	-	9	-

Reorganization in Adsorbed Layer. Immediately after the early stage of adsorption, each macromolecule has the same conformation as in solution. Then, its conformation is expected to be modified as thermodynamic equilibrium is approached with an increased number of segments in contact with the solid surface for both flexible and rigid polymers. As polymer rigidity increases, and particularly for polyelectrolytes, the adsorbed segment trains will be increasingly aligned. This reorganization inside the adsorbed layer occurs even if there is no free molecule on top of the adsorbed

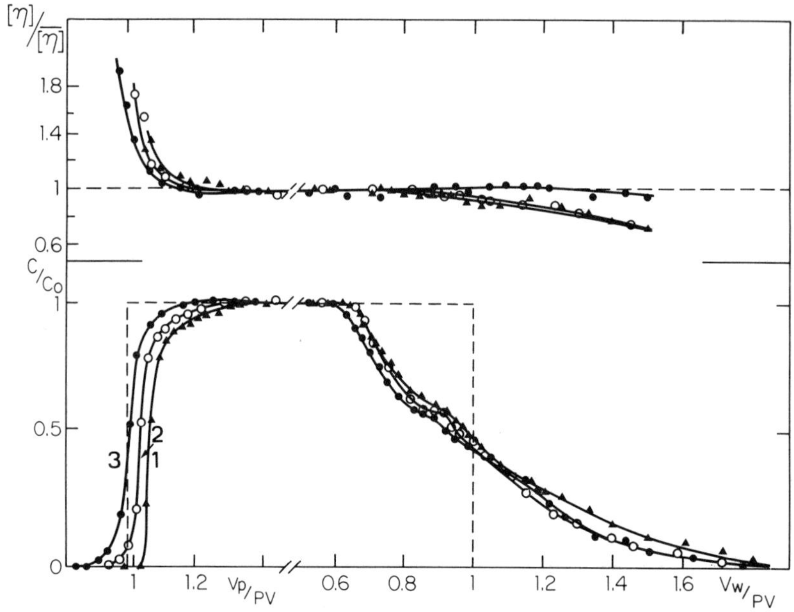

Fig. 3. Injection of xanthan slugs (233 ppm, 50 g/l NaCl) through sand pack IV at 0.3 m/d

layer and if the polymer is monodisperse. However, the presence of free molecules is thought to decrease the reorganization time t_R. Indeed this reorganization implies train desorption and readsorption of these trains in other locations and the desorption step which is the slower process is accelerated if free molecules compete to decrease the probability of re-adsorption at the same location. This explains why the exchange rate increases with polymer concentration, even for monodisperse polymers (28). The effects of these exchanges for polydisperse polymers are discussed below.

Since the reorganization time t_R is thought to be mainly determined by the time of desorption of trains and weakly dependent on loop reorganization, t_R is expected to increase both with segmental adsorption energy and the average number of segments in a train. Under our experimental conditions, t_R should increase as salinity increases for polymer adsorbed on negatively charged sands (Fig. 2 and 3), and for the same salinity, should be higher when adsorption occurs on kaolinite (Fig. 1). Indeed, the adsorption energy is expected to be higher on the positive sites which still remain at neutral pH on kaolinite lateral surfaces. The experimental results (Fig. 1 to 3 and Table III) suggest such tendencies although without verification because experiments have not been designed to test such τ_r dependence, but it is clear that the reorganization time is very long, more than 48 hours.

During this time of reorganization, if the instantaneous adsorption is higher than the quasi-irreversible one at thermodynamic equilibrium (a condition which is fulfilled in our experiments, see Table III), the part of this instantaneous adsorption which is weakly adsorbed is expected to be desorbed as soon as free polymer concentration decreases. This is realized at the trailing edge of the slug, leading to a decrease in its velocity. This reversible adsorption A_{rev} increases with the time available for adsorbed layer reorganization in presence of the polymer. Under the present

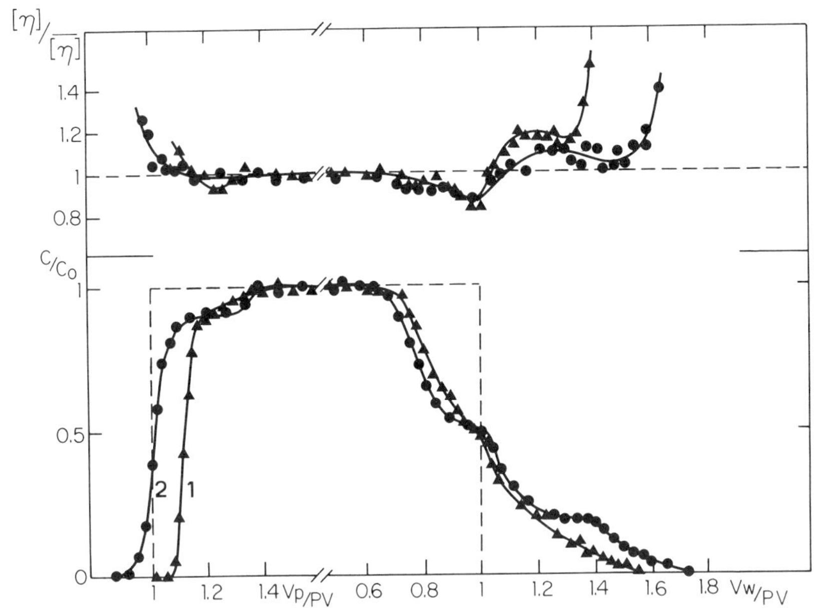

Fig. 4. Injection of polyacrylamide slugs (240 ppm, 20 g/l NaCl) through sand pack III at 0.3 m/d

experimental conditions this time is equal to residence time $t_R = L/v_f$, where L is porous media length and v_f is frontal velocity in absence of adsorption. This analysis also explains why the amount of reversible adsorption decreases for successive slugs. However, the value of A_{rev} is not expected to become zero even after several slugs have been injected since reversible adsorption may exist even after pseudo-thermodynamic equilibrium (see discussion below).

Exchanges between Free and Adsorbed Polymer. The exchange of macromolecules between solution and adsorbed layer occurs even with a strictly monodisperse polymer. This exchange requires penetration of free molecules through adsorbed layer overcoming osmotic pressure due to higher polymer concentration inside adsorbed layer. Although penetration implies a reptation process expected to be slow (27), this time is thought to be negligible compared to train desorption time. However, the rate of exchange is molecular weight dependent and is higher when high molecular weight molecules replace lower molecular weight ones in the adsorbed layer, leading to a preferential adsorption of high molecular weight molecules. In other words, the proportion of high molecular weight molecules is expected to be higher in adsorbed layer than in free polymer solution at thermodynamic equilibrium.

Since the amount of adsorbed molecules increases with molecular weight (2, 3, 27-31), the overall irreversible adsorption increases during reorganization time, i.e. the irreversible adsorption A_{irr} due to the injection of each slug is always positive or zero for the last slugs (Tables III and IV).

An injection of xanthan solution at 20 g/l NaCl into a sand + kaolinite pack at 0.3 m/d i.e. under experimental conditions very similar to those for which the results are shown in Fig. 1 and Table III, but with a much larger slug size (3.5 PV), has been carried out. The residence time t_r was thus much longer and the velocity of the trailing edge was found to be much slower than that in Fig. 1. This is expected since the reversible part of adsorption was higher due to a longer actual reorganization time. In addition, the concentration profile of the second slug injected was found to be very similar to that plotted in Fig. 1 for the third slug.

As soon as the polymer has been adsorbed at the leading edge of each slug, the reorganization in adsorbed layer in contact with a polydisperse polymer solution promotes the adsorption of the highest molecular weight macromolecules and thus the desorption of the lowest. Due to this exchange mechanism, the first samples of polymer solution which moves at the leading edge of the slugs are expected to contain mainly low molecular weight, and they should have an average relative intrinsic viscosity $[\eta]_r$ smaller than one. Such a situation occurs when the adsorption level is high like for the XCPS solution moving through the sand + kaolinite pack (Fig. 1) but also for the HPAM solution in a sand pack (Fig. 4) but to a less extent. However, some high molecular weight species are also desorbed and move faster due to the adsorbed layer exclusion chromatography effect. This last mechanism explains why the first samples arriving at the porous medium outlet may have intrinsic viscosities higher than 1, mainly when adsorption level is small, i.e. both for experiments carried out in sand packs and experiments in sand + kaolinite pack for the last slugs.

As a practical application, the polymer solution fractions moving ahead of slug leading edge in a reservoir containing highly adsorbing materials such as clays is expected to contain low molecular weight polymer having low viscosity.

It is to be noted that the water flow time between two successive polymer injections was around two days in all the experiments reported in this paper, but this time is not a determining parameter since reorganization and exchange processes are very slow in absence of free polymer (28).

As a practical consequence of the above analysis of reorganization and exchange processes, the determination of the overall reversible part of polymer adsorption due to adsorbed layer reorganization requires the injection of a concentrated slug of large size at very low flow rate, in order to ensure that the residence time is actually longer than the reorganization time. In practice, since this reorganization cannot be predicted under complex conditions prevalent in natural brines and porous media, it is safer to inject successive slugs until two identical successive concentration profiles at the outlet are observed.

Adsorption and Desorption in Pseudo-Stagnant Zones. At the leading edge of the two first polyacrylamide slugs (Fig. IV) a pseudo-plateau or tailing is observed between 1.1 and 1.35 PV at a relative concentration C/C_0 equal to 0.92 before reaching 1 for $V_p/PV > 1.4$. Since the conditions to create hydrodynamic retention in our experimental conditions are not realized (see section below on hydrodynamic retention), these pseudo-plateaus cannot be attributed to hydrodynamic retention. Moreover, the third slug injected not plotted in Fig. 4 for clarification) did not display any pseudo-plateau, showing that the effects of slow adsorption kinetics are implied in the formation of these pseudo-plateaus. Three successive injections of polyacrylamide slugs under identical conditions showed identical shapes, showing the good reproducibility of the experiments. Consequently, these pseudo-plateaus are attributed to very slow convection and/or adsorption in pseudo-stagnant zones of the pore structure very similar to those where hydrodynamic retention is expected at higher flow rates.

The pseudo plateau observed at low polymer concentration ($C/C_0 \simeq 0.2$) in the trailing edge is due to a similar phenomenon. It must be noted that these pseudo-plateaus have a completely different origin than those observed at high flow rates in very dispersive but weakly adsorbent sandstones and which were found to be due to viscous fingering (20). As soon as the polymer concentration in the flowing zones decreases, the molecules which are desorbed during adsorbed layer reorganization diffuse from these pseudo-stagnant zones towards flowing zones. This diffusion which is governed by a competition between osmotic pressure and hydrodynamic forces due to solvent flow is a very slow process. This explains why this pseudo - plateau in the trailing edge can be observed only at very low flow rates, and could even be a shoulder if the flow was stopped before to be resumed. In addition, as expected from the above analysis, no pseudo-plateau was observed neither in the third slug concentration profile nor when the flow rate was higher (for γ_{PT} = 8.4, 16.8 and 33.6 s^{-1}).

Since this osmotically driven diffusion towards flowing zones is slower for high molecular weight macromolecules, a pseudo-plateau followed by an upturn in the $[\eta]_r$ versus V_w/PV curve is observed during this desorption. Since the magnitude of this desorption due to adsorbed layer reorganization is greater in the first slug, the $[\eta]_r$ vs V_w/PV curve is above that corresponding to the second slug (Fig. 4)

Similar pseudo plateau and upturn can be observed in $[\eta]_r$ vs V_w/PV curve for higher flow rate xanthan slug injection (Fig. 5) for V_w/PV ~ 1.2, showing that this molecular weight dependant desorption may be observed for xanthan as well as for polyacrylamide but the effect is smaller in the case of xanthan.

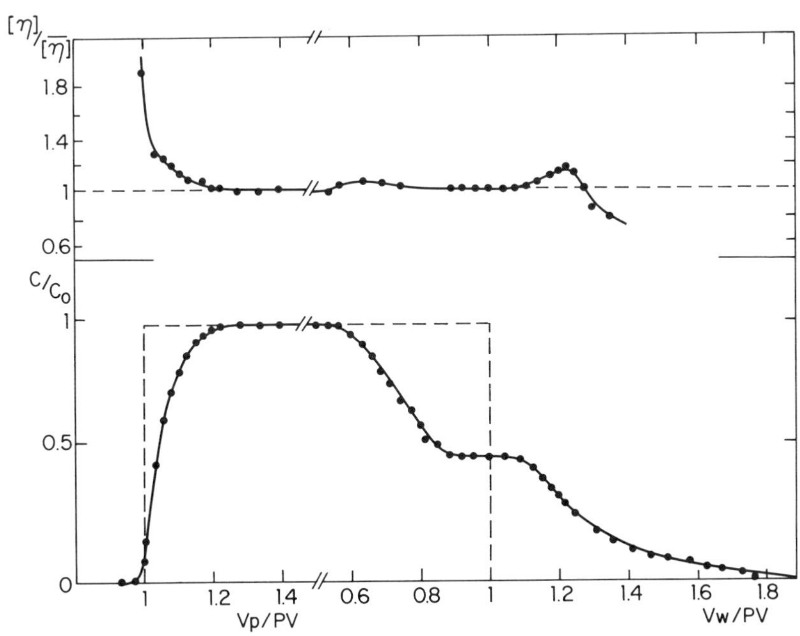

Fig. 5. Injection of a xanthan slug (NaCl 50 g/l - 233 ppm) through sand pack V at 0.6 m/d

Adsorption and Desorption at Thermodynamic Equilibrium

After complete reorganization in the adsorbed layer including the effects of polydispersity, i.e. at thermodynamic equilibrium, the adsorbed layer contains mainly high molecular weight fractions. This molecular weight distribution MWD inside the adsorbed layer could be calculated from MWD of the polymer in solution and equilibrium constants for each molecular weight. After displacement of the polymer inside the porous media by water, the true thermodynamic equilibrium without free molecule in solution may be very long to be realized (some months under polymer flooding conditions) and would mean a complete desorption. However, in the experiments described here, i.e. after some days of displacement by water, a small part of adsorbed layer is expected to be desorbed. The permeability reduction R_k which is defined as the ratio of permeability before and after polymer adsorption is strictly equal to 1 for XCPS due to a negligible hydrodynamic thickness which is expected to be of the order of macromolecules diameter, i.e. 2 nm (21, 32). For HPAM, R_k is found to be equal to 1.14 which is due to a thickness of adsorbed layer of 0.35 μm (2, 24, 32).

Under these pseudo equilibrium conditions, the injection of a new polymer slug may lead to both adsorption at the leading edge and desorption of an equal amount at the trailing edge since this adsorption is entirely reversible. This situation is realized in sand packs (see Tables III, IV and Fig. 1, 2, 4) because, due to electrostatic repulsions, only weakly adsorbing sites may exist. It should be noticed that a similar situation, i.e. reversible adsorption may occur also without adsorbed layer when adsorption energy is very weak.

When the last slugs are injected (Fig. 1, 3 and 4), the equilibrium is almost reached. The mean molecular weight of the polymer samples are higher than the mean value at the leading edge and lower at the trailing edge. Thus, a chromatographic effect, which is quite similar to surface exclusion chromatography under non adsorbing conditions is clearly displayed. This adsorbed layer exclusion chromatography is expected to occur only if there is a net repulsion between free and adsorbed molecules as discussed above.

The adsorption at the leading edge is expected to increase strongly with polymer concentration and to occur mainly when free polymer concentration becomes higher than that in the adsorbed layer since osmotic pressure becomes favorable for adsorbed layer penetration.

At the trailing edge, desorption occurs as soon as free polymer concentration decreases, but this desorption is expected to be very large when polymer concentration becomes lower than the critical concentration C_c at which the isotherm reaches a pseudo-plateau. If the shape of the isotherm displays a linear increase of adsorption with polymer concentration, a plateau at a value equal to C_c is expected at the trailing edge of the slug. Such a plateau is actually observed in Fig. 2 to 5. The polymer concentration at the plateau is found to be 140 ppm in Fig. 2 (XCPS at 20 g/1 NaCl, sand pack n° I). The plateau is also observed at 140 ppm in Fig. 3 (XCPS at 50 g/1, sand pack n° IV).

It should be noticed that the concentration plateau at the trailing edge is more clearly seen with the second slug because instantaneous desorption due to adsorbed layer reorganization which conceals the plateau is smaller during the second slug injection (Tables III and IV).

In Fig. 5, a longer plateau is observed in the trailing edge concentration profile at a concentration of 105 ppm. The main reason that this plateau is clearly observed even in the first slug concentration profile is that the polymer velocity was two times larger (v = 0.6 m/j), which implies a shorter time for reorganization and thus smaller instantaneous desorption. Under similar experimental conditions (XCPS 50 g/l and with the sand used for packing porous media n° V), adsorption experiments in test tubes carried out according to a procedure previously described (2, 3), showed that the isotherm plateau was actually reached at a concentration between 100 and 110 pm, thus corresponding to the desorption plateau.

Such a plateau due to desorption at thermodynamic equilibrium is also observed for polyacrylamide in a sand pack at a concentration of 120 ppm. This result is expected since weak energy type isotherms have also been found for the adsorption of this polymer on negatively charged surface (2 ,3).

Very weak adsorption type isotherms are very unusual for polymers because the adsorption energy per macromolecule is exponentially increasing with the number of segments per molecule actually adsorbed in trains. However, under our experimental conditions, the adsorption energy per adsorbed segment is expected to be very small due to the electrostatic repulsions between anionic polymer molecules and highly negatively charged sand surface.

It should be noticed that a desorption is also expected from kaolinite, but since the isotherm is of high energy type (2) due to some positive charges (-AlOH groups), this desorption is expected to be very slow and would occur at too low polymer concentration (a few ppm, see Ref. 28) to be detected in our experiments.

Hydrodynamic Retention

The overall retention of polymers during their propagation through porous media includes both adsorption which is partly reversible as discussed above and a flow-dependent retention, which is usually called hydrodynamic retention and which was found to increase with flow rate and to be reversible if flow rate is decreased (9, 18, 19). However, the presence of microgels could be partly responsible for the hydrodynamic retentions previously observed.

A more precise description of hydrodynamic retention can be proposed since new results concerning the hydrodynamic behavior of macromolecules in small calibrated membranes in weak deformation regime are available (24).

Hydrodynamic retention in porous media occurs only when i) macromolecules in the MWD have a radius of gyration larger than the half of the dimension d_r of some restrictions in pore structure ii) the Peclet number corresponding to polymer flow through these small openings is high enough ($P_e \gg 1$) to prevent downstream diffusion of the macromolecules towards zones of large restrictions ($d_r \gg 2R_G$) iii) hydrodynamic stresses are too small to decrease the energy barrier to be overcome to force the macromolecules through the restrictions. Under these conditions, the hydrodynamically entrapped macromolecules prevent the passage of even smaller ones, leading to the accumulation of polymer upstream to pore restrictions up to reach an equilibrium between hydrodynamic forces and osmotic pressure between these concentrated zones and those were polymer solution flows.

For spherical particle packs, the first of these conditions is fulfilled only for very small diameter spheres ($d_{sphere} \leqq 5\ R_G$). However, this first condition can be realized in sand packs and sandstones, even if permeability or particle size is large, due to very complex structure which can contain very small openings ($d_r \ll 2\ R_G$). Thus in our experiments the first condition can be realized but the second i.e. a Peclet number higher than one in the small openings is not because of too low velocity. Hydrodynamic retention at much higher velocity (30 m/d) has been observed and will be described in a forthcoming paper on hydrodynamic retention.

It must be noted that such hydrodynamic retention is less likely for semi-rigid polymers like xanthan since their conformation makes more probable their hydrodynamic orientation parallel to the opening axis. The hydrodynamic retention previously observed in Berea sandstones (18) was probably induced by the presence of microgels in the xanthan sample used.

CONCLUSIONS

1) The leading edge velocity of polymer slugs in adsorbent porous media is determined by competition between molecular weight-independent instantaneous adsorption and molecular weight-dependent adsorbed layer exclusion chromatography. The instantaneous adsorption should be concentration dependent and may be lower or higher than that at thermodynamic equilibrium.

2) Immediately after this first stage of instantaneous adsorption, a reorganization inside the adsorbed layer begins, even for monodisperse polymers, to reach thermodynamic equilibrium. The reorganization time which may be very long in the case of high energy adsorption, decreases as free polymer concentration in solution increases. During this reorganization, the hydrodynamic thickness of adsorbed layer is expected to decrease since the number of segments in adsorbed trains increases.

3) The trailing edge mean velocity is governed by competition between exclusion chromatography and desorption, the latter of which occurs as soon as free polymer concentration decreases. Both exclusion chromatography and desorption due to adsorbed layer reorganization are molecular weight dependent.

4) For polydisperse polymers, the exchanges between macromolecules in the solution and the adsorbed layer lead to a preferential desorption of low molecular weight macromolecules previously adsorbed. Therefore, at thermodynamic equilibrium, the molecular weight distribution in adsorbed layer will contain much higher molecular weight fractions than in free polymer solution.

5) An adsorbed layer exclusion chromatography which should be restricted to good solvent conditions is effective not only after thermodynamic equilibrium is reached but also during the reorganization process. For high permeability porous media with large particles like natural sands, the true inaccessible pore volume has a negligible effect on polymer velocity compared to that of surface or adsorbed layer exclusion phenomena.

6) Even after the thermodynamic equilibrium has been reached with free polymer concentration equal to that injected, the polymer which is reversibly adsorbed will be desorbed as soon as free polymer concentration decreases below the critical concentration C_c, above which a plateau is observed in the isotherm. This desorption will create a plateau at a concentration equal to C_c in the trailing edge concentration profile.

7) The phenomena of adsorption and desorption in pseudo-stagnant zones of porous media create pseudo-plateau or tailing of the leading edges of concentration profiles and shoulders in the trailing edges.

8) The conditions under which hydrodynamic retention can occur in flow of polymer solutions through porous media have been analyzed, leading to the conclusion that Peclet numbers corresponding to potential retention zones were too low to induce this type of retention.

9) The overall dispersion of very dilute polymer solution slugs is determined, even at low flow rates such as those prevailing in oil reservoirs far from the wells, by hydrodynamic dispersion because of the extremely small diffusion coefficients expected for high molecular weight polymers in a confined situation.

10) The overall dispersion of the trailing edges of semi-dilute solution slugs such as those used in enhanced oil recovery operations is governed by viscous fingering which is only very slightly reduced by transversal diffusion, a process which is negligible for high molecular weight polymers.

11) As practical applications i) the conditions for determining instantaneous and irreversible adsorption in laboratory experiments have been given ii) the polymer molecules moving ahead of slug leading edge in oil reservoirs containing highly adsorbing minerals such as clays are expected to have low molecular weight and thus to induce low viscosity of the polymer solution.

Acknowledgments

The authors wish to acknowledge ARTEP (Association for Research on Petroleum Drilling and Production Techniques) for financial support and permission to publish this work. They are indebted to Ph. Delaplace and M. Nitabah both for carrying out very careful experiments and for fruitful discussions.

References

1. CHAUVETEAU, G. : "Fundamental Criteria in Polymer Flow Through Porous Media", in "Water-Soluble Polymers", Glass J.E. Ed., Adv. in Chem. Ser., 1986, 213.
2. LECOURTIER, J. and CHAUVETEAU, G.: "Adsorption des polyacrylamides et du xanthane sur les surfaces minérales", 3ème Colloque Européen RAP, Rome, Avril 1985, 187
3. LECOURTIER, J., LEE, L.T. and CHAUVETEAU, G.: "Adsorption of EOR Polymers on Mineral Surfaces", AICHE Mtg., Houston, Apr. 1987.
4. BRIGHAM, W.E., REED, P.W. and DEW, J.N.: "Experiments on Mixing During Miscible Displacement in Porous Media", SPEJ, 1961, 1
5. KOVAL, E.J.: "A Method for Predicting the Performance of Unstable Miscible Displacement in Heterogeneous Media", SPEJ, 1963, 145
6. FAYER, F.J.: "An Approximate Model With Physically Interpretable Parameters for Representing Miscible Viscous FIngering", SPE 13166, Houston, Sept. 1984
7. VOSSOUGHI, S., SMITH, J.E., GREEN, D.W. and WILLHITE, G.P.: "A New Method to Simulate the Effects of Viscous Fingering on Miscible Displacement Processes in Porous Media", SPEJ, 1984, 56
8. DAWSON, R, and LANTZ, R.B.: "Inaccessible Pore Volume in Polymer Flooding", SPEJ, 1972, 448
9. WILLHITE, G.P. and DOMINGUEZ, J.G.: "Mechanisms of Polymer Retention in Porous Media" in Improved Oil Recovery by Surfactant and Polymer Flooding", S.O. Shah and T.S. Schechter, Acad. Press. Inc., New York, 1977, 511

10. LIAUH, W.C., DUDA, J.O. and KLAUS, E.E.: "An Investigation of the Inaccessible Pore Volume Phenomenon", AICHE Symp. Ser. 212, 1982, 78, 70
11. LOTSCH, T., MULLER, T. and PUSH, G.: "The Effect of Inaccessible Pore Volume on Polymer Core Flood Experiments", SPE 13590, Arizona, April 1985
12. LECOURTIER, J. and CHAUVETEAU, G.: "Xanthan Fractionation by Surface Exclusion Chromatography", Macromolecules, 1984, 17, 1340,
13. LECOURTIER, J. and CHAUVETEAU, G.: "Propagation of Polymer Slugs through Porous Media", SPE Paper n°13034, Dallas, Sept. 17-19, 1984
14. GUPTA, S.P. and TRUSHENSKI, S.P.: "Micellar Flooding - The Propagation of the Polymer Mobility Buffer Bank", SPEJ, 1978, 5,
15. SZABO, M.T.: "Some Aspects of Polymer Retention in Porous Media Using a C^{14}-Tagged Hydrolyzed Polyacrylamide", SPEJ, 15, 1975, 323
16. MUNGAN, N.: "Rheology and Adsorption of Aqueous Polymer Solutions", Technology, 1965, 45
17. COHEN, Y. and CHRIST, F.R.: "Polymer Retention and Adsorption in the Flow of Polymer Solutions through Porous Media", SPEJ, 1985, 25
18. MAERKER, J.M.: "Dependence of Polymer Retention on Flow Rate", J. Pet. Techn., 1973, 25, 1307
19. CHAUVETEAU, G. and KOHLER, N.: "Polymer Flooding: the Essential Elements for Laboratory Evaluation", SPE, Paper n°4745, 2nd Improved Oil Recovery Symposium, Tulsa, April, 1974
20. SORBIE, K.S., PARKER, A. and CLIFFORD, P.J.: "Experimental and Theoretical Study of Polymer Flow in Porous Media", SPE Paper n° 14231, Las Vegas, Sept. 1985
21. MULLER, G., ANRHOURRACHE, N., LECOURTIER, J. and CHAUVETEAU, G.: "Salt Dependence of the Conformation of a Single Stranded Xanthan", Int. J. of Biol. Macrom., 1986, 8, 167
22. LECOURTIER, J., CHAUVETEAU, G. and MULLER, G.: "Salt-Induced Extension and Dissociation of a Native Double Stranded Xanthan", Int. J. of Biol. Macrom., 1986, 8, 307
23. MULLER, G.: "Propriétés statiques et dynamiques de polymères utilisés en récupération assistée du pétrole", 2ème Colloque Européen RAP, Paris, Nov. 1982, 163
24. BAGASSI, M., Thesis: "Comportement hydrodynamique des macromolecules dans les milieux poreux fins en régime de déformation faible", University of Brest, 1986, France
25. CHAUVETEAU, G. and KOHLER, N.: "Influence of Microgels in Xanthan Polysaccharide Solutions on their Flow through Various Porous Media", SPEJ., 1984, 24, 361,
26. CHAUVETEAU, G.: "Rod-Like Polymer Solution Flow through Fine Pores: Influence of Pore Size on Rheological Behavior", J. of Rheol., 1982, 26 (2), 111
27. De GENNES, P.G.: "Scaling Concepts in Polymer Physics", Ithaca, New York Cornell Univ. Press, 1979
28. PEFFERKORN, E., CARROY, A. and VAROQUI, R.: "Dynamic Behavior of Flexible Polymers at a Solid/Liquid Interface", J. Polym. Sci., 1985, 23 (10), 1957
29. TAKUHASHI, A. and KAWAGUCHI, M.: "The Structure of Macromolecules Adsorbed on Interfaces", Adv. Polym. Sci., 1982, 46, 1
30. COHEN STUART, M.A., COSGROVE, T. and VINCENT, B.: "Experimental aspects of Polymer Adsorption at Solid/Solution Interfaces", 1986, 24, 143
31. HESSELINK, F. Th.: "On the Theory of Polyelectrolyte Adsorption", J. Coll. Int. Sci., 1977, 60, 448
32. CHAUVETEAU, G., TIRREL, M. and OMARI, A.: "Concentration Dependence of the Effective viscosity of Polymer Solutions in Small Pores with Repulsive or Attractive Walls", J. Coll. Int. Sc., 1984, 1, 100

THE SIMULATION OF POLYMER FLOW IN HETEROGENEOUS POROUS MEDIA

K.S. Sorbie and P.J. Clifford

AEE Winfrith
Dorset, DT2 8DH
England

ABSTRACT

In this paper, some recent progress is described with a programme of study on polymer flow in heterogeneous (layered) porous media. This work has used numerical simulation, including models for all the main polymer phenomena, in order to establish recovery mechanisms in polymer flooding. Fluid crossflow is central to the recovery mechanisms in stratified systems with communicating layers. Scaling theory has been used to establish the relative magnitude of the various crossflow effects at both the field and laboratory scale. A combination of simulation and scaling theory has enabled us to define appropriate laboratory experiments, which reproduce the main reservoir features in a correctly scaled way. Some experimental results are presented and analysed by computer simulation.

INTRODUCTION

Polymer flooding has been applied in a wide range of reservoirs throughout the world over the last twenty years. In many cases, this treatment has been applied when the reservoir oil has a high viscosity, and waterflooding efficiency is low due to an unfavourable mobility ratio. In other cases, polymers have been used in heterogeneous reservoirs to remedy problems of water channelling, and hence early breakthrough, in high permeability zones. This latter problem can be very serious even when the local water-oil mobility ratio within a layer is quite favourable. The most common type of macroscopic heterogeneity found in oil reservoirs is layering where adjacent layers may have widely differing permeabilities. Many complex effects are observed when polymers flow in such systems. The polymer reduces the overall mobility of the aqueous phase, and hence changes the flow patterns within layers and between layers of different permeability in the reservoir. The polymer slug may also disperse, adsorb onto the rock matrix, chemically degrade, and show other behaviour such as viscous fingering, excluded volume effects, non-Newtonian flow etc. The aim of this paper is to outline our approach to the problem of understanding and developing quantitative models for describing polymer flooding processes in macroscopically layered systems.

In order to understand how polymers operate in a heterogeneous reservoir the following research strategy was adopted:

(i) Developing polymer simulation models in order to establish recovery mechanisms in heterogeneous systems with communicating layers.

(ii) Applying scaling theory in order to estimate the magnitude of various effects at both the laboratory and field scales.

(iii) Using results from scaling theory to define laboratory experiments, which illustrate correctly scaled polymer flow mechanisms for heterogeneous field systems.

(iv) Performing experiments and use the results to improve the theoretical description of polymer processes, for application in reservoir simulation.

Some of our progress with both the theoretical and experimental aspects of the above programme is described in this paper.

POLYMER SIMULATION MODELS

A number of publications have appeared in the literature describing simulation models for polymer flooding (1-8). The transport properties of the polymer, the interaction with the rock matrix, possible chemical reactions and the non-Newtonian rheological behaviour of polymeric liquids combine to make polymer flow through porous media very complex. The main features which should be included in a polymer simulator for both field and laboratory use are as follows:

(i) Polymer apparent viscosity in porous media: it is known that the viscosity of a polymer solution is strongly dependent on its concentration (9) and on its flow rate (10). In porous medium flow, polymers may show shear-thinning behaviour (11-13). In addition, polyacrylamide exhibits complex viscoelastic behaviour at higher flow rates (14).

(ii) Adsorption-Retention: virtually all types of polymer used in EOR are adsorbed/retained in the porous medium (6, 9, 10, 15). This phenomenon can have a very important effect on the outcome of a polymer flooding process. It may reduce the permeability of rock to the aqueous phase, an effect which remains long after the rock has been post-flushed with brine and is often described by a residual resistance factor (9).

(iii) Transport properties: polymer molecules show dispersion behaviour in porous media which may be different from tracer molecules (16). Polymers also show excluded or inaccessible pore volume effects which may cause the polymer molecules to travel faster through the porous medium than tracer species (13, 16-18).

(iv) Chemical reactions: polymers will gradually degrade chemically in reservoir environments, causing a corresponding loss of solution properties such as viscosity (19-21). Polymers may also react chemically with other species such as trivalent metal ions to cross link and form gel-like materials in the reservoir (22, 23).

(v) Thermal effects: Polymer properties such as viscosity, degradation rate etc are dependent on temperature. If large amounts of cool water are injected into a hot reservoir there may be significant temperature changes and calculations should take account of this (24, 25).

(vi) Sensitivities: most of the above effects are sensitive to factors such as local salinity etc. A simulation model should include the generalised sensitivities of such properties on these other factors.

Polymer is invariably found in the aqueous phase, and its transport equation is usually taken to be a generalised convection dispersion equation of the following form:

$$\nabla\cdot\left[\frac{k\,k_{rw}\rho_w}{\eta_w}C_i(\nabla P_w - \rho_w g\nabla h)\right] + \nabla\cdot\left[\phi\rho_w S_w D_i \nabla C_i\right] \tag{1}$$

$$+ \phi S_w R_i + q_i = \frac{\partial}{\partial t}(\phi\tilde{m}_i)$$

where

$$\tilde{m}_i = \rho_w S_w C_i + \rho_R \Gamma_i/\phi \tag{2}$$

All symbols have their usual meaning (see Nomenclature) and only the more important ones are defined here: C_i is the concentration of component i in the water (eg polymer, Cl etc). The viscosity of the aqueous phase, η_w, may depend on polymer or ion concentration, temperature etc; D_i is the dispersion of component i in the aqueous phase; R_i and q_i are the source/sink terms for component i through chemical reaction and injection/production respectively. Polymer adsorption, as described by the Γ_i term in equation (2), may feed back onto the mobility term in equation (1) through permeability reduction. In addition to the polymer/tracer transport equation above, a pressure equation must be solved (5-8), in order to find the velocity fields for each of the phases present; ie aqueous, oleic and micellar (if there is a surfactant present). If thermal effects are also to be included, then a heat balance equation is also required. The SCORPIO simulator (26, 27), which is used in our studies allows for all of these effects.

As an example of applying our simulator to model a specific effect in a polymer flood, we consider chemical degradation of polymer and its dependence on temperature. Many laboratory studies have appeared in the literature on the bulk phase stability of EOR polymers (eg 19-21). Although the process appears to be quite complex, the ultimate result is that a desirable property (eg viscosity) is eroded with time. If we assume a simple first order reaction for the polymer degradation term, R_i, then we may define the polymer half-life at a given temperature as:

$$\tau_{\frac{1}{2}} = \frac{\ln 2}{\gamma} \tag{3}$$

where γ is the first order rate constant; this assumption is not necessary in our simulation model where any more general treatment is possible but it does simplify the discussion here. If we now define polymer viscosity to be linearly dependent on concentration then η_w also undergoes exponential decay with time. Using the above approach, we may calculate the effect of chemical degradation on a polymer flood. Figure 1 shows the cumulative oil/time profiles for a waterflood, stable polymer flood and a polymer flood with $\tau_{\frac{1}{2}}$ = 6 months in a simple isothermal two layer reservoir system with the properties given in Table 1.

In a non-isothermal waterflood in the same 2 layer system in which the injected water is at 70°F and the reservoir is at 200°F, Figure 2 shows the temperature distribution after 4 years of waterflooding (thermal properties in Table 2). Note that there is considerable cooling in the streak and we might expect that the chemical degradation rate for the polymer may be much lower in this region. If the degradation rate is characterised by an activation energy E_a, then we assume that its rate depends on the absolute temperature, T, according to the familiar Arrhenius equation:

$$\left(\frac{dC}{dt}\right)_{deg} = - f(C) \exp (E_a/RT) \tag{4}$$

where f(C)is a function of polymer concentration (proportional to C for a first order reaction) and R is the gas constant. For an estimated activation energy of 100 kJ mol^{-1} and a temperature of 200°F, the reaction rate will approximately halve for each 18°F (10°C) drop in temperature. With this assumption, we have repeated the calculation for the stable and degrading polymer floods in the 2 layer system. The half-life is again taken to be 6 months at 200°F but this now increases (ie degradation rate decreases) at lower temperatures. Figure 3 shows the calculated cumulative oil recovery for this case along with some other cases for comparison. It is clearly seen that the recovery efficiency for this case where the degradation rate depends on temperature is very close to that for the stable polymer (28).

In order to understand the effects of chemical degradation, it has been necessary to use simulation models to establish the mechanisms of oil recovery in stratified systems. This is discussed in the following section.

FLOW MECHANISMS IN POLYMER FLOODING OF LAYERED SYSTEMS

In layered reservoirs, such as many candidates for polymer flooding in the North Sea and the United States, polymer acts by the mechanism of vertical flow diversion, and it is this mechanism alone which we discuss in this section. Polymer behaviour in a layered system is often described as follows: an injected polymer slug predominantly enters a high permeability layer, which it "blocks" due to its high viscosity, so diverting chase water into layers of lower permeability. The reality is more complicated, particularly in reservoirs where the layers are in good vertical communication (28, 29). The situation is pictured in Figure 4. The polymer slug does act as a "plug" in the high permeability layer, but its effect on

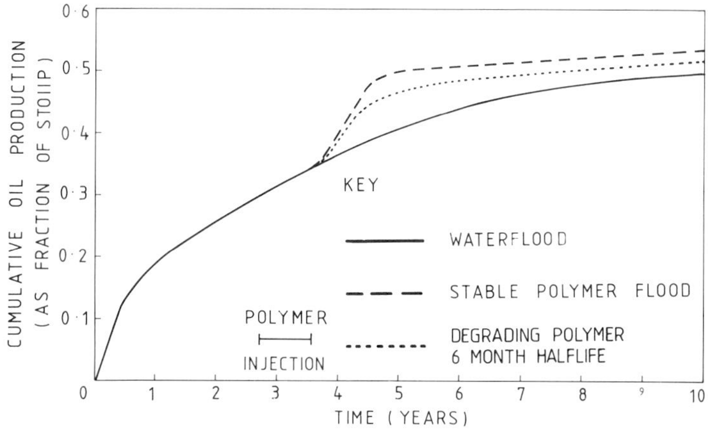

FIG 1 EFFECT OF POLYMER DEGRADATION WITH A 6 MONTH HALF LIFE ON CUMULATIVE OIL PRODUCTION IN THE BASE CASE TWO LAYER MODEL

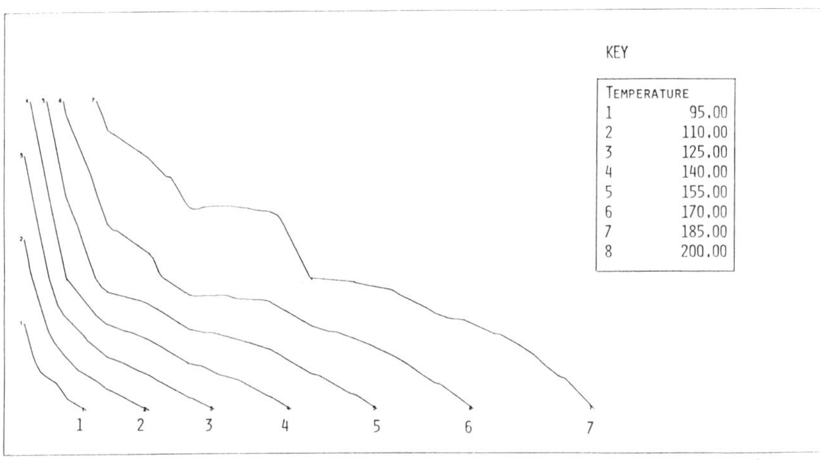

FIG 2 TEMPERATURE DISTRIBUTION AFTER 4 YEARS IN THE BASE CASE MODEL USING THE THERMAL PROPERTIES IN TABLE 2

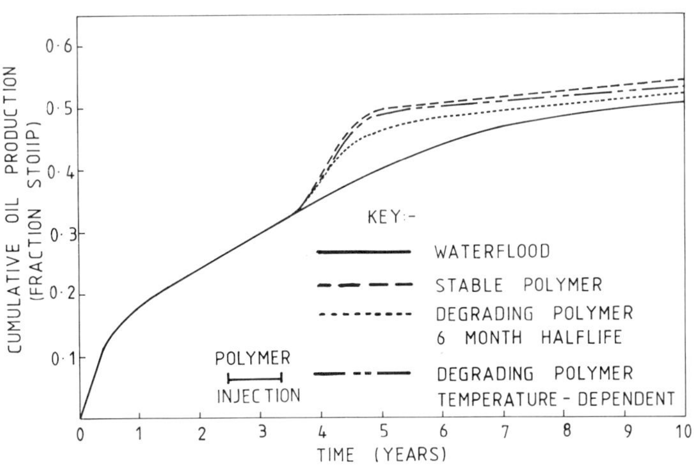

FIG 3 EFFECT OF POLYMER DEGRADATION ON CUMULATIVE OIL PRODUCTION WHEN DEGRADATION RATE DEPENDS ON THE LOCAL TEMPERATURE (FIG 2)

TABLE 1

THE BASE CASE TWO-LAYER MODEL

	LAYER 1	LAYER 2
LENGTH	3000 ft	3000 ft
WIDTH	1000 ft	1000 ft
THICKNESS	20 ft	80 ft
HORIZONTAL PERMEABILITY	2000 md	100 md
VERTICAL PERMEABILITY	200 md	10 md

FOR BOTH LAYERS,

Porosity = 0.20
Connate water S_{WC} = 0.25
Residual oil S_{OR} = 0.22

$$\text{Relative permeability to water} = 0.3 \times \left(\frac{S_W - S_{WC}}{1 - S_{OR} - S_{WC}} \right)^2$$

$$\text{Relative permeability to oil} = 0.9 \times \left(\frac{S_O - S_{OR}}{1 - S_{OR} - S_{WC}} \right)^3$$

FLUID PROPERTIES

oil = 1.046cp
water = 0.307cp

Polymer solution viscosity varies with concentration as:

C/ppm	/cp
0	0.307
100	0.82
200	1.52
300	2.42
400	3.62
500	5.25

water = 61.45 $lbsft^{-3}$
oil = 48.24 $lbsft^{-3}$

TABLE 2

RESERVOIR AND FLUID THERMAL PROPERTIES USED IN THE CALCULATIONS IN SECTION 2

THERMAL CAPACITY OF RESERVOIR ROCK = 35.0 BTU/ft^3 - ^{o}F
WATER = 62.4 BTU/ft^3 - ^{o}F
OIL = 31.2 BTU/ft^3 - ^{o}F

THERMAL CAPACITY OF UNDERBURDEN AND OVERBURDEN
= 35.0 BTU/ft^3 - ^{o}F

THERMAL CONDUCTIVITY OF RESERVOIR AND SURROUNDINGS
= 25.0 BTU/ft-day-^{o}F

flow is localised rather than global. The fluid flow ratio in the high to low permeability layers is altered only in the neighbourhood of the slug. Chase water is diverted out of the high into the low permeability layer at the rear of the slug, carying some polymer with it. Flow in the low permeability layer adjacent to the slug is enhanced, and this additional flow crosses back into the high permeability layer ahead of the slug. Therefore, in the time before the polymer has approached the production well, the oil which it sweeps from the low permeability layer actually emerges from the high permeability layer. It should also be noted that there will be a delay between the start of polymer injection and the first appearance of incremental oil. The delay period is the time taken for crossflowed oil to flow from injector to producer, ahead of the slug, along the high permeability layer.

The outcome of polymer flooding in a reservoir with layers in vertical communication clearly differs from the case in a reservoir with non-communicating layers. In the latter case, the polymer causes a gross change in the flow ratio between the layers. The fluid diversion, and hence incremental oil production, begins as soon as polymer is injected and the incremental oil is produced directly from the low permeability layers.

Most layered reservoirs, without an impermeable rock barrier between layers, approximate to a case of "infinite" vertical communication, in which the pressure gradients are identical along the length of the different layers. This approximation, as used for example by Zapata and Lake (30), is a consequence of the thinness of a typical pay zone compared to the well-to-well distance. It is not possible, a priori, to say that a vertically communicating reservoir is better or worse for polymer application than a non-communicating reservoir. In the case with vertical communication, the slug will be lost from the high permeability layer by progressive crossflow of its rear end. However, while it remains intact, it will cause a greater rate of incremental oil flow than a similar slug in a non-communicating reservoir. Also, the initial slug emplacement, and the effects of polymer adsorption and chemical degradation, will differ between the two cases.

To illustrate some of these features, a set of simulations was carried out in a simplified reservoir vertical cross-section. The (20x4) block section represented a model of the type shown in Figure 4, with a 60 ft layer of high permeability underlying a 120 ft layer of low permeability in a reservoir of length 3000 ft. The oil/water mobility ratio was close to unity. The system was flooded at a constant rate of 0.2 PV per year. An initial waterflood of 0.6 PV (at the end of which a watercut in excess of 80% was reached) was followed by 0.2 PV of polymer solution, and then by chase water. It is recognised that a polymer flood in a vertically communicating system will be characterised by large flows, and rapid concentration changes, in small regions close to slug boundaries. Therefore a simple simulation of this kind will not always give an accurate quantitative calculation of polymer slug emplacement and crossflow. However, it is adequate to display the mechanisms which are occurring during the polymer process.

Certain features of the reservoir and fluids were varied in different calculations. Figure 5 shows the way in which polymer crossflow depends on the vertical to horizontal permeability ratio, (k_z/k_x), of the reservoir rock. The injected polymer viscosity is here ten times water viscosity, and for most calculations, the high permeability layer has ten times the

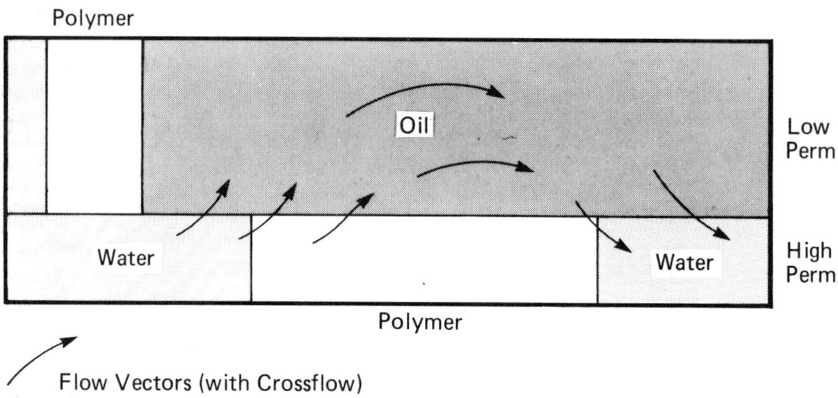

FIG 4 OIL RECOVERY MECHANISM FOR POLYMER FLOODING OF HETEROGENEOUS SYSTEMS

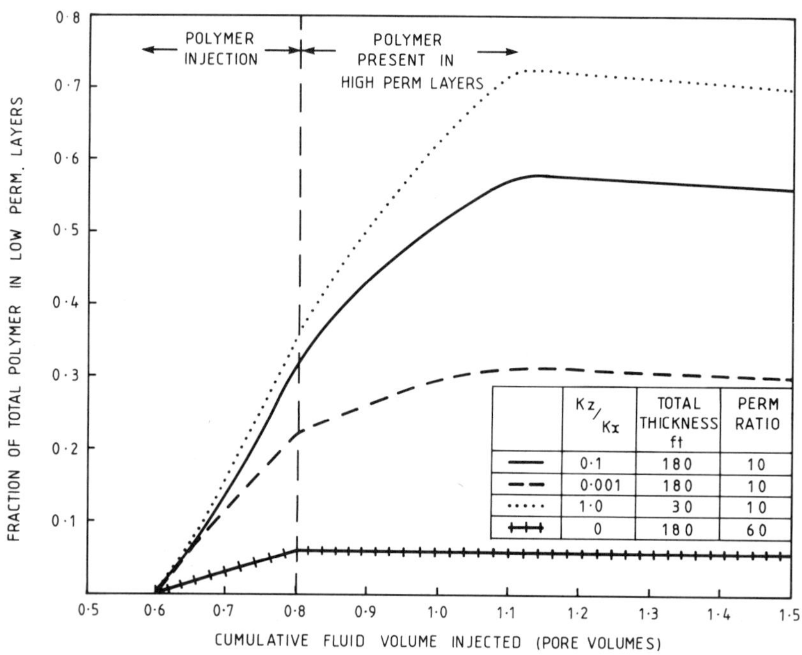

FIG 5 DEPENDENCE OF POLYMER CROSSFLOW ON VERTICAL PERMEABILITY OF 2-LAYER RESERVOIR

permeability of the low permeability layer. Even for a modest vertical permeability (k_z/k_x) = 0.1) in a thick pay zone (180 ft total), it is seen that the quantity of polymer in the low permeability layer does not remain constant after the end of polymer injection. In fact, it increases from 30% to 60% of the total polymer, due to polymer crossflow from the rear of the slug in the high permeability layer. A thinner reservoir of 30 ft thickness, with $k_z/k_x = 0.1$, will have as much as 70% of the total polymer ending up in the low permeability layer. It is necessary to reduce k_z/k_x to a value as small as 10^{-3} before the polymer flow even approaches the case for a non- communicating reservoir ($k_z/k_x = 0$), in which the quantity of polymer in the low permeability layer remains fixed after the end of polymer injection.

The oil crossflow in a vertically communicating reservoir is illustrated by Figure 6. It is seen that for the system with good vertical communication ($k_z/k_x = 0.1$), polymer initially produces an increase in the oil content of the high permeability layer, compared to a waterflood. This is due to the crossflow of oil into the high permeability layer ahead of the advancing slug, from where it is eventually produced. This feature is much reduced in the case of poor vertical communication. Figure 7 illustrates the timing of incremental oil production in the different cases. A reservoir with $k_z/k_x = 0$ shows incremental production starting as soon as polymer injection begins, reaching a maximum rate following the end of polymer injection, with an eventual decline in the rate as polymer is produced from the high permeability layer. A communicating case (k_z/k_x = 0.1), on the contrary, shows no incremental production until after the end of polymer injection, followed by a very sharp pulse of oil flow as polymer approaches the producer. By the time of the maximum incremental oil production rate, most of the polymer has long since left the high permeability layer, having already completed its task.

One consequence of these flow mechanisms is that, since much of the polymer in a vertically communicating reservoir leaves the high permeability layer in a time significantly less than the well-to-well transit period, it is not necessary for the polymer to survive chemically for all of that period. Therefore, polymer chemical degradation at a given rate has a smaller effect than might be expected. Figure 1 shows that the effect of a six month degradation half-life is relatively small, even though incremental production occurs over a period of about 2 years (28, 29).

The results presented here are for only a narrow range of cases. However, experience shows that the same mechanisms apply across a wide range of geological structures and fluid properties, including three-dimensional layered reservoirs and partially adsorbing polymers.

THE SCALING OF POLYMER FLOODING PROCESSES IN HETEROGENEOUS SYSTEMS

The importance of fluid crossflow between layers, and the enhanced viscous crossflow effects of the polymer in the reservoir have been discussed. In order to study this crossflow behaviour in the laboratory, we must be able to design experiments that show these effects in a correctly scaled way. However, it is well known that there are other crossflow effects which occur, both in the field and in experimental systems, and it is important to consider how these relate in magnitude to the viscous crossflow. All crossflow effects are summarised in Figure 8 and are described below:

(i) Viscous crossflow: pressure gradients develop between layers, since fluids of different mobilities move at different rates within each of the layers (30).

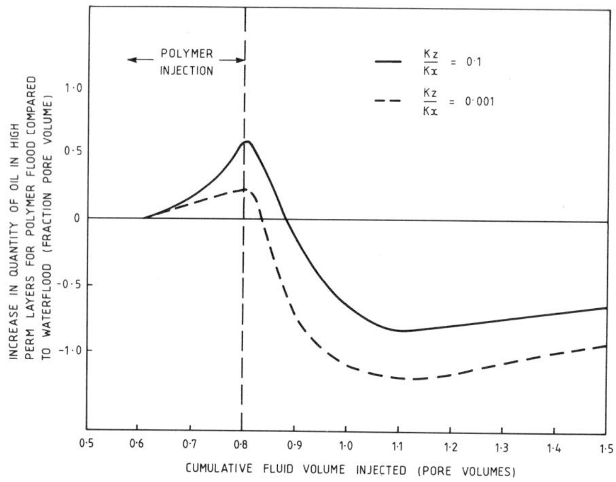

FIG 6 CROSSFLOW OF OIL IN 2-LAYER RESERVOIRS OF DIFFERENT VERTICAL PERMEABILITY

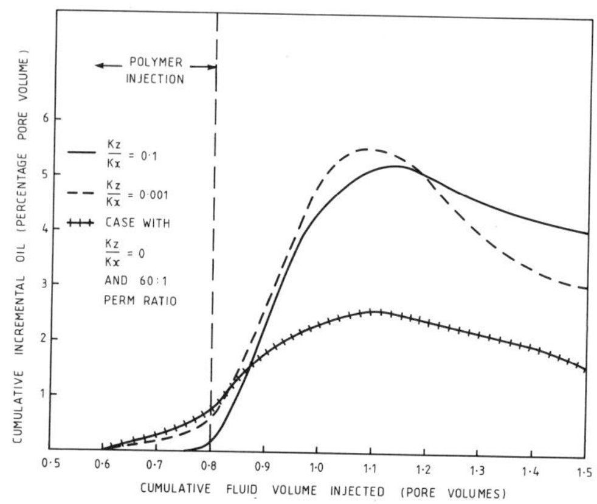

FIG 7 TIMING OF INCREMENTAL OIL PRODUCTION AS A FUNCTION OF VERTICAL PERMEABILITY OF 2-LAYER RESERVOIR

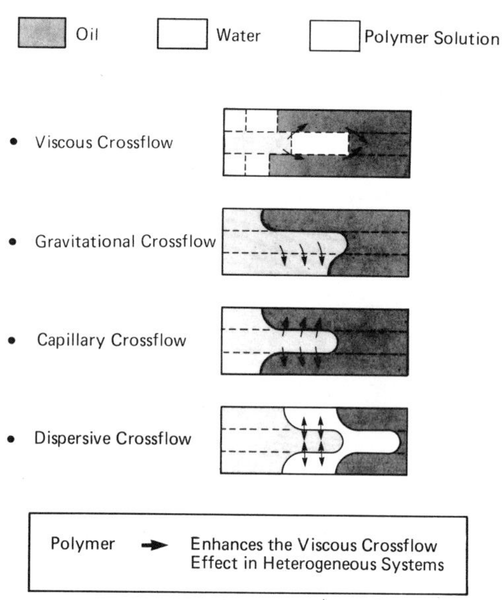

FIG 8 CROSSFLOW EFFECTS IN HETEROGENEOUS SYSTEMS

(ii) Capillary crossflow: an advancing waterfront in one layer invades an adjacent oil rich zone (in a water wet system), thus modifying the relative positions of the waterfronts in each layer, and generally improving oil recovery efficiency (31).

(iii) Gravitational crossflow: the aqueous phase tends to migrate into lower layers of the reservoir due to oil-water density differences (32, 33).

(iv) Dispersion: concentration fronts of transported components within a phase are spread out, both longitudinally and between layers, due to a combination of local porous medium heterogeneity and molecular diffusion (34, 35).

We have recently studied the relative magnitudes, and scaling, of viscous, gravitational, capillary and dispersive forces in polymer floods in layered systems (36). The study of dimensionless groups governing the scaling of results from two phase flow experiments to the reservoir were initiated by Leverett et al (37), over forty years ago. Both theoretical and experimental work has been presented on this topic on numerous occasions since that time (32, 33, 38-44). However, until recently (36), no one had applied results from scaling theory specifically to the study of polymer flooding processes.

Any reservoir or laboratory system may be described by the set of dimensionless dependent and independent variables shown in Figure 9. In order to ensure that the same independent variable values give the same dependent variable values for the laboratory and field system, a set of dimensionless groups must be held constant. These groups, known as similarity groups, describe the ratio of the various forces to each other. We have extended the approach of Rapoport (38) to include non-isotropic rock permeabilities (36). This allows us to investigate the scaling of crossflow effects in relation to the vertical to horizontal permeability ratio. Four similarity groups are derived in this manner, and these are summarised in Table 3; dispersion groups are also presented, where the dispersion in the flow equation has been taken to be a linear function of horizontal velocity for both longitudinal (x or y) and transverse (z) flow and α is the dispersivity.

The validity of these scaling groups in describing polymer flooding processes has been confirmed by numerical simulation (36). An example of such a calculation is presented in Figure 10 for a polymer flooding calculation in a simple, scaled, three layer system, both at the field and laboratory scale (see Table 4). From the values of the scaling groups alone, it would be impossible to gauge the contribution of gravity or the effect of polymer etc on oil recovery; in order to do this, numerical simulation - or a correctly scaled experiment - is required.

An important point to note from the similarity groups in Table 3, is that the ratio of gravity to viscous effects may be represented by the group:

$$G_{gv} = \frac{k_x \Delta\rho g \Delta z^2 \Delta y}{q \Delta x \mu_o} \tag{5}$$

and in a linearly scaled system this group will remain a constant. However, capillary pressure effects scale in a more complex way (36), in that the quantity (dP_c/dS_w) must scale with the linear dimension of the system over

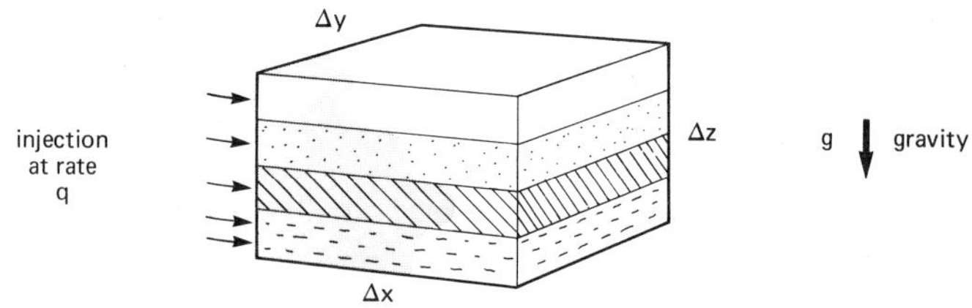

INDEPENDENT GROUPS = $\frac{x}{\Delta x}, \frac{y}{\Delta y}, \frac{z}{\Delta z}, \frac{t}{\Delta t}$

DEPENDENT GROUPS = S_w water saturation at a point

C_i polymer/tracer concentration

$\frac{P}{\Delta P}$ scaled pressure

$\frac{\text{cumulative oil}}{\text{original oil in place}}$

GIVEN INDEPENDENT VARIABLES → SAME DEPENDENT VARIABLES
IF SIMILARITY GROUPS HELD FIXED

FIG 9 INDEPENDENT AND DEPENDENT VARIABLES IN THE SCALING GROUP CALCULATIONS

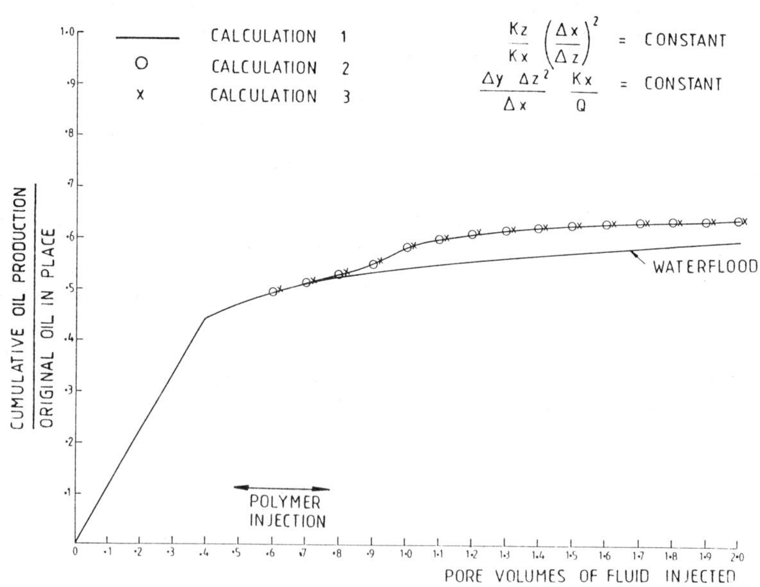

FIG 10 NUMERICAL CONFIRMATION OF THE SCALING GROUPS IN WATER AND POLYMER FLOODS FOR THE SCALED SYSTEMS IN TABLE 4

TABLE 3

SIMILARITY GROUPS USED TO SCALE VISCOUS, CAPILLARY, GRAVITY AND DISPERSIVE EFFECTS IN WATER AND POLYMER FLOODING

These express the relative magnitudes of the viscous, gravitational, capillary and dispersive effects

$$\frac{\mu_o}{\mu_w} = \text{oil/water viscosity ratio}$$

$$\frac{q\,\Delta x\,\mu_o}{k_x\,\Delta y\,\Delta z\left(\frac{dPc}{dSw}\right)} = \frac{\text{Viscous Forces}}{\text{Capillary Forces}}$$

$$\frac{\Delta\rho\,g\,\Delta z}{\left(\frac{dPc}{dSw}\right)} = \frac{\text{Gravitational Forces}}{\text{Capillary Forces}}$$

$$\frac{k_x}{k_y}\left(\frac{\Delta y}{\Delta x}\right)^2 \quad \text{and} \quad \frac{k_x}{k_z}\left(\frac{\Delta z}{\Delta x}\right)^2$$

DISPERSION

$$\frac{\alpha_x}{\phi\,\Delta x} = \frac{\text{Longitudinal Dispersion}}{\text{Longitudinal Convection}}$$

$$\frac{\alpha_z\,\Delta x}{\phi\,\Delta z^2} = \frac{\text{Transverse Dispersion}}{\text{Transverse Convection}}$$

NB These Groups apply to water and polymer floods and can include the effects of both adsorption and chemical reaction.

TABLE 4

CORRECTLY SCALED 3 LAYER SYSTEMS AT THE EXPERIMENTAL AND RESERVOIR SCALE (NO CAPILLARY PRESSURE)

CALCULATION	1	2	3	4	UNITS
SCALE	EXPERIMENT	RESERVOIR	RESERVOIR	RESERVOIR	
ΔX	1.640	1000	1000	1000	ft
ΔY	0.328	200	200	200	ft
ΔZ	0.295	180	216	180	ft
Kx (Low)	500	500	416.67	250	mD
Kx (High)	10000	10000	8333.33	5000	mD
K_V/K_H	0.1	0.1	0.144	0.1	-
Ø (Low)	0.2	0.2	0.2	0.2	-
Ø (High)	0.38	0.38	0.38	0.38	-
μ_O	2.0	2.0	2.0	2.0	cp
μ_W	1.0	1.0	1.0	1.0	cp
$\Delta\rho$	13.21	13.21	13.21	13.21	lbs/ft^3
V_D*	10	10	10	5	ft/day

All calculations were performed with a three layered system, the middle layer having the high permeability.

The calculations 1 → 3 were performed first for a waterflood and then for a 20% pore volume polymer slug injected after 0. pore volumes of water. Calculation 4 was only performed for waterflood.

* V_D is defined as the initial horizontal Darcy velocity in the high permeability streak.

the whole saturation range. Suppose that we have a given capillary pressure curve that is found (from calculation) to be quite significant at the experimental scale. We may then assess how significant the same capillary pressure curve is when the model is linearly scaled up by a large factor, F, to the reservoir scale keeping all fluids and rock properties the same. If the Darcy velocity is also kept constant, then the values of the scaling groups describing viscous/capillary and gravity/capillary forces increase by a factor F as we go to the reservoir scale whereas the group describing viscous/gravity forces remains the same. However, care must be exercised in interpreting this fact; by essentially removing the effect of capillary pressure, other complex couplings between these forces are also affected and it is not possible to say that the relative contributions of viscous and gravitational effects to the oil recovery mechanism are identical in both systems.

The above discussion indicates, that, in order to reproduce correctly scaled viscous crossflow phenomena in both experimental waterfloods and polymer floods, the effects of capillary pressure at the laboratory scale should be minimised. This may be done in oil-water displacement experiments by using high flow rates and fluids of higher viscosity; the scaling groups and numerical simulation can be used in order to estimate how large a reduction in capillary effect can be achieved in this way. Alternatively, it may be more appropriate to work in single phase and use polymer (or glycerol thickened water) to displace an in situ tracer or brine solution. In this case, however, care must be taken that dispersive effects do not dominate the flow patterns. Experimental work will be presented in the next section for both two phase and single phase displacement experiments which illustrate polymer flooding displacement mechanisms.

LABORATORY EVALUATION OF POLYMER DISPLACEMENT MECHANISMS IN LAYERED SYSTEMS

In order to reproduce the appropriate crossflow phenomena in polymer flooding in layered systems, the laboratory core assemblies must have some known degree of heterogeneity with communication between layers. Fluid flow in such systems for waterflooding and single phase flow has been studied by several workers (32-35, 44-55). These studies have emphasised various effects such as that of gravity segregation (32), flow rate (46), transverse dispersion (34, 35) and slug breakdown (51, 52). In some cases the effects of polymer have been demonstrated in layered systems (53, 54, 55). Szabo (55) has reported results for polyacrylamide floods in a three layer stratified sand pack, in which polymer adsorption is present. Szabo gives an approximate qualitative analysis of the oil recovery mechanism which is consistent with the ideas presented this work.

At this laboratory, we are investigating stratified cores with known levels of heterogeneity of the following types:

(i) Cored cores which have the centre removed from a cylindrical core of Clashach (16, 56) sandstone, and filled with high permeability ballotini, as shown in Figure 11 (a).

(ii) Rectilinear layered cores, with two rectangular slabs of Clashach sandstone separated by a streak of ballotini, as shown in Figure 11 (b).

A design feature of these systems is separate effluent ports for each layer. This allows us to demonstrate one of the key features of a non-unit mobility flood; namely the change in the local flow rates in each of the layers, and the altered effluent profiles from the layers.

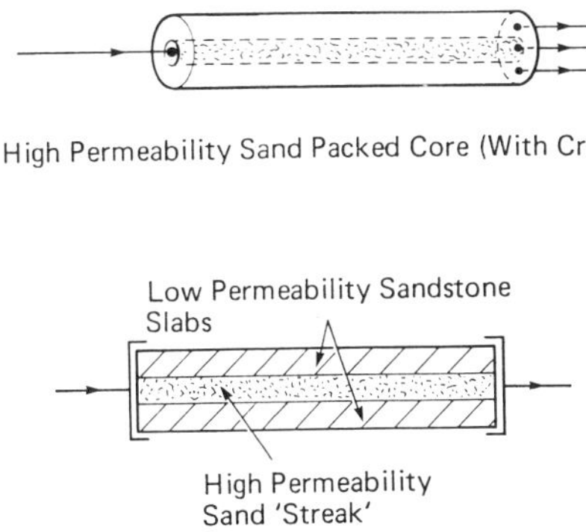

(a) High Permeability Sand Packed Core (With Crossflow)

(b) Layered Sand/Rock Pack (With Crossflow)

FIG 11 TWO DIMENSIONAL EXPERIMENTAL CORE ARRANGEMENTS FOR POLYMER FLOODS

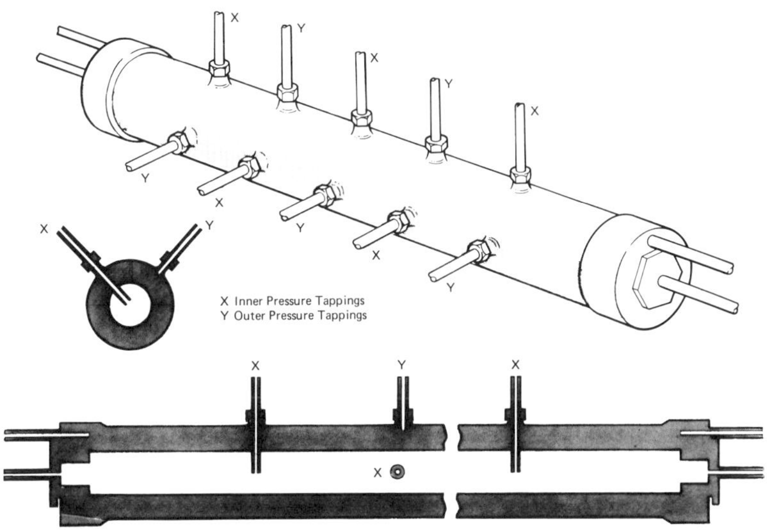

FIG 12 CORED CORE DUAL PERMEABILITY EXPERIMENT

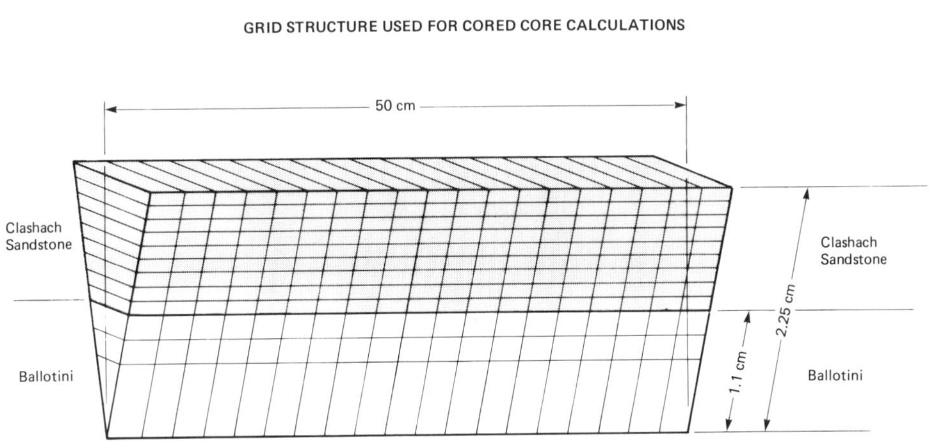

FIG 13 CURVILINEAR GRID USED IN THE MODELLING OF THE TWO PHASE DISPLACEMENT EXPERIMENTS IN THE CORED CORE ASSEMBLY

At present, our work is more advanced with the cored core, which is shown in more detail in Figure 12, and experimental results are only presented for this system in this paper. A simple system with this geometry has been used previously for demonstrating sweep improvement in a qualitative way in polymer flooding (54). We have modelled this system by solving the flow equations in a curvilinear grid based on the lengthwise pie-shaped slice shown in Figure 13. Both two phase and single phase experiments have been carried out, and numerical simulation has been used to analyse these. In all floods, injection is at constant rate into the ballotini higher permeability layer, and production is usually from both layers.

Experimental details of the core and fluids used in the two phase displacement experiments are presented in Figure 14. The flood sequence used in this core assembly is shown in Figure 15. The polymer used was a slow filtered (16, 57) xanthan polysaccharide manufactured by Pfizer. During these floods, measurements were made of cumulative oil, watercuts and polymer concentrations in the effluent water. The experimentally determined cumulative oil and fractional flow of water (watercut) for both the waterflood and the xanthan polymer flood are shown in Figure 16. The fact that the polymer flood improves oil recovery compared with the waterflood, is an indication that viscous crossflow has been significantly enhanced, if our proposed mechanisms are correct. Since we have no direct measurement of the quantities of oil in the high and low permeability layers during these floods, we must resort to numerical simulation in order to clarify what is happening. Using the measured fluid, rock and polymer properties, simulations were performed using the grid in Figure 13. The calculated profiles for cumulative oil recovery, watercut, and polymer concentration are compared with experimental results in Figures 17, 18 and 19. The theoretical results are in good agreement for cumulative oil and watercut; the calculated polymer profile is in reasonable agreement with experiment, although there is some uncertainty in the experimental measurements of polymer concentration. Using the simulation, we have obtained the predictions of the total oil in place in each layer as a function of time (pv) for both the waterflood, and the polymer floods, as plotted in Figure 20. The time of water breakthrough for both the water and polymer floods is shown on this figure, and the following points should be noted:

(i) In both the waterflood and the polymer flood all of the recoverable oil is removed from the high permeability ballotini streak by 0.6 pv.

(ii) All of the incremental oil is obtained from the low permeability Clashach annulus.

(iii) Before breakthrough oil leaves the low permeability layer and enters the high permeability layer.

The conclusion from the calculations is quite clear: polymer is causing some of the incremental oil to crossflow from the lower to the higher permeability layer, from which it is later produced, in accordance with the mechanisms discussed in Section 3. We note that supporting calculations indicate that gravity and capillary effects are not very important in this system, in spite of its small scale.

More detailed experiments have been performed in single phase at a range of favourable viscosity ratios in another, better characterised, cored core assembly (58). Details of this system are presented in Table 5. In these experiments, glycerol was used as the viscosifier in place of polymer in order to simplify the modelling of the transport and to emphasise only the viscous effects. Extensive use of chlorine-36 labelled brine and

DIMENSIONS

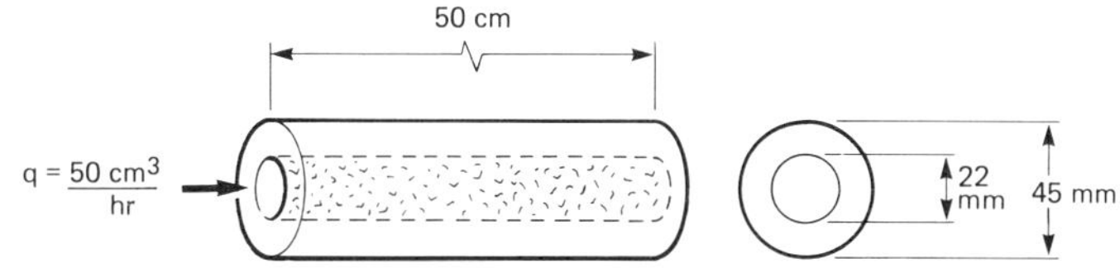

PERMEABILITIES	Ballotini central streak	= 6.2 darcy
	Clashach sandstone annulus	= 1.3 darcy

FLUIDS	Viscosities oil (dodecane)	μ_o	= 1.26 cp
	water (artificial sea water)	μ_w	= 0.87 cp

POLYMER Xanthan polysaccharide (Pfizer FLOCON 4800)
Very slow filtered
800 ppm stock has η_p = 24 cp

FIG 14 EXPERIMENTAL DETAILS OF THE CORE AND FLUIDS USED IN THE TWO PHASE DISPLACEMENT EXPERIMENTS

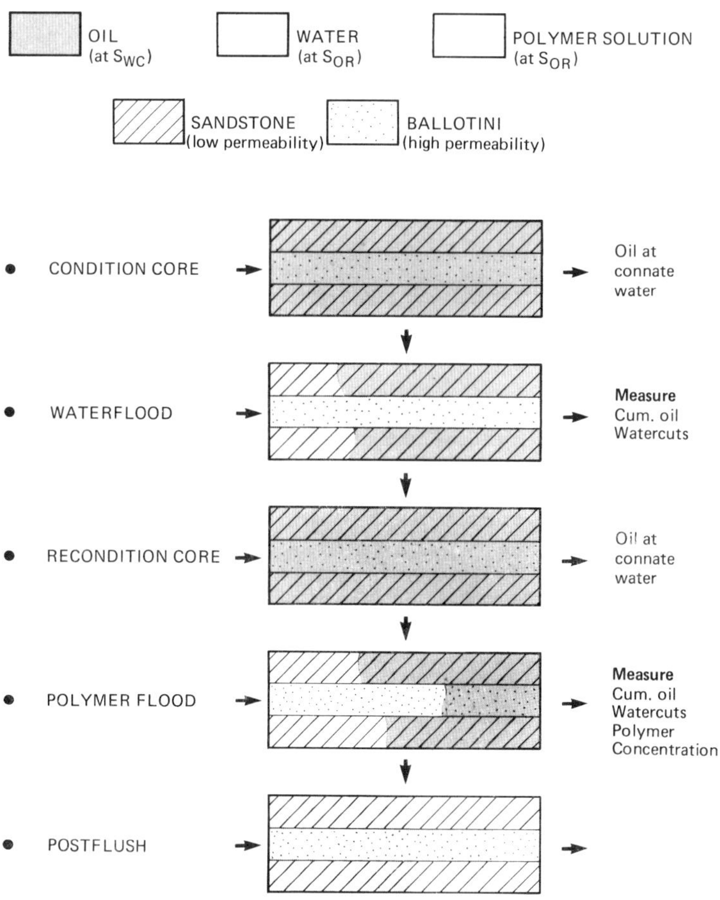

FIG 15 FLOODING SEQUENCE IN CORED CORE OIL DISPLACEMENT EXPERIMENTS

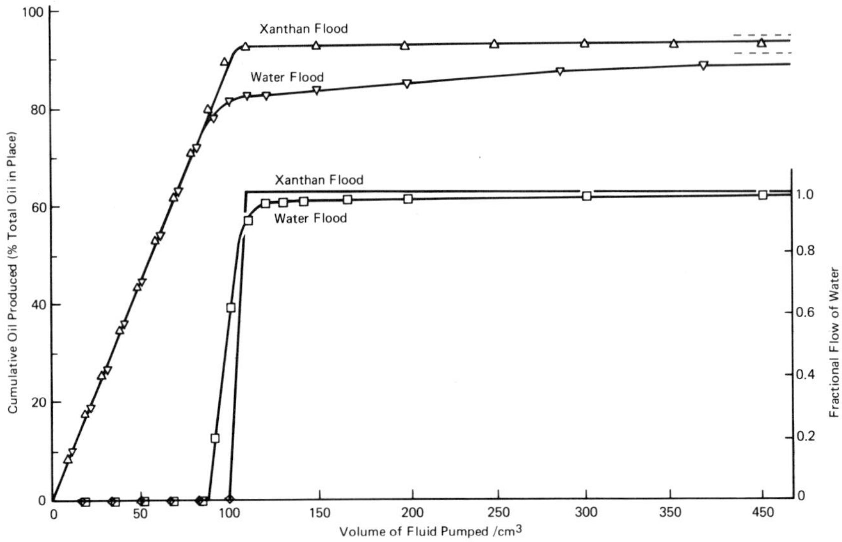

FIG 16 CUMULATIVE OIL AND WATERCUT PROFILES FOR THE TWO PHASE WATER AND POLYMER FLOODS IN THE CORED CORE GEOMETRY

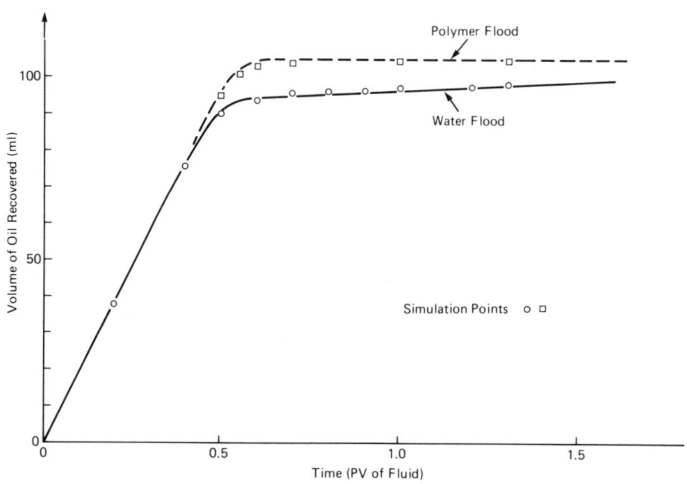

FIG 17 CUMULATIVE OIL PROFILE FOR WATER AND POLYMER FLOODING OF CORED CORE: EXPERIMENTAL AND SIMULATION RESULTS

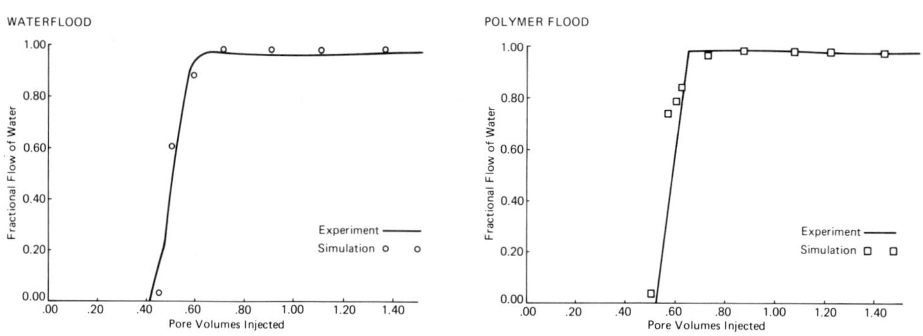

FIG 18 TOTAL FRACTIONAL FLOW OF WATER AT OUTLET FOR CORED CORE ASSEMBLY: EXPERIMENTAL AND SIMULATION RESULTS

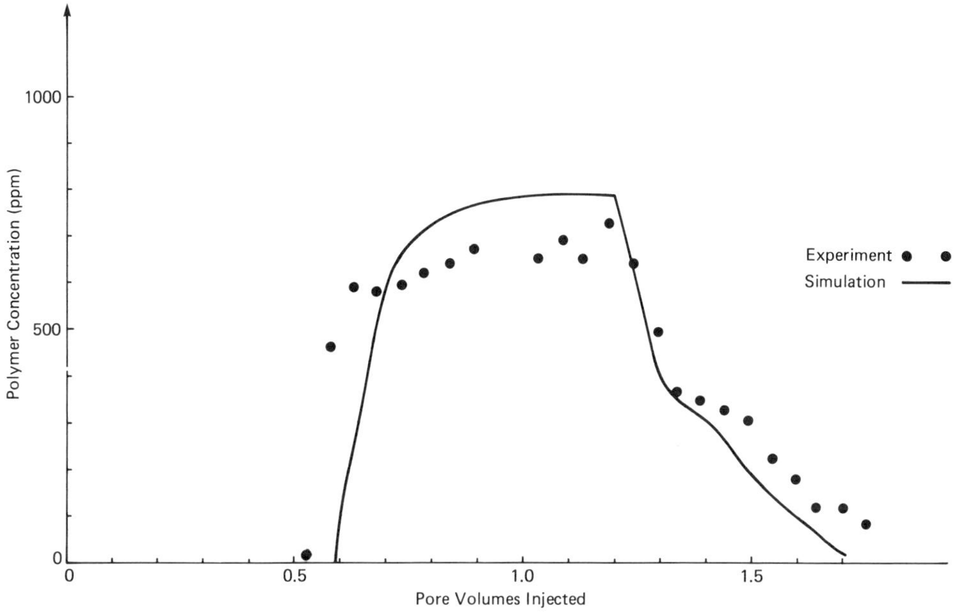

FIG 19 TOTAL EFFLUENT CONCENTRATION OF XANTHAN IN CORED CORE POLYMER FLOOD: EXPERIMENTAL AND SIMULATION RESULTS

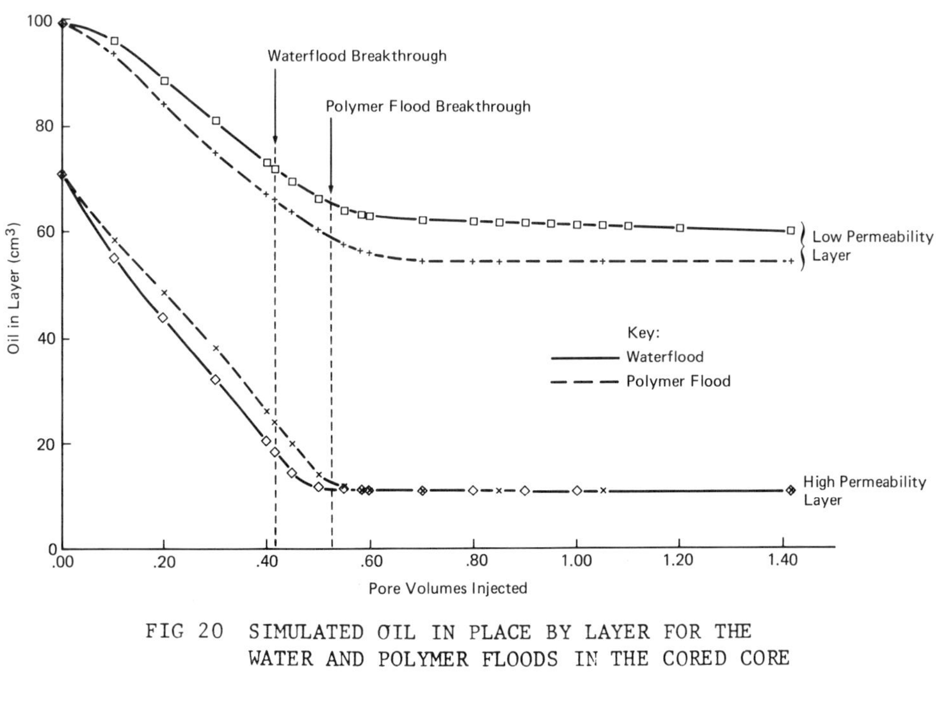

FIG 20 SIMULATED OIL IN PLACE BY LAYER FOR THE WATER AND POLYMER FLOODS IN THE CORED CORE

carbon-14 labelled glycerol (both β-emitters) was made, in order to fully elucidate the crossflow mechanisms. We present results for a flood sequence consisting of a preflush with unlabelled brine, into a core full of chlorine-36 labelled brine with injection and production in the ballotini only, followed by a 10:1 viscosity ratio glycerol flood where injection is into the ballotini, and production is from both layers. Note that the preflush was intended to sweep the chlorine-36 from the ballotini streak, leaving all remaining labelled chloride in the low permeability layer; the recovery of some of this material from the high permeability streak would be a clear demonstration of fluid crossflow. All effluent profiles are measured, along with the fluid flow rates from each layer (Q_{BL} and Q_{CL}), and pressures at all tappings along the core.

Figure 21 shows the observed effluent profiles of chlorine 36 and carbon-14 in both layers, and Figure 22 shows the corresponding flow ratio, (Q_{BL}/Q_{CL}). In Figure 21, we note that there is a small peak in the chlorine-36 effluent from the ballotini, at about the time that the viscous glycerol breaks through. This is a small but important effect and it signals the crossflow of material as anticipated above. Figure 22 shows that, after about 0.5 pv injection, there is a sharp drop in the flow ratio, (Q_{BL}/Q_{CL}), to a minimum value at about 0.8 pv, when the flow from the high and low permeability sections has almost equalised. This ratio gradually climbs back up to its original value (about 5) over the next 4 pv of injection, as we would expect as the core now fills with 10 cp material. This flow ratio variation is also a consequence of the fluid crossflow occurring in the system.

The most convincing demonstration of crossflow is obtained when we compare results for the "cumulative recovery" from displacements at unit, 2:1 and 10:1 viscosity ratio in this system. The cumulative recovery is defined as the fraction of the total amount of fluid originally in the system which has been displaced; this is calculated from the normalised effluent profiles, and the total flow rates in each layer. Total cumulative recovery profiles are shown for these viscosity ratios in Figure 23. As we expect, the most viscous flood leads to the "highest" recovery in the sense that it brings the recovery forward in time, in exactly the same way that a polymer flood might do. However, it is very instructive to look at <u>where</u> this increased cumulative production comes from. The short answer to this is, of course, from the low permeability Clashach layer, but this conceals the details of the mechanism involved. The respective pore volumes of the ballotini are Clashach and 75 cm^3 and 105 cm^3 and hence they contain 0.42 and 0.58 respectively of the total material. By looking more closely at where this material is produced from, a very interesting observation emerges. Figure 24 presents the cumulative recovery profiles for each of the low permeability Clashach, and higher permeability ballotini layers for the unit and non-unit mobility floods. For the unit-mobility flood, all the material in the ballotini, 0.42 of the total, was produced by about 2 pv as we might expect. However, only 0.3 of the total material in the system has been produced from the Clashach at this stage and this quantity rises only to 0.41 after 5 pv. The results for the 10:1 mobility flood show that, by 2 pv, 0.7 of the total material has been produced from the ballotini although it only contained 0.42 of this material initially; 0.26 has been produced from the Clashach by this time. In other words, <u>less</u> material is produced directly from the Clashach in the viscous flood compared to the unit mobility displacement although the source of the "incremental" recovery is the low permeability Clashach layer itself. The incremental production occurs largely by crossflow of material into the high permeability ballotini layer from which it is subsequently produced. These results give a very clear experimental demonstration of the predicted crossflow effects.

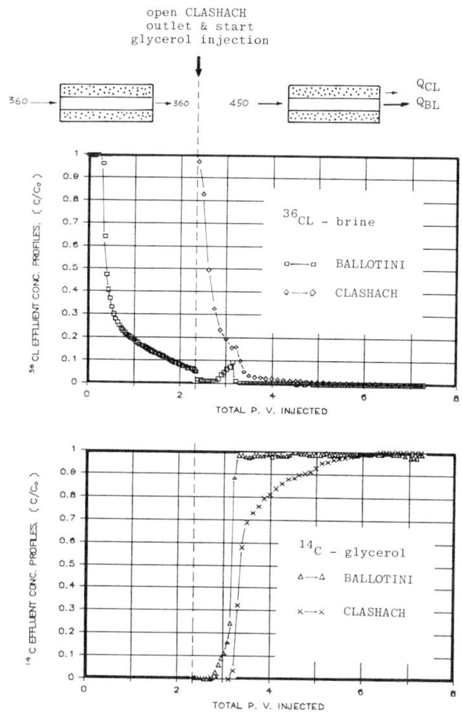

FIG 21 CHLORINE-36 AND CARBON-14 EFFLUENT PROFILES FROM BOTH LAYERS

TABLE 5 CORE DIMENSIONS AND FLUID PROPERTIES

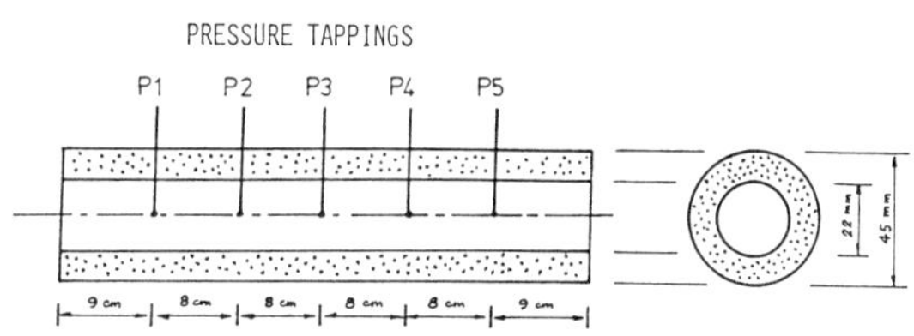

Outer diameter: 4.5 cm

Inner diameter: 2.2 cm

Outer x' sectional area: 12.10 cm^2 (A_1)

Inner x' sectional area: 3.80 cm^2 (A_2)

Ratio of A_1/A_2: 3.18

Length of core: 50 cm

pore volume of Clashach (outer): 105 cm^3 ; porosity = 0.174

pore volume of Ballotini (inner): 75 cm^3 ; porosity = 0.395

	Clashach permeability (darcies)	Ballotini permeability (darcies)
$k_{INLET\ 1}$	0.7569	12.75
k_{12}	0.7569	12.75
k_{23}	0.7130	11.37
k_{34}	0.6743	11.78
k_{45}	0.7612	12.66
$k_{5\ OUTLET}$	0.6214	12.63

Fluid Properties

A.S.W. (brine)	10 cP Glycerol	20 cP Glycerol
viscosity= 0.87 cP @ 30°C	viscosity = 9.89 cP	viscosity = 21.44 cP

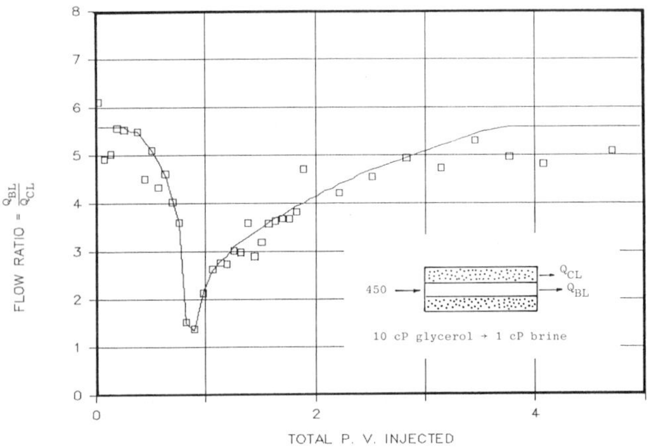

FIG 22 BALLOTINI/CLASHACH FLOW RATIO DURING THE 10:1 VISCOSITY RATIO FLOOD

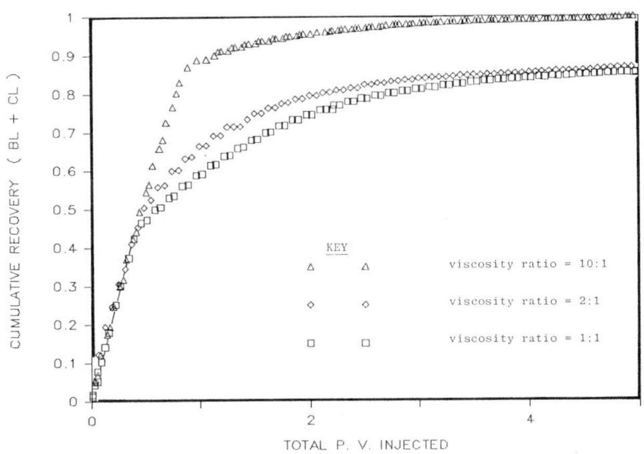

FIG 23 TOTAL CUMULATIVE RECOVERY PROFILES FOR THE UNIT, 10:1 AND 2:1 VISCOSITY RATIO DISPLACEMENT

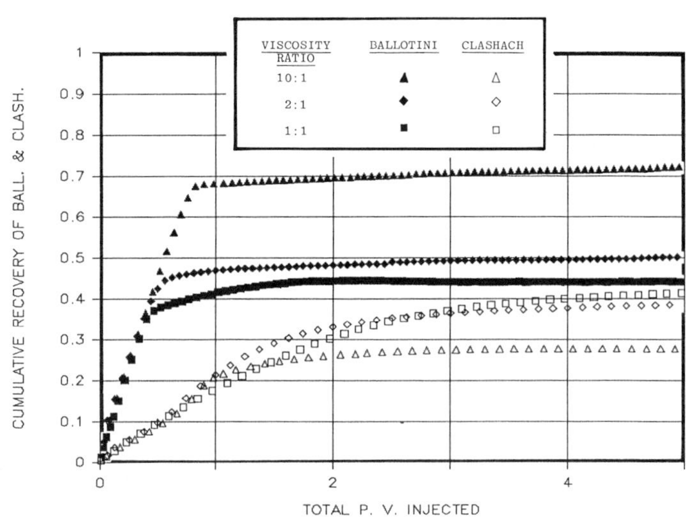

FIG 24 CUMULATIVE RECOVERY PROFILES FOR EACH LAYER

These effects have been simulated and good agreement has been found for these single phase displacements in the cored core (58). Figure 25 shows a comparison between the experimental and calculated effluent profiles in the ballotini and Clashach for the 10:1 displacement discussed earlier. Overall, results are quite well reproduced. Figure 26 shows that there is very good agreement between the calculated and experimentally observed flow ratios, (Q_{BL}/Q_{CL}), for the 10:1 and 2:1 viscosity ratio floods. We can combine these, and other results of our findings into a picture of viscous crossflow, which is shown schematically in Figure 27. This figure indicates why the flow ratio, (Q_{BL}/Q_{CL}), does not drop immediately in the initial part of the viscous flood, and illustrates some aspects of the observed effluents and pressures between layers in the core. The pressure behaviour is not discussed here. A more detailed account of this work is presented elsewhere (58).

We now consider the scaling of results from the two phase and single phase floods presented above, neglecting differences between core and reservoir geometry. In the two phase water and polymer floods, scaling is straightforward since capillary and gravity effects are negligible at the rates at which the experiment was performed. Therefore, if we scale the system linearly, or in accordance with the dimensional scaling groups, results will reflect the field scale case exactly at the correct rates and fluid properties. In the single phase system there is, of course, no capillary pressure and the only processes operating are viscous flow and dispersion. In fact, in the simulation results it was found that transverse dispersion, D_T, had to be included in the calculation in order to obtain the correct effluents at unit mobility ratio; however, in the higher mobility ratio floods the effect of D_T was much less significant (59). In the calculations a constant value for D_T was taken within a given flood and the values 0.0, 1×10^{-4} and 1×10^{-3} $cm^2.s^{-1}$ were examined. From the dispersion groups in Table 3, it can be seen that to obtain an identical contribution of dispersion relative to viscous effects (in systems of identical shape), the longitudinal and transverse dispersivity of each system must scale in proportion to its length. Certain studies (59) indicate that this is valid, at least for longitudinal dispersivity, for reservoir rocks over a length scale ranging from 0.1 to 1000 m, in which case experiments will give a valid picture of a single-phase viscous flood on the reservoir scale. However, if reservoir rocks can be found with a different dependence of dispersivity on length, experiments could either overstate or understate the role of dispersion in the reservoir. Noting that the dispersivity measured in our experiments is about a factor of 1000 less than values obtained for reservoirs, we suggest that the relative contributions of viscous and dispersive crossflow in the experiments may be a good approximation to the reservoir scale.

CONCLUSIONS

In this paper, numerical simulation, scaling theory and experiments have been used in order to study the recovery mechanism operating in the polymer flooding of macroscopically layered systems. All of the work has centred on vertical sweep improvement in which the local water/oil mobility ratio has been close to unity. The main conclusions are as follows:

(i) A versatile simulator, including models for all of the main polymer flow phenomena in porous media, is required in order to establish flow mechanisms in the polymer flooding of stratified systems and to examine sensitivities such as the effect of slug size, polymer degradation, thermal effects etc.

(ii) In stratified systems, with some degree of communication between the layers, the polymer operates through a localised mechanism

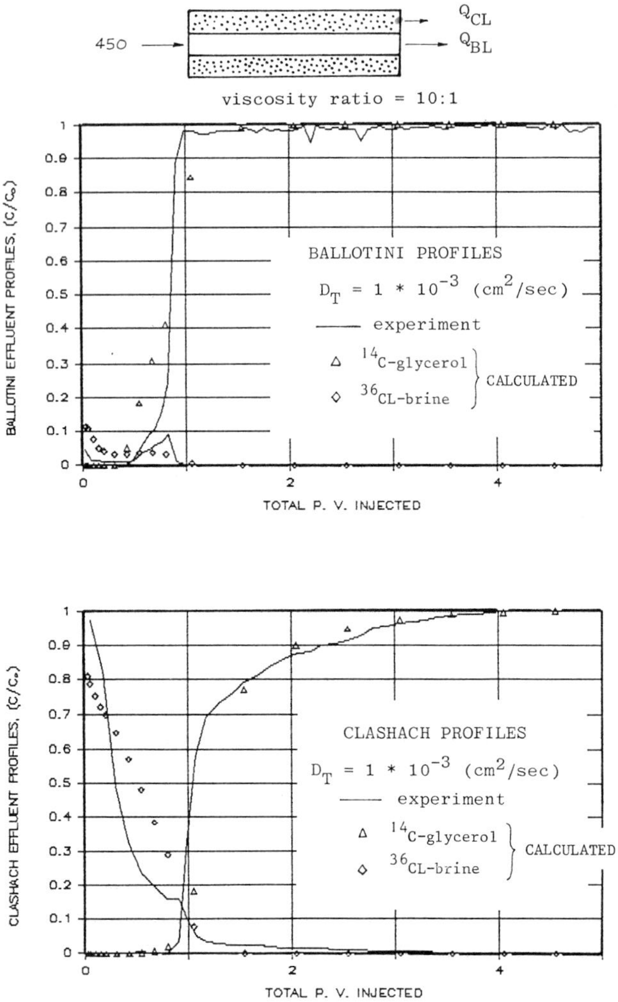

FIG 25 CALCULATED ($D_T = 1 \times 10^{-3}$ cm^2/s) AND EXPERIMENTAL EFFLUENT PROFILES FOR THE 10:1 VISCOSITY RATIO FLOOD

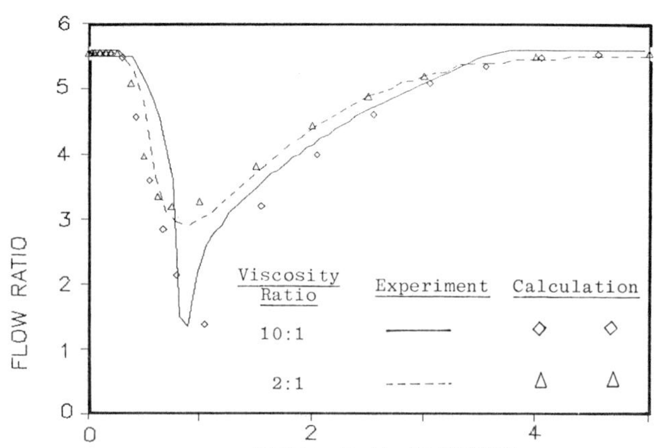

FIG 26 CALCULATED AND EXPERIMENTAL BALLOTINI/CLASHACH FLOW RATIOS FOR 10:1 AND 2:1 VISCOSITY RATIO FLOODS

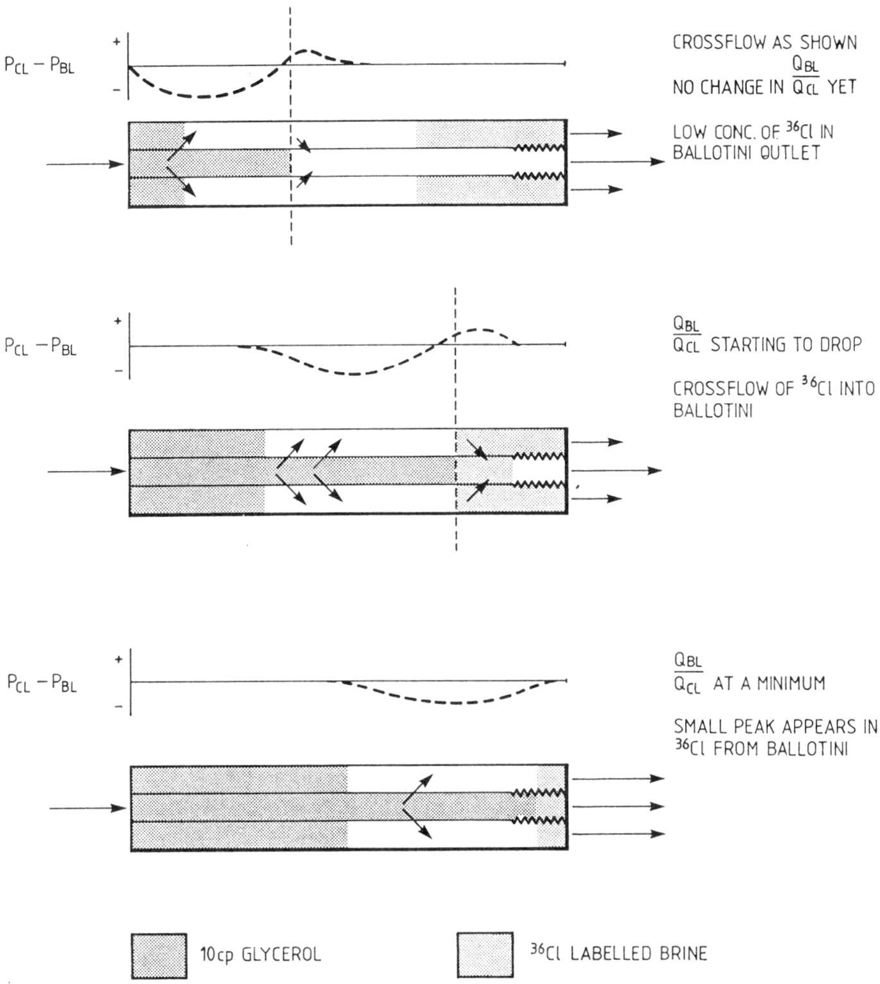

FIG 27 SCHEMATIC DIAGRAM OF CROSSFLOW EFFECT IN LAYERED CORE OR RESERVOIR

of enhanced viscous crossflow; this remains important down to very low values of vertical to horizontal permeability ratio (k_z/k_x).

(iii) A combination of scaling theory and numerical simulation may be used in order to estimate the relative importance of viscous crossflow compared with capillary, gravitational and dispersive crossflow effects.

(iv) Capillary effects may be relatively more important in two phase laboratory floods than at the field scale and these should be reduced by using more viscous fluids, working at higher rates or by performing single phase floods. In single phase floods, the magnitude of dispersive effects should be examined in order to ensure that they contribute at an "appropriate" level.

(v) Both two phase and single phase experiments in well characterised heterogeneous cores were presented, along with theoretical analysis. Crossflow effects were clearly demonstrated in the experiments and scaling was discussed.

NOMENCLATURE

C_i	concentration of polymer or tracer in the aqueous phase
D_i	dispersion of component i in the aqueous phase
D_T	refers specifically to transverse dispersion
E_a	activation energy for polymer degradation reaction
f(c)	concentration function in Arrhenius rate expression for degradation
g	gravitational constant
h	vertical height above a datum plane
k	absolute rock permeability
k_x, k_z	rock permeability in the horizontal (x) and vertical (z) directions
k_{rw}, k_{ro}	relative permeabilities of water and oil
$\tilde{m}_i$	mass density of component i
P_w	pressure in the aqueous phase
P_c	capillary pressure
Q_{BL}, Q_{CL}	flow rates from the Ballotini and Clashach layers in cored core assembly
q_i	source/sink term in the transport equation for component i due to wells
R	the gas constant
R_i	source/sink term for component i due to chemical reaction

S_w	water saturation
S_{OR}, S_{WC}	residual oil and connate water saturations
T	temperature
Δx, Δy, Δz	system overall dimensions in scaling groups
α_x, α_z	horizontal and vertical (transverse dispersivities)
η_w	aqueous phase viscosity; may depend on polymer concentraction, shear rate etc
μ_w, μ_o	water and oil viscosities
ρ_w, ρ_R	density of water and rock
ϕ	porosity
$\tau_{\frac{1}{2}}$	half-life for first order chemical degradation of polymer
γ	first order rate constant in polymer degradation reaction
Γ_i	adsorption level of component i

ACKNOWLEDGEMENTS

We would like to thank the following colleagues: Bob Hawes, Rex Wat and Lawrence Roberts for many helpful discussions and useful advice on the work reported in this paper; Karen Rolf and Hazel Crofts for their assistance with computing and Janet Miller for typing the manuscript. This work is funded by the UK Department of Energy.

REFERENCES

1. ZEITO, G. A., "Three Dimensional Numerical Simulation of Polymer Flooding in Homogeneous and Heterogeneous Systems", SPE 2186, 43rd Ann. Fall Conf. of SPE, Houston, Texas, Sept. 27-Oct. 2, 1968.

2. SLATER, G. E. and FAROUQ-ALI, S. M., "Two Dimensional Polymer Flood Simulation", SPE 3033, 45th Ann. Fall Conf. of SPE, Houston, Texas, 4-7 Oct., 1970.

3. JEWETT, R. L. and SHURZ, A. F., "Polymer Flooding: A Current Appraisal", J. Pet. Tech., p 675-683, June 1970.

4. PATTON, J. T., COATS, K. H. and COLEGROVE, G. T., "Prediction of Polymer Flood Performance", Soc. Pet. Eng. J., p 72-84, March 1971.

5. BONDOR, P. L., HIRASAKI, G. J. and THAM, M. J., "Mathematical Simulations of Polymer Flooding in Complex Reservoirs", Soc. Pet. Eng. J., p 369-382, October 1972.

6. VELA, S., PEACEMAN, D. W. and SANDVIK, E. I., "Evaluation of Polymer Flooding in a Layered Reservoir with Crossflow, Retention and Degradation", SPE 5102, 49th Ann. Fall Conf. of SPE, Houston, Texas, 6-9 Oct., 1974.

7. TODD, M. R., and CHASE, C. A., "A Numerical Simulator for Predicting Chemical Flood Performance", SPE 7689, 5th Symp. on Reservoir Simulation of SPE, Denver, Colorado, 1979.

8. NAIKI, M., "Numerical Simulation of Polymer Flooding Including the Effects of Salinity", PhD dissertation, University of Texas at Austin, August 1979.

9. VAN POOLEN, H. K., "Fundamentals of EOR", Chapter 5, Penwell Books, Tulsa, 1980.

10. SANDVIK, E. I. and MAERKER, J. M., "Application of Xanthan Gum for EOR", ACS Symposium Series No 45, Extracellular Microbial Polyaccharides, Eds P. A. Sandford and A. Laskin, 1977.

11. HEEMSKERK, J., JANSSEN van ROSMALEN, R., HOLTSLAG, R. J. and TEEW, D., "Quantification of Viscoelastic Effects of Polyacrylamide Solutions", SPE/DOE 12652, presented at SPE/DOE 4th Symp. on EOR, Tulsa, Oklahoma, 15-18 April, 1984.

12. CHAUVETEAU, G. and ZAITOUN, A., "Basic Rheological Behaviour of Xanthan Polysaccharide Solutions in Porous Media: Effects of Pore Size and Polymer Concentration", Proc. of 1st European Symp. on EOR, Bournemouth, England, September 1981.

13. CHAUVETEAU, G., "Rodlike Polymer Solution Flow Through Fine Pores; Influence of Pore Size on Rheological Behaviour", J. Rheol., 26, p111, 1982.

14. SOUTHWICK, J. G. and MANKE, C. W., "Molecular Degradation, Injectivity and Elastic Properties of Polymer Solutions", SPE 15652, presented at 61st SPE Annual Fall Conference, New Orleans, 5-8 October, 1986.

15. SZABO, M. T., "An Evaluation of Water Soluble Polymers for Secondary Oil Recovery Parts I and II", J. Pet. Tech., pp 553-570, May 1979.

16. SORBIE, K. S., PARKER, A. and CLIFFORD, P. J., "Experimental and Theoretical Study of Polymer Flow in Porous Media", SPE 14231, presented at 60th Annual Fall Conference of SPE, Las Vegas, 22-25 September, 1985.

17. DAWSON, R. and LANTZ, R. B., "Inaccessible Pore Volume in Polymer Flooding", Soc. Pet. Eng. J., pp448-452, October 1972.

18. SHAH, B. N., WILLHITE, G. P. and GREEN, D. W., "The Effect of Inaccessible Pore Volume on the Flow of Polymer and Solvent Through Porous Media", SPE 7586, presented to 53rd SPE Ann. Fall Conf., Houston, Texas, October 1978.

19. DAVISON, P. and MENTZER, E., "Polymer Flooding in North Sea Oil Reservoirs", SPE 9300, presented at the 55th Annual Fall Conference of SPE-AIME, Dallas, Texas, 1980.

20. AKSTINAT, M. H., "Polymers for EOR in Reservoirs of Extremely High Salinities and Temperatures", SPE 8979, SPE Int. Symp. on Oilfield and Geothermal Chemistry, Stanford, California, 28-30 May, 1980.

21. ASH, S. G., CLARKE-STURMAN, A. J., CALVERT, R. and NISBET, T.M., "Chemical Stability of Biopolymer Solutions", SPE 12085, 58th Ann. Fall Conf. of SPE, San Francisco, California, 5-8 October, 1983.

22. BATYCKY, J. P., MAINI, B. B. and MILOSZ, G., "A Study of the Application of Polymeric Gels in Porous Media", SPE 10620, presented at SPE Int. Symp. on Oilfield and Geothermal Chemistry, Dallas, Texas, January 1982.

23. NAVRATIL, M., SOVAK, M. and MITCHELL, M. S., "Diverting Agents for Sweep Improvements in Flooding Operations - Laboratory Studies", SPE 10621, presented at SPE Int. Symp. on Oilfield and Geothermal Chemistry, Dallas, Texas, January 1982.

24. BROWN, C. E., CHRISTIE, M. A. and GATT, A. M., "Reservoir Temperature Distributions Around Injection Wells; Effect on EOR Schemes", presented at 2nd European Symposium on EOR, Paris, 8-10 November, 1982.

25. SORBIE, K. S., ROBERTS, L. J. and FOULSER, R. W. S., "Polymer Flooding Calculations for Highly Stratified Brent Sands in the North Sea", presented at 2nd European Symposium on EOR, Paris, 8-10 November, 1982.

26. SCOTT, T., ROBERTS, L. J., SHARPE, S. R., CLIFFORD, P. J. and SORBIE, K. S., "In Situ Gel Calculations in Complex Reservoir Systems Using a New Chemical Flood Simulator", SPE 14234, presented at 60th Annual Fall Conference of SPE, Las Vegas, 22-25 September, 1985.

27. SCOTT, T., SHARPE, S. R., SORBIE, K. S., CLIFFORD, P. J., ROBERTS, L. J., FOULSER, R. W. S. and OAKES, J., "A General Purpose Chemical Flood Simulator", SPE 16029, presented at SPE Numerical Simulation Conference, San Antonio, Texas, 2-4 February, 1987.

28. CLIFFORD, P. J. and SORBIE, K. S., "The Effects of Chemical Degradation on Polymer Flooding", SPE 13586, presented at SPE Int. Symp. on Oilfield and Geothermal Chemistry, Phoenix, Arizona, 9-11 April, 1985.

29. CLIFFORD, P. J. and SORBIE, K. S., "Polymer Flooding in Stratified Systems: Recovery Mechanisms and the Effects of Chemical Degradation", presented to IEA Collaborative Project on EOR, Trondheim, Norway, 4-5 October, 1984.

30. ZAPATA, V. J. and LAKE, L. W., "A Theoretical Analysis of Viscous Crossflow", SPE 10111, 56th Ann. Fall Conf. of SPE, San Antonio, Texas, 5-7 October, 1981.

31. YOKOYAMA, and LAKE, L. W., "The Effects of Capillary Pressure on Immiscible Displacements in Stratified Permeable Media", SPE 10109, presented at 56th Annual Fall Conference of SPE, San Antonio, Texas, 5-7 October, 1981.

32. CRAIG, F. F., SANDERLIN, J. L. and MOORE, D. W. "A Laboratory Study of Gravity Segregation in Frontal Drives", Trans. AIME 210, p275, 1957.

33. GODDIN, C. S., CRAIG, F. F., WILKES, J. O. and TEK, M. R., "A Numerical Study of Waterflood Performance in a Stratified System with Crossflow", J. Pet. Tech., 18, p765, 1966.

34. KOONCE, K. T. and BLACKWELL, R. J., "Idealised Behaviour of Solvent Banks in Stratified Reservoirs", Soc. Pet. Eng. J., 1, pp 318-328, December, 1965: Trans. AIME, 234, p318, 1965.

35. WHEAT, M. R. and DAWE, R. A., "Transverse Dispersion in Slug Mode Chemical EOR Processes in Stratified Porous Media", SPE 14890,

presented at 5th SPE/DOE Symposium on EOR, Tulsa, Oklahoma, 20-23 April, 1986.

36. EVANS, A. J., SORBIE, K. S. and CLIFFORD, P. J., "Scaling Relations for Water and Polymer Floods in Stratified Systems", AEE Winfrith M2143, January 1985.

37. LEVERETT, M. C., LEWIS, W. B. and TRUE, M. E., "Dimensional Model Studies of Oil-Field Behaviour", Trans. AIME 146, p175, 1942.

38. RAPOPORT, L. A., "Scaling Laws for Use in Design and Operation of Water - Oil Flow Models", Trans. AIME 204, p143, 1955.

39. GEERTSMA, J., CROES, G. A. and SCHWARZ, N., "Theory of Dimensionally Scaled Models of Petroleum Reservoirs", Trans. AIME, 207, p118, 1956.

40. ENGELBERTS, W. F. and KLINKENBERG, L. J., "Laboratory Experiments on the Displacement of Oil by Water, from Packs of Granular Material", Proc. Third World Petr. Congress, The Hague, Section II, 544, 1951.

41. BENTSEN, R. G., "Scaled Fluid-Flow Models with Permeabilities Differing from that of the Prototype", J. Can. Pet. Tech., 15, No. 3, p46, 1976.

42. CROES, G. A. and SCHWARTZ, N., "Dimensionally Scaled Experiments and the Theories on the Water Drive Process", Trans. AIME, 204, p35, 1955.

43. PERKINS, F. M. and COLLINS, R. E., "Scaling Laws for Laboratory Flow Models of Oil Reservoirs", Trans. AIME, 219, p383, 1960.

44. GAUCHER, D. H. and LINDLEY, D. C., "Waterflood Performance in a Stratified Five-Spot Reservoir - A Scaled Model Study", Trans. AIME, 219, p208, 1960.

45. VAN MEURS, P., "The User of Transparent Three Dimensional Models for Studying the Mechanism of Flow Processes in Oil Reservoirs", Trans. AIME, 210, p295, 1957.

46. RICHARDSON, J. G. and PERKINS, F. M., "A Laboratory Investigation on the Effect of Rate on Recovery of Oil by Waterflooding", Trans. AIME, 210, p114, 1957.

47. CARPENTER, C. W., BAIL, P. T. and BOBECK, J. E., "A Verification of Waterflood Scaling in Heterogeneous Communicating Flow Models", Soc. Pet. Eng. J., p9, 1962.

48. GRIFFITH, J. D. and FITCH, R. A., "Experimental and Calculated Performance of Miscible Floods in Stratified Reservoirs", J. Pet. Tech., p1289, November 1964.

49. ROOT, P. J. and SKIBA, F. F., "Crossflow Effects During an Idealised Displacement Process in a Stratified Reservoir", Trans. AIME, 234, p229, 1965.

50. CRAIG, F. F., "The Reservoir Engineering Aspects of Waterflooding", Vol. 3. H. L. Doherty Series, SPE of AIME, New York, 1971.

51. WRIGHT, R. J., WHEAT, M. R. and DAWE, R. A., "Slug Size and Mobility Requirements for Chemically Enhanced Oil Recovery within Heterogeneous Reservoirs", SPE 13704, presented at SPE Middle East Oil Technical Conference, Bahrain, 11-14 March, 1985.

52. WRIGHT, R. J. and DAWE, R. A., "Fluid Displacement Efficiency in Layered Porous Media; Mobility Ratio Influence", Rev. Inst. Fr. du Pétrole, 38, pp 455-474, July 1983.

53. NOVOSAD, J, "The Effect of Rock Heterogeneity and Wettability on Chemical Flooding", presented at 5th Annual Meeting of IEA on EOR, Trondheim, Norway, 4-5 October, 1984.

54. SANDIFORD, B. B., "Polymer Selection for Mobility and Sweep Improvement Processes", presented at the ACS Meeting, New York City, August, 1981.

55. SZABO, M. T., "Laboratory Investigations of Factors Influencing Polymer Flood Performance", Soc. Pet. Eng. J., pp 338-346, August 1975.

56. LEVER, A. and DAWE, R. A., "Water Sensitivity and Migration of Fines in the Hopeman Sandstone", J. Petrol Geol., 7, No 1, pp 97-107, Jan. 1984.

57. CHAUVETEAU, G. and KOHLER, N., "Influence of Microgels in Xanthan Polysaccharide Solutions on their Flow Through Various Porous Media", SPE 9295, presented at SPE 55th Ann. Fall Conf., Dallas, 21-24 September, 1980.

58. SORBIE, K. S., WAT, R. M. S., ROWE, T. and CLIFFORD, P. J., "Core Floods in Well Characterised Heterogeneous Systems: Experimental and Simulation Results", SPE 16275, presented at SPE Int. Symp. on Oilfield Chemistry, San Antonio, Texas, 4-6 Feb., 1987.

59. ARYA, A. HEWITT, T. A., LARSON, R. and LAKE, L. W., "Dispersion and Reservoir Heterogeneity", SPE 14364, presented at 60th Annual Fall Conference on SPE, Las Vegas, Nevada, 22-25 September, 1985.

CORRELATION OF THE FLOW OF FLOCON 4800 BIOPOLYMER WITH POLYMER CONCENTRATION AND ROCK PROPERTIES IN BEREA SANDSTONE

G. Paul Willhite* and J. T. Uhl

*Dept. of Chemical and Petroleum Engineering
University of Kansas, Lawrence, Kansas
Chevron Oil Field Research Company
La Habra, California

SUMMARY

An experimental program was carried out to determine the mobility of Flocon 4800 biopolymer in Berea sandstone cores. Experimental work conducted at 25C included rocks with brine permeabilities ranging from 15.5 md to 848 md for polymer concentrations of 500-1500 ppm. Frontal advance rates varied from 0.1 ft/d to 117 ft/d. Flocon 4800 was found to follow a power-law model during flow through porous rock. Correlations were developed between polymer mobility, polymer concentration, and brine permeability of the rock after contact with polymer. Using these correlations, it is possible to estimate the mobility of Flocon 4800 in chemical flooding processes conducted in Berea core material within the range of polymer concentrations studied. Mathematical models derived from capillary bundle approaches and rheological parameters derived from steady shear measurements produced poor predictions of polymer mobility in Berea sandstone cores.

INTRODUCTION

Polymers are used for mobility control in chemical flooding processes such as micellar-polymer and caustic-polymer flooding and in polymer augmented waterflooding. Selection of a polymer for mobility control is a complex process because it is not possible to predict the behavior of a polymer in porous rock from rheological measurements such as viscosity/ shear rate curves. Polymers used for mobility control are non-Newtonian fluids. Flow characteristics are controlled by the shear field to which the polymer is subjected. Properties of polymers can be measured under steady shear in rheometers. However, in porous rock, it is difficult to define the shear environment a polymer experiences as it flows through tortuous pores.

This paper describes an experimental program to characterize the flow properties of biopolymers in porous rock. An objective of the program was to develop a correlation between polymer flow in porous media and properties of the porous rock. Experimental data were obtained for the flow of biopolymer in fired Berea core material over a wide range of permeabilities.

BACKGROUND

Rheology

Polymers used for mobility control in oil reservoirs are non-Newtonian fluids. Thus, the viscosity of the fluid varies with shear rate. These fluids can be characterized using standard rheological measurements and models. Most biopolymers exhibit the behavior shown in Figure 1[1] when the shear rate is varied from near zero to rates as high as 10,000 sec^{-1}. At low shear rates, the fluid behaves as a Newtonian fluid and thus the viscosity does not vary with shear rate. As shear rate increases, the polymer molecules deform and the viscosity decreases. This is also referred to as shear-thinning. At extremely high shear rates, a second Newtonian region appears. The viscosity of the solution approaches the viscosity of the solvent in the second Newtonian region. Most rheological studies do not extend into the second Newtonian region because of experimental difficulties and because it is believed to be beyond the range of practical interest.

In the shear-thinning region, the viscosity is closely approximated by the power-law model given by Equation 1. Power-law parameters, K and n can be determined readily from analysis of data from standard rheometers. Typical values of n range between 0.2 and 0.9 for shear thinning fluids.

$$\mu = K\dot{\gamma}^{(n-1)} \qquad (1)$$

where

μ = viscosity, cp

$\dot{\gamma}$ = shear rate, sec^{-1}
K = power law constant, $mPa.s^{n}$
n = power law exponent

Rheological models have also been developed to describe fluid behavior over the shear rate range which includes Newtonian behavior at low and high shear rates. The Carreau Model A[2] has been found to fit polymer data satisfactorily. Equation 2 is the Carreau Model A. In Equation 2, μ_o is the Newtonian viscosity in the low shear region, μ_∞ is the Newtonian viscosity in the high shear regions, and μ is the shear rate. The parameter n is the power-law exponent and τ_r is a characteristic time constant. All parameters are determined by fitting experimental data.

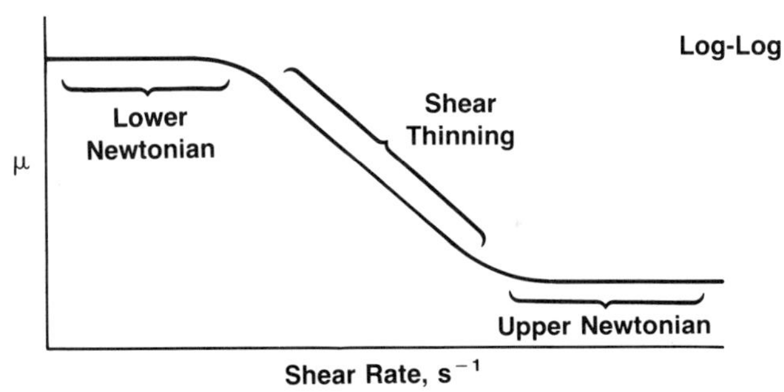

Fig. 1. Rheology of a shear thinning fluid.

It is necessary to have shear rate data over a wide range of rates in order to use Equation 2. The parameter τ_r is found by extrapolating the Newtonian and shear-thinning regions to a common intersection point.[3] The parameter n is the same in both the Carreau and power-law models.

$$\mu - \mu_\infty = \frac{\mu_o - \mu_\infty}{[1 + (\tau_r \dot{\gamma})^2]^{(1-n)/2}} \tag{2}$$

Each parameter varies with polymer type, concentration, solvent, amount, and species of dissolved ions and temperature. All of these parameters can be determined from analysis of steady shear data.[4] Parameters may then be correlated empirically as done in Figure 2 for the power law model for Flocon 4800 ME at 25C.

Polymer Flow in Porous Rocks

Mathematical Models. Mathematical models of polymer flow in porous rocks are usually based on capillary bundle models of flow in porous media. The rock is visualized to be a bundle of tortuous, non-connected capillaries which have uniform radii, R. This mathematical model represents flow in the tortuous capillaries which have effective length (L_e) by equivalent straight capillary tubes with length L. An approximation attributed to Carman[5] assumes that a particle of fluid entering a tortuous capillary must travel L_e/L faster than a particle of

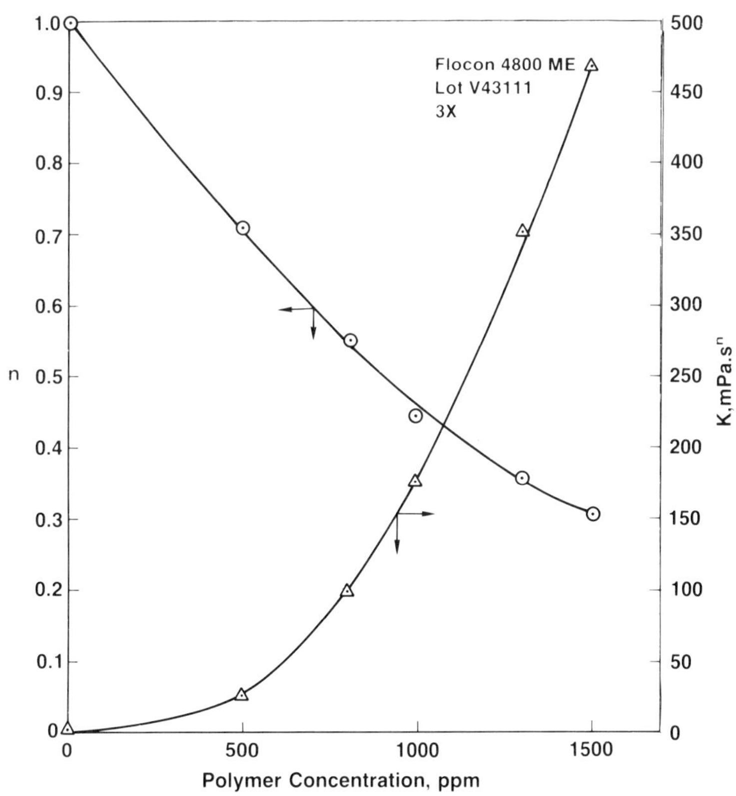

Fig. 2. Variation of power-law parameters with polymer concentration, Flocon 4800ME.

fluid flowing through the equivalent straight capillary in order to reach the end of the capillaries at the same time.

Using this model, the corresponding Darcy velocity for flow through porous rock is given by Equation 3.

$$u = \frac{\phi R^2 \Delta P}{8 \mu L} \left(\frac{L}{L_e}\right)^2 \qquad (3)$$

where

u = Darcy velocity in the direction parallel to length L volumetric flow rate/cross sectional area
μ = viscosity of the fluid
ϕ = the porosity.
ΔP = pressure drop in the direction parallel to length L

The Darcy velocity is also given by Equation 4 for the flow of Newtonian fluids in porous rock.

$$u = \frac{k \, \Delta P}{\mu \, L} \qquad (4)$$

where
k = permeability.

The pore radius and effective length obtained from Equations 3 and 4 are given by Equation 5.

$$\left(R \frac{L}{L_e}\right)^2 = \frac{8k}{\phi} \qquad (5)$$

The term $(L_e/L)^2$, is the tortuosity. When R is known, the tortuosity is an adjustable parameter chosen to obtain agreement between experimental and predicted data.

An equivalent shear rate for flow in porous media can also be derived from the capillary bundle model. For Newtonian fluids, the shear rate in the porous media is given by Equation 6

$$\dot{\gamma} = \frac{4 \, u}{\sqrt{8k\phi}} \qquad (6)$$

Although many research personnel consider capillary bundle models to be a gross simplification of flow in porous media, they provide a conceptual framework to estimate effects of non-Newtonian properties on flow in porous media.

Non-Newtonian characteristics are introduced by expressing the wall shear in the capillary tube as an equivalent shear derived from a rheological model such as the power-law model (Equation 1) or the Carreau Model A (Equation 2). Derivations of polymer flow models based upon power-law and Carreau Model A are found in references 6 and 7. Equation 7

$$u = \left(\frac{\phi n}{3n+1}\right) \left(\frac{8k}{\phi}\right)^{\frac{n+1}{2n}} \left(\frac{\Delta P}{2KL}\right)^{1/n} \qquad (7)$$

is the Blake-Kozeny model for the flow of a non-Newtonian fluid in a porous rock which has porosity ϕ and permeability k. Polymer flow rates may be calculated from rheological properties K and n which are easily measured for a polymer using a rheometer and properties of the rock.

The corresponding shear rate at the wall of the capillary tube is given by Equation 8.

$$\dot{\gamma}_w = \left(\frac{3n + 1}{n}\right) \frac{u}{\sqrt{8k\phi}} \tag{8}$$

Equation 7 can also be written in the form of Darcy's law as in Equation 9.

$$u = \frac{k}{\mu_a} \left(\frac{\Delta P}{L}\right) \tag{9}$$

where

μ_a = apparent viscosity of the polymer when the Darcy velocity is u.

The apparent viscosity is defined by Equation 10.

$$\frac{1}{\mu_a} = \frac{4}{Ku^{n-1}} \left(\frac{n}{3n+1}\right)^n (8k\phi)^{\frac{n-1}{2}} \tag{10}$$

An apparent shear rate can be defined for the porous media when the polymer is in the shear-thinning region. This apparent shear rate is given by Equation 11. Note the apparent shear rate in the rock is not the same as the shear rate at the capillary wall because shear rate is not constant across the capillary cross section.

$$\dot{\gamma}_a = \left(\frac{3n+1}{4n}\right)^{\frac{n}{n-1}} \frac{4u}{\sqrt{8k\phi}} \tag{11}$$

There are several versions of the Blake-Kozeny model to improve agreement between experimental and predicted results. For example, Equation 12 is the modified Blake-Kozeny model developed by Christopher and Middleman[8]. The tortuosity was assumed to be 25/12 in this model.

$$u^n = \frac{k}{H} \frac{\Delta P}{L} \tag{12}$$

where

$$H = \frac{K}{12} \left(\frac{9n + 3}{n}\right)^n (150\ k\phi)^{\frac{1-n}{2}} \tag{13}$$

The Blake-Kozeny model assumes that polymer retention on the rock does not alter the permeability of the rock. Modifications of the Blake-Kozeny model have been proposed[9] to account for permeability reduction by correlating the radius of the capillary with the amount of permeability reduction.

The Blake-Kozeny and modified Blake-Kozeny models have been verified by comparing observed pressure drops with computed pressure drops for flow of non-Newtonian fluids through packed beds. Agreement between predicted and measured pressure drops is within 20%. Thus, it is possible to predict flow behavior through some porous materials from rheological properties and characteristics of the porous media.

Although these models are widely used in numerical simulators of chemical flooding processes, there have been few published studies which show how well these models represent polymer flow in reservoir rock. Of the papers found, two references report agreement between flow properties observed in reservoir rock and predicted properties. Hirasaki and Pope[9] found good agreement between predicted and experimental apparent viscosities for dilute (200-300 ppm) solutions of a biopolymer flowing through reservoir rock at residual oil saturation (S_{or}). In these cores, permeability to brine ranged from 7.7 to 22.8 md. Vogel and Pusch[7] report good agreement between apparent viscosities predicted from a capillary bundle model based upon Carreau Model A and apparent viscosities measured during the flow of biopolymer and polyacrylamide in sandpacks. Permeabilities of the sandpacks were about 5 darcies. However, Teeuw, et al.[6] found that the predicted apparent viscosities were a factor of two higher than observed in core flow experiments. Castagno, et al.[10] also found that the predicted viscosity for flow of biopolymer through Frannie reservoir rock was significantly less than expected from polymer properties. In contrast, Duda, et al.[11] measured pressure drops that were 200% greater than predicted pressure drops for concentrated xanthan solutions. The difference between experimental and computed pressure drops was attributed to regions of expansion and contraction in the pore space where excess pressure drops occurred.

Empirical Models

Chauveteau[3] also studied flow of biopolymers in porous rock. By using well filtered biopolymer solutions, he determined apparent viscosities as a function of Darcy velocity in Fountainbleau sandstone over permeabilities ranging from 3.3 md to 256 md. Polymer retention was low and it was possible to restore the permeability of the rock to its prepolymer value after each polymer flow experiment. Apparent viscosities were fitted with the Carreau Model A. Analysis of the experimental data yields pairs of apparent viscosity and Darcy velocity. Conversion of Darcy velocity to apparent shear rate in the porous rock was done using Equation 14.

$$\dot{\gamma}_a = \frac{4\alpha u}{\sqrt{8k\phi}} \qquad (14)$$

The parameter α was obtained empirically by assuming the shear rate for the onset of shear-thinning behavior in the viscometric data is equal to the shear rate at the velocity where the onset of shear-thinning is observed in the rock. A critical shear rate is found for each polymer system from the intersection of tangents to the viscosity shear rate curve in the lower Newtonian and shear-thinning regions. This critical shear, γ_c, is $1/\tau_r$ from Equation 2. A similar procedure is used to find the critical velocity, u_c, from apparent viscosity-Darcy velocity data.

Chauveteau's research produced an important result. Apparent viscosities in Fountainbleau sandstone were significantly less than found in rheological measurements. This is particularly evident in the lower Newtonian region where apparent viscosities range from 17% to 38% lower than solution viscosities. The largest differences were observed at the lowest permeabilities. These results were explained using a model which assumes that the layer of fluid next to the pore wall is depleted of polymer because the polymer molecule is too large to get close to the wall. Consequently, polymer molecules flow faster than the solvent they were injected with. The model has two adjustable parameters which are determined from molecular size arguments and fitting experimental data.

Gogarty[12] studied the flow of polyacrylamides in Berea sandstone. A shear-thinning region was found where the apparent viscosity decreased linearly with frontal advance rate when the data were plotted on log-log paper. Gogarty found that the power-law exponent for polymer flow in porous rock (n_c) was greater than the power-law exponent (n) determined from rheometer data. An empirical correlation was developed between the apparent viscosity in the shear thinning region and the frontal advance rate. This correlation is defined in Equations 15-17.

$$\mu_a = F\left(\frac{u}{\phi}\right)^{(n_c-1)} \tag{15}$$

where F and n_c are found from the experimental data. A correlation of apparent viscosity with rheological data was obtained by defining an apparent shear rate in the core as in Equation 16

$$\dot{\gamma}_a = \left[\frac{B\,(u/\phi)}{f(k)\,\sqrt{k_{wp}/\phi}}\right]^{\frac{(n_c-1)}{(n-1)}} \tag{16}$$

where

$$f(k) = m \log\left(\frac{k_{wp}}{k_r}\right) + p \tag{17}$$

The parameters B, m, k_r, and p are found empirically. Using Equation 16, the apparent viscosity in the core is computed from Equation 18

$$\mu_a = K\,\dot{\gamma}_a^{\,n-1} \tag{18}$$

The term k_{wp} in Equations 16 and 17 is the brine permeability after mobile polymer has been displaced from the rock. Polyacrylamides usually cause large reductions in brine permeability due to retention of polymer by adsorption and mechanical entrapment.

The Gogarty correlation was developed because it was not possible to predict the apparent viscosity of polyacrylamide solutions in situ. Gogarty demonstrated that it is possible to correlate data using rheological models. However, it is necessary to have good experimental data for the polymer/rock system of interest to develop correlations needed for design and simulation.

EXPERIMENTAL

Flow experiments were conducted using the Pfizer Flocon 4800 series biopolymer. This polymer was selected because initial experiments indicated that permeability reduction due to polymer retention could be held to 15-20% by carefully filtering the polymer solutions. This allowed study of the effects of polymer concentration on polymer flow without contending with permeability changes. All experiments were conducted in Berea core material at a constant injection rate. Pressure drop was measured continuously to insure steady-state flow existed prior to changing the flow rate.

Polymer solutions were prepared following the standard procedure recommended by Pfizer. Nominal polymer concentrations were 500, 1000, and 1500 ppm. Formaldehyde (1500 ppm) was added to the solutions to prevent microbial degradation. The Pfizer polymer is supplied as a broth with a percent activity for each lot. The activity is determined from a viscosity/concentration calibration curve provided by Pfizer.[13] This activity was not used to prepare solutions. Instead, polymer solutions were made up on the basis of percent by weight using the weight fraction of solids precipitated from the broth by isopropanol. Lot numbers, activities and percent solids by weight are summarized in Table 1 for the polymers used. Polymer solutions were filtered twice through a 0.3 micron Whatman cartridge filter and once through a 0.45 micron Millipore filter under a maximum pressure drop of 10 psi. Solutions were clear after filtration. Viscosity shear rate data were obtained using a Haake CV100 visometer. Power-law parameters K and n were determined for each solution.

Lot to lot variations in polymer properties were observed. Because of this, there were differences in rheological properties between fluids which had the same nominal concentration. It was necessary to determine viscosity-shear rate curves for several concentrations in order to match polymer solution characteristics of previous runs.

Fired Berea sandstone cores from the same block were selected to cover a wide range of permeabilities. All cores were 2" in diameter by about 12" long and were mounted in an aluminum sleeve using epoxy. Cores were fitted with end caps and were saturated with 3% KCl following evacuation. Porosity was determined gravimetrically. Permeability to brine was determined at room temperature at three water injection rates.

Polymer injection rates were selected to yield interstitial velocities over a range of 0.02 ft/d to about 117 ft/d. Two constant rate injection pumps were used. Flow rates from 2 cc/min to 10 cc/min were obtained using a Cheminert pump with manual rate selection. Injection rates between 0.002 cc/min and 2 cc/min were supplied by a Cheminert pump which was equipped with a computer interface. The computer control system was designed to permit specification of flow rate as an operating parameter. The computer also automatically stepped through a sequence of flow rates when the pressure drop stabilized. Criteria for stabilized flow was that the average of four successive pressure measurements must be within 0.5%-1.5%, depending upon the range of the transducer. This system enabled us to obtain a large amount of flow data in a reasonable period of time.

Pressure was measured using a set of Validyne pressure transducers which were monitored by the HP computer system. Transducer ranges available were 200 psi, 25 psi, and 2 psi. Transducers were

Table 1. Properties of Polymer Broths

Polymer Type	Lot Number	Activity wt%	Solids IPA-ppt wt%	Cores
4800	V49081	4.7	3.21	826A,776A
4800ME	V43111	4.5	3.15	793A,793B,796B
4800	10912-101-8	18.9	11.71	828D

cross-calibrated to determine ranges where accurate data could be obtained. Transducers were switched manually at appropriate points in the runs. A schematic of the experimental system is shown in Figure 3. All polymer flow experiments were conducted at room temperature which was nominally 25C.

Flow experiments in most runs began with the highest polymer concentration to minimize variation in permeability between runs made with polymer solutions having different concentrations. In the first run, a 1500 ppm polymer solution was injected into the core until the pressure drop across the core stabilized. This normally took about 5 PV of polymer. This experiment was done at the highest rate which was possible and still remain within the pressure limitations of the apparatus (200 psi). In some runs, the viscosity of the effluent was determined at selected points in order to verify that the effluent was essentially the same as the injected material.

Flow rate experiments were done from the largest to smallest rate for a given concentration. The flow rate range was 0.002-0.009 cc/min to 2 cc/min for most 1000 ppm and 500 ppm runs. Polymer was flushed from the core with 3% KCl following the rate test at a given concentration. Permeability to brine was determined at 1-3 flow rates after at least 10 PV of brine had been injected. Values of these permeabilities are presented in Table 2.

In two cores, permeability was reduced after the first series of polymer flow experiments by oilflooding to interstitial water saturation and waterflooding to residual oil saturation. Permeability to brine was determined at residual oil saturation prior to polymer flow experiments. Polymer flow experiments were done following the sequence described above. Small amounts of oil were produced during some experiments. This usually occurred during the highest flow rate of a run. Displaced oil was estimated from the volume of oil collected in a graduated cylinder. In one

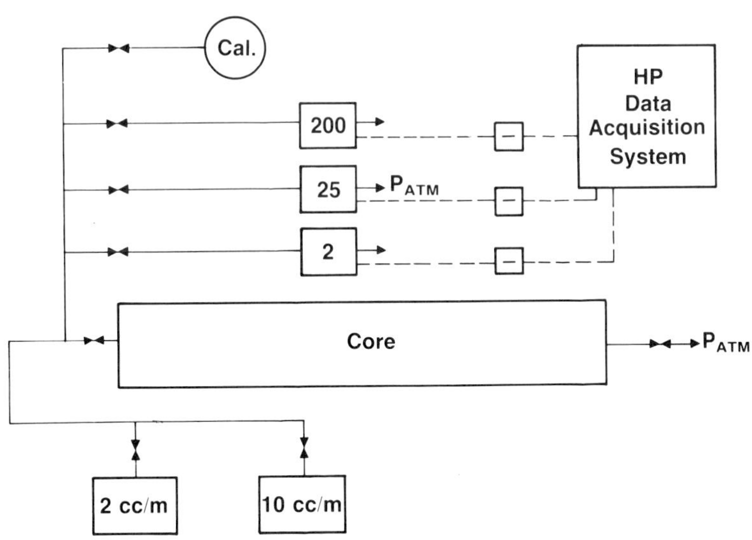

Fig. 3. Schematic of polymer flow equipment.

Table 2. Permeability Reduction Following Exposure of Berea Core to 1,500 ppm Flocon 4800 3X

Core	k_a md	k_w md	k_{wp} md	k_{wp}/k_w
826A	913	909	818-847	0.9-0.93
828D	916	876	668	0.76
793A	447	412	352	0.85
793B	474	425	352	0.83
796A	224	177	153	0.86
796B		192	155	0.80

or two runs, a change in oil saturation occurred in the middle of a concentration sequence. The entire concentration sequence was rerun in this case.

A third sequence of experiments was done on one core following the flow experiments at residual oil saturation. The core was heated to 60C and polymer flooded at maximum rate using the 1500 ppm solution to reduce the oil saturation and increase brine permeability. Oil displaced by this polymer flood caused the brine permeability to increase from 10 md to about 90 md. Thus, it was possible to conduct polymer flow experiments on a core with the same internal pore configuration but with a different permeability.

Analysis of Data

Experimental data collected from each run consisted of an overall pressure drop corresponding to the imposed flow rate. Data were analyzed assuming a power-law model. In the power-law model, the volumetric flow rate and pressure drop are related by Equation 19.

$$\left(\frac{Q}{A}\right)^{n_c} = \lambda_p^* \left(\frac{\Delta P}{\Delta L}\right) \tag{19}$$

where

Q = volumetric flow rate
A = cross-sectional area of core
λ_p^* = polymer mobility constant
ΔP = pressure drop
ΔL = length
n_c = power-law exponent for polymer flow in porous rock

When the power-law model applies, a plot of pressure drop versus volumetric flow rate is a straight line on log-log paper as shown in Figure 4 for five runs in Core BE793A 03. Rearranging Equation 19 we have

$$\Delta P = \left[\frac{\Delta L}{\lambda_p^* A^{n_c}}\right] Q^{n_c} \tag{20}$$

which is in the form of $\Delta P = aQ^{n_c}$. The parameters a and n_c were obtained from the experimental data by the method of least squares. The value of λ_p^* was computed from the value of a using Equation 21.

$$\lambda_p^* = \frac{(2.3309)^{n_c} \Delta L}{(6.3283 \times 10^{-3}) a} \tag{21}$$

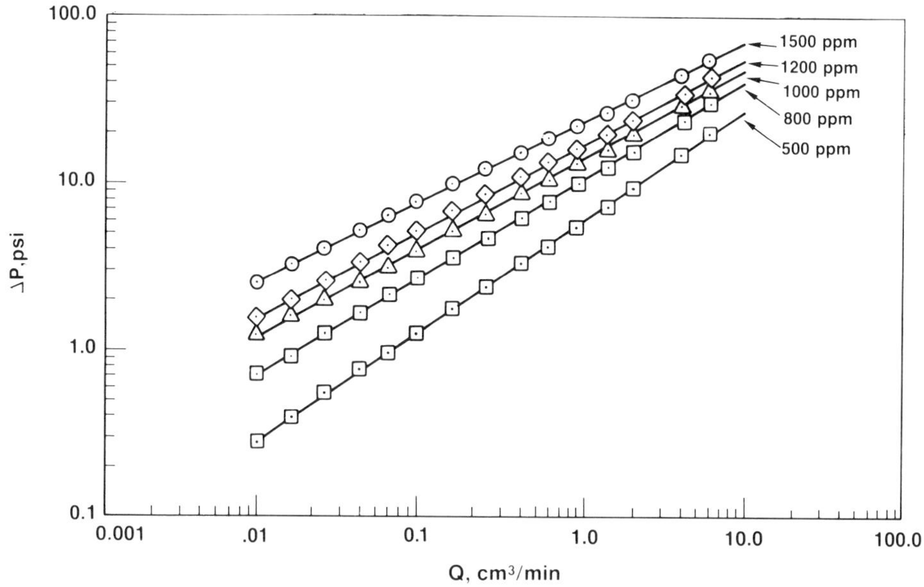

Fig. 4. Pressure drop-flowrate data for several polymer concentrations in Core BE 793A03.

The units of λ_p^* are md/cp/(ft/d)$^{1-n_c}$ when Q, A, ΔP, and ΔL have units of cc/min, cm^2, psi and feet. The constant 2.3309 accounts for the cross-sectional area of the core and unit conversion. The constant 6.3282 x 10^{-3} is a conversion factor which has the units of (ft/d)/((psi/ft)(md/cp)). Note that the polymer mobility constant, λ_p^*, is the polymer mobility at a Darcy velocity of 1 ft/d.

Equation 19 may also be written as Equation 22.

$$u^{n_c} = \lambda_p^* \left(\frac{\Delta P}{\Delta L}\right) \tag{22}$$

When ΔP and ΔL are in psi and feet respectively, the Darcy velocity can be computed from Equation 22 in the units of ft/d by correcting for units as in Equation 23.

$$u^{n_c} = (6.3283 \times 10^{-3}) \lambda_p^* \left(\frac{\Delta P}{\Delta L}\right) \tag{23}$$

The equivalent expression in the form of Darcy's law is Equation 24.

$$u = (6.3283 \times 10^{-3}) \lambda_p \left(\frac{\Delta P}{\Delta L}\right) \tag{24}$$

where

$$\lambda_p = \frac{\lambda_p^*}{u^{n_c - 1}} \tag{25}$$

λ_p = polymer mobility at Darcy velocity, u, in md/cp

The mobility of the polymer increases as the Darcy velocity increases.

RESULTS

Flow experiments were completed on six Berea cores using polymer concentrations of 500, 1000 and 1500 ppm. The experimental data were fitted by the power-law model for most of the flow rate range as shown in Figure 4. Departure from the power-law model was observed at flow rates less than 0.009 cc/min (frontal advance rate - 0.1 ft/d) in several runs. This was expected since flow behavior should become Newtonian at low frontal advance rates. When Newtonian flow occurs, the power-law constant is 1 which means that the slope of the graph of ΔP versus Q on log-log paper is 45 degrees. Newtonian flow was not attained at the lowest frontal advance rate (0.022 ft/d).

Values of n_c varied with concentration as expected from viscometric data. Figure 5 shows the correlation of n_c and λ_p^* determined from viscometric data. Values of n_c for a given polymer concentration varied slightly with brine permeability after polymer treatment. A power-law model was used to correlate n_c with k_{wp}.

Inspection of the data for each polymer concentration indicated that λ_p^* varied with brine permeability after polymer treatment. Theoretical and empirical models discussed in the Background Section were reviewed to identify possible ways of correlating the data. We found that values of

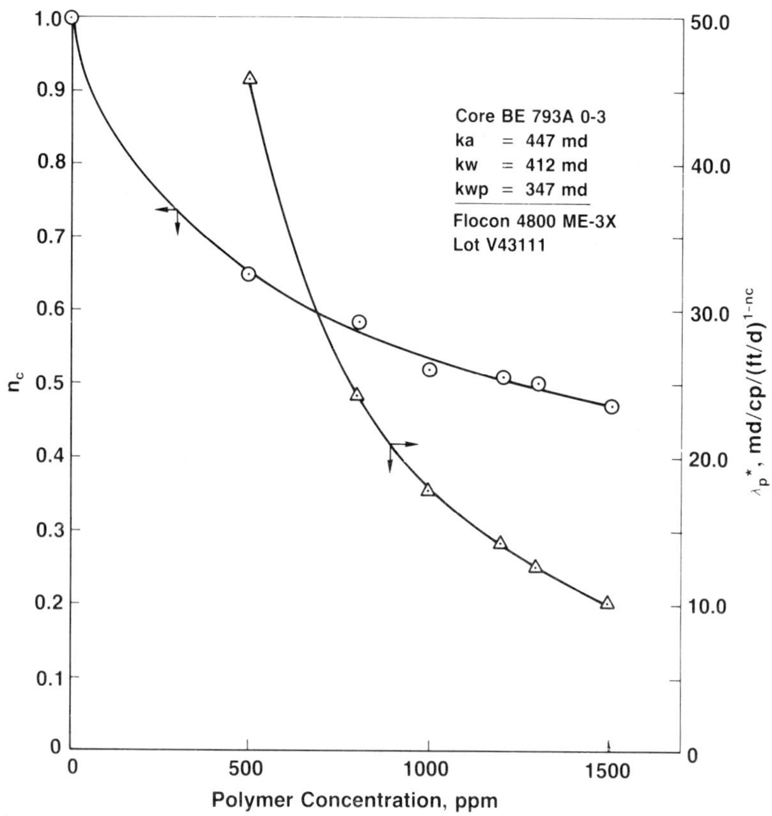

Fig. 5. Correlation of n_c and λ_p^* with polymer concentration for Core BE 793A 03.

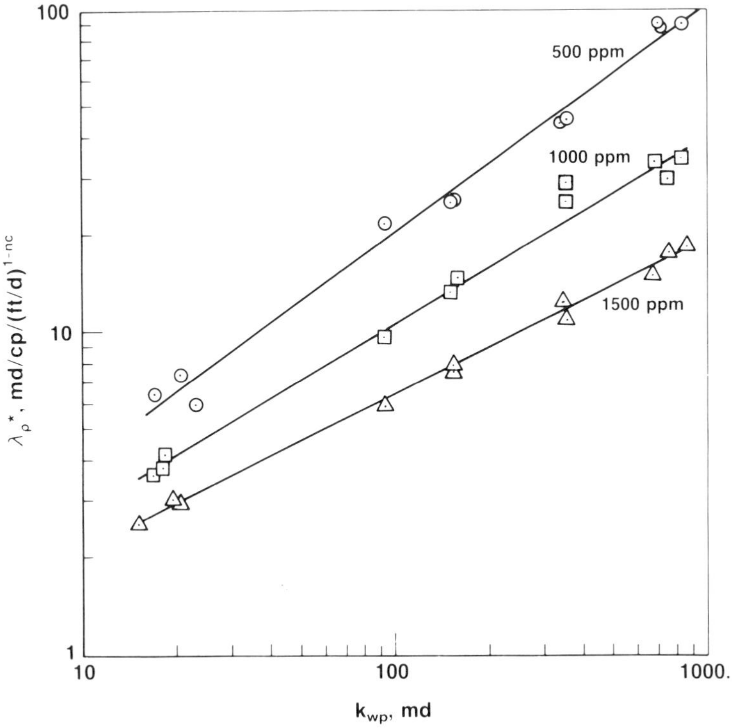

Fig. 6. Correlation of λ_p^* with k_{wp} for polymer concentrations of 500,1000 and 1500 ppm.

λ_p^* could be correlated with brine permeability after contact with polymer as shown in Figure 6. There is some scatter in the data for a particular concentration. Most of this scatter is believed to be caused by differences in solution properties between different lots of polymers as discussed in the Experimental Section. The data in Figure 6 were fitted with a least squares program to obtain the correlations for each concentration given in Equations 26, 28, and 31.

500 ppm

$$\lambda_p^* = 0.783\ (k_{wp})^{0.708} \tag{26}$$

$$n_c = 0.618\ (k_{wp})^{0.009} \tag{27}$$

1000 ppm

$$\lambda_p^* = 0.746\ (k_{wp})^{0.578} \tag{28}$$

$$n_c = 0.659\ (k_{wp})^{-0.035} \tag{29}$$

1500 ppm

$$\lambda_p^* = 0.679\ (k_{wp})^{0.488} \tag{30}$$

$$n_c = 0.710\ (k_{wp})^{-0.073} \tag{31}$$

DISCUSSION

Comparisons were made between polymer mobilities computed from the Blake-Kozeny model (Equation 7), the modified Blake-Kozeny model (Equation 13) and the actual data. Equation 32 is the expression for λ_p^* derived from the Blake-Kozeny model.

$$\lambda_p^* = \frac{1}{2K} \left(\frac{\phi n}{3n+1}\right)^n \left(\frac{8k}{\phi}\right)^{\frac{n+1}{2}} \tag{32}$$

Typical results are shown in Tables 3-5 for the Blake-Kozeny model where polymer mobilities are compared at a frontal advance rate of 1 ft/d. For the conditions of this study, the polymer mobility was underestimated by factors ranging from 1.3 to 6.7. Best agreement was observed at 500 ppm for high permeability cores. Poorest agreement occurs at 1500 ppm and 15.5 md cores. The average reduction in polymer mobility was 0.42 for the Blake-Kozeny model and 0.36 for the modified Blake-Kozeny model. Capillary bundle models consistently predict lower polymer mobilities in porous rocks than observed experimentally.

Capillary bundle models assume the power-law exponent, n, determined from steady shear measurements is identical to the power-law exponent, n_c, observed during flow in porous rocks. A comparison of n and n_c is shown in Figure 7. In our experiments, the value of n_c was greater than n for polymer concentrations above 500 ppm. A review of Chauveteau's data also reveals that the power-law exponent for flow in porous rocks is larger than the power-law index measured from viscometric data.

The comparison of predicted and experimental polymer mobilities indicates that capillary bundle models which rely on rheological parameters

Table 3. Comparison of Predicted and Experimental Polymer Mobilities, Flocon 4800 in Berea Sandstone Cores.

Concentration 500 ppm
Frontal Advance Rate = 1 ft/d

k_{wp} md	ϕ	λ_p Blake-Kozeny md/cp	λ_p Experimental md/cp
818	0.209	39.0	52.0
818	0.226	39.5	53.2
703	0.226	34.8	51.8
695	0.209	34.0	52.5
348	0.212	19.3	26.4
354	0.208	19.4	26.3
149	0.155	9.0	13.2
157	0.19	9.8	14.4
92.	0.209	6.4	12.5
23.	0.226	2.06	3.5
20.5	0.226	1.9	4.4
16.7	0.209	1.56	3.7

$K = 0.0267$ Pa.s^n)
$n = 0.655$) Rheological Parameters (averaged)
$n_c = 0.647$)

Table 4. Comparison of Predicted and Experimental Polymer Mobilities, Flocon 4800 in Berea Sandstone Cores.

Concentration 1,000 ppm
Frontal Advance Rate = 1 ft/d

k_{wp} md	ϕ	λ_p Blake-Kozeny md/cp	λ_p Experimental md/cp
818	0.209	7.4	17.2
743	0.226	7.04	15.4
727	0.226	6.93	15.5
684	0.209	6.49	16.8
344	0.212	3.94	12.1
350	0.208	3.97	12.6
147	0.155	1.94	5.8
157	0.19	2.15	7.0
92.	0.209	1.49	4.8
18.3	0.226	0.47	2.1
18.	0.226	0.46	1.98
16.6	0.209	0.43	1.8

$K = 0.162$ Pa.s^n)
$n = 0.464$) Rheological Parameters (averaged)
$n_c = 0.553$)

Table 5. Comparison of Predicted and Experimental Polymer Mobilities, Flocon 4800 in Berea Sandstone Cores.

Concentration 1,500 ppm
Frontal Advance Rate = 1 ft/d

k_{wp} md	ϕ	λ_p Blake-Kozeny md/cp	λ_p Experimental md/cp
847	0.209	3.27	8.5
748	0.209	3.01	8.0
668	0.226	2.87	8.4
342	0.212	1.81	5.7
352	0.208	1.84	5.5
153	0.155	0.96	3.1
155	0.19	1.04	3.3
90.	0.209	0.75	2.7
20.5	0.226	0.29	1.4
19.5	0.226	0.29	1.4
15.	0.209	0.23	1.2

$K = 0.426$ Pa.s^n)
$n = 0.307$) Rheological Parameters (averaged)
$n_c = 0.496$)

K and n obtained from steady shear experiments do not describe flow in porous rocks. This finding has significant impact on simulation of polymer injection rates in potential polymer floods and permeability modification projects. Injection rates predicted using Blake-Kozeny models will be less than can be attained in practice. Therefore, it is necessary to determine

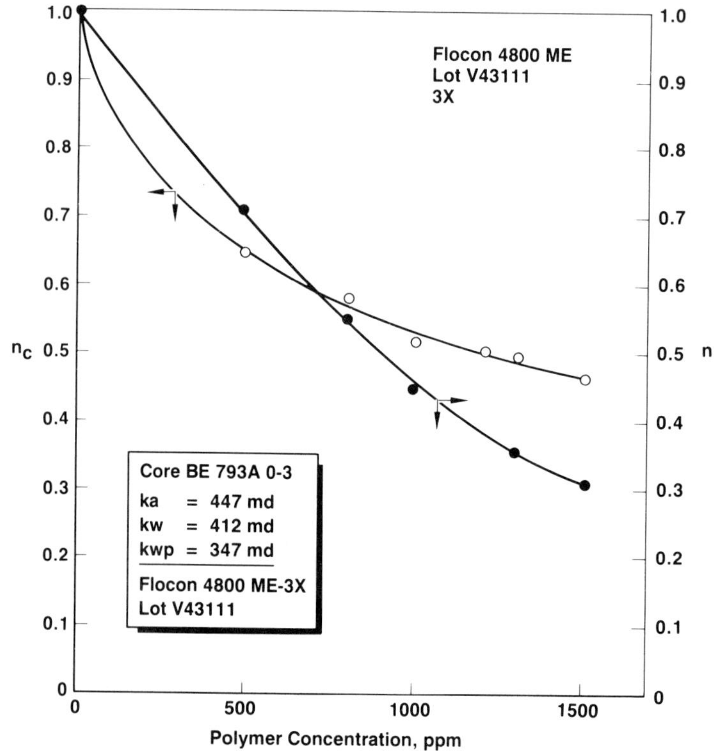

Fig. 7. Comparison of n_c and n for polymer concentrations of 0-1500 ppm.

polymer mobilities from laboratory measurements on reservoir rocks or porous rocks with comparable characteristics.

The correlation developed between the mobility of Flocon biopolymer and brine permeability following polymer contact in Berea cores has implications for mobility control in chemical flooding processes. Because the mobility of a polymer solution is not a linear function of the rock permeability, a polymer solution that provides mobility control in one rock may not maintain mobility control in another rock of the same type which has a significantly lower permeability.

The development of a correlation between water permeability following polymer treatment, polymer concentration, and polymer mobility, opens several opportunities for improvement of mobility control design in laboratory experiments. By using the correlation, it is possible to estimate the polymer concentration required to obtain a given mobility in the displacement of biopolymer through Berea core. Laboratory corefloods are usually run at a frontal advance rate of 1 ft/d. The mobility of biopolymer at a frontal advance rate of 1 ft/d is given by Equation 33.

$$\lambda_p = \lambda_p^* \phi^{1-n_c} \tag{33}$$

If the design mobility (λ_p) is known, the required polymer concentration can be estimated using data from Figure 6. To do this we compute the value of λ_p^* from Equation 33. The brine permeability following polymer flow must be estimated. For straight polymer flood applications using filtered polymer, this will be 15-20% less than brine permeability before contact with polymer. Otherwise, polymer flow tests on cores will be necessary to determine the permeability reduction following exposure to polymer. In chemical flood applications, it is necessary to estimate the permeability to water at residual oil saturation attained by chemical flood before reducing the permeability to account for polymer retention. For fired Berea core material, it is possible to determine the polymer concentration which is necessary to obtain a given mobility. The concentration range was limited to 500-1500 ppm in this study. It is possible that a similar correlation will be found at higher polymer concentrations.

This method of estimating polymer concentration replaces the approach of estimating a shear rate which is thought to exist in situ and measuring fluid properties at this shear rate. It is possible to define such a shear rate in terms of the correlation developed in this study for use in evaluation of solutions. For the biopolymer studied, the apparent shear rate in the porous rock is given by Equation 34.

$$\dot{\gamma}_a = \left[\frac{K \lambda_p^*}{k_{wp}}\right]^{\left(\frac{1}{1-n}\right)} u^{\left(\frac{1-n_c}{1-n}\right)} \tag{34}$$

At a frontal advance rate of 1 ft/d, the apparent shear rate is computed from Equation 35

$$\dot{\gamma}_a = \left[\frac{K \lambda_p^*}{k_{wp}}\right]^{\left(\frac{1}{1-n}\right)} \phi^{\left(\frac{1-n_c}{1-n}\right)} \tag{35}$$

Apparent shear rates estimated from the experimental data using Equation 35 are always higher than those estimated from Equation 11 which are based on the capillary bundle model. The mobility of the polymer at the apparent shear rate is given by Equation 36.

$$\lambda_p = \frac{k_{wp}}{K (\dot{\gamma}_a)^{n-1}} \tag{36}$$

Laboratory work did not include reservoir core material. Since the mobility of polymer solutions depends upon the shear field within the porous rock, we expect different rocks to yield different values of n_c and λ_p^*. However, results from Berea cores suggest that similar correlations might be found from core material of the same geologic origin.

Another use of the correlation between λ_p^* and k_{wp} is extrapolation of laboratory data to the field. We have shown that it is in general not possible to predict polymer flow in porous rocks from rheological properties. Thus, it is necessary to conduct some laboratory experiments to determine the values of n_c and λ_p^* for a given rock/polymer system. A limited number of core samples can be run. While these samples may be chosen as representative, the permeabilities of the samples may not be equal to the average permeabilities of the interval. The correlation developed for biopolymer flow in Berea sandstones provides a method for estimating polymer flow properties in samples which have a different permeability than the samples on which the measurements were made. Runs at the same polymer concentration in rocks with two different permeabilities are required. Extrapolation would be based upon fitting the data to the relationship

$$\lambda_p^* = a\, k_{wp}^{\;b} \tag{37}$$

where a and b are determined by fitting the data. With such a correlation it is possible to convert laboratory measurements to field parameters needed to estimate injection rates or interpret the pressure response of an injection well.

CONCLUSIONS

The following conclusions are based on the flow of Flocon 4800 biopolymer through Berea sandstone cores over the range of parameters studied.

1. Flow of Flocon 4800 polymer solutions through Berea cores is described by a power-law model at frontal advance rates ranging from 0.1 ft/d to 117 ft/d. The upper rate is the maximum frontal advance rate studied and is not a limit on power-law behavior in the rock.

2. The power-law exponent for flow through Berea core material was larger than the power-law exponent determined from steady shear measurements when polymer concentrations exceed 500 ppm.

3. Correlations were developed between polymer concentration, polymer mobility, and permeability of Berea sandstone to brine following contact with polymer.

4. Polymer mobility can be estimated from these correlations for application in other chemical recovery processes.

5. Polymer mobilities are underestimated by a factor of two or more when predicted using capillary bundle models and rheological parameters obtained from steady shear measurements on the polymer solutions.

6. Capillary bundle models are inadequate for prediction of in situ shear rates for polymer flow in porous rocks.

ACKNOWLEDGEMENTS

The authors thank Mr. R. L. Ulrich for his laboratory assistance and Chevron Oil Field Research Company for permission to publish this paper.

REFERENCES

1. Crocket, M. J., Davies, A. R., and Walters, K.: "Numerical Simulation of Non-Newtonian Flows", Elsevier, New York (1984), 3.

2. Carreau, P. J.: "Rheological Equations from Molecular Network Theories", Transactions of the Society of Rheology, 16:1, (1972), 99-127.

3. Chauveteau, G. and Zaitoun, A.: "Basic Rheological Behavior of Xanthan Polysaccharide Solutions in Porous Media: Effects of Pore Size and Polymer Concentration", European Symposium on Enhanced Oil Recovery, Bournemouth (1981), 197-212.

4. Walters, K.: "Rheometry", Chapman and Hall, Ltd., London (1975).

5. Carman, P. C.: "Fluid Flow Through Granular Beds", Trans. Inst. Chem. Eng., 15 (1937), 150-166.

6. Teeuw, D. and Hesselink, F.: "Power-Law Flow and Hydrodynamic Behavior of Biopolymer Solutions in Porous Media", SPE 8982 presented at the Fifth SPE International Symposium on Oil Field and Geothermal Chemistry, Stanford, California, May 28-30, 1980.

7. Vogel, P. and Pusch, G.: "Some Aspects of the Injectivity of Non-Newtonian Fluids in Porous Media", European Symposium on Enhanced Oil Recovery, Bournemouth, (1981), 179-195.

8. Christopher, R. H. and Middleman, S.: "Power-Law Flow Through A Packed Tube", I & EC Fundamentals (November 1965), 422-426.

9. Hirasaki, G. J. and Pope, G. A.: "Analysis of Factors Influencing Mobility and Adsorption in the Flow of Polymer Solutions Through Porous Media", SPEJ, (August 1974), 337-346.

10. Castagno, R. E., Shupe, R. D., Gregory, M. D. and Lescarboura, J. A.: "A Method for Laboratory and Field Evaluation of a Proposed Polymer Flood", SPE 13124, presented at the 59th Annual Technical Conference and Exhibition, Houston, Texas, September 16-19, 1984.

11. Duda, J. L., Hong, S. A. and Klaus, E. E.: "Flow of Polymer Solutions in Porous Media: Inadequacy of the Capillary Model", Ind. Eng. Chem. Fundam., 22, (1983), 299-305.

12. Gogarty, W. B.: "Mobility Control With Polymer Solutions", SPEJ, (June 1967), 161-173.

13. Letter from J. Tarlton to G. P. Willhite, June 5, 1985.

HIGH TEMPERATURE AND HARDNESS STABLE COPOLYMERS OF VINYLPYRROLIDONE AND ACRYLAMIDE

G. A. Stahl[1], A. Moradi-Araghi, and P. H. Doe

Phillips Petroleum Company
Research and Development
Bartlesville, Oklahoma 74004

ABSTRACT

The preparation and solution properties of poly(vinylpyrrolidone-co-acrylamide), synthetic water-soluble copolymer which has non-precipitating behavior in hard brines at high temperatures, is discussed. One sample of the copolymer has remained in synthetic seawater for over six years at 121°C (250°F) without precipitation. In that time the viscocity of its seawater solution has remained constant. It is recognized that while polyacrylamide is a highly effective water thickener, it rapidly hydrolyzes at high temperatures. In the presence of divalent cations, such as Ca++ or Mg++, hydrolyzed polyacrylamide will precipitate and lose viscocity. The usefulness of polyacrylamide is restricted to applications in which temperatures of less than about 75°C are encountered.

Polyvinylpyrrolidone, on the other hand, has been shown to resist hydrolysis at elevated temperatures. Unfortunately, polyvinylpyrrolidone lacks the viscosifying power to be economically useful in enhanced oil recovery processes.

We have found that copolymers of vinylpyrrolidone and acrylamide can be effective polymers for oil recovery applications. We anticipate the greatest use will be in operations in which high temperatures are encountered. These conditions are now commonly called hostile environments. Vinylpyrrolidone-acrylamide copolymers, on aging in hard brines at high temperatures, will undergo some hydrolysis of the acrylamide moieties, but the degree of hydrolysis is limited. The presence of vinylpyrrolidone in the copolymer apparently restricts the level of hydrolysis.

An equimolar copolymer of these monomers will hydrolyze about 40% of the available acrylamide in months of aging at 121°C. Aged at the same conditions, polyacrylamide will quantitatively hydrolyze. The vinylpyrrolidone must be incorporated in the copolymer to protect acrylamide as a mixture of homopolymers will precipate on aging.

Non-quantitative hydrolysis and viscosity producing power of the copolymers described herein permits design of a copolymer based on the application temperature. Lower temperature applications will require less vinylpyrrolidone in the copolymer. The preparation and general characteristics of vinylpyrrolidone-acrylamide copolymers are discussed. Particular attention is given the relationship of vinylpyrrolidone level and the equilibrium level of acrylamide hydrolysis at 121°, and 150°C.

INTRODUCTION

In recent years oil production has been moving into deeper and consequentially hotter reservoirs. This trend has produced a challenge for the manufacturers of the water-soluble polymers employed in the various oil recovery processes. The processes employing water-soluble polymers include drilling, cementing, stimulation, and enhanced oil recovery (EOR). EOR may have the most stringent requirements for polymers owing to the long application times. An EOR project may require retention of desirable polymer properties for one, two, or more years at reservoir conditions.

The polymers used in EOR may be employed in treatment processes such as mobility control, injection well reprofiling, production well water shut-off, or steam diversion. At temperatures greater than 70-75°C (160-170°F),most water-soluble polymers undergo extensive hydrolysis. As a result, the thickened brine solutions lose viscosity (or in the case of crosslinked gels, they lose gel strength) and are rendered useless[2/3]. The mechanism of failure is usually precipitation by the reaction of the hydrolyzed polymer with naturally occuring polyvalent ions.

Polysaccharides such as the xanthan gums have been used in recovery operations, however they are probably not suitable for applications in reservoirs with temperatures in excess of 93°C (200°). Like polyacrylamide, the natural polymers hydrolyze at high temperatures. The use of stabilizer additive packages might raise this limit[4/5], but the possibility that these packages will protect the polymer from hydrolysis as the polymer moves away from the well bore is remote. The package and the polymer will probably seperate during movement because of differing partition coefficents with the reservoir.

In a widely read paper, Davison and Mentzer[6] reported testing some 140 different polymers for applicability in EOR in seawater at 90°C (194°F). Almost all except scleroglucan and polyvinylpyrrolidone failed by precipitation in a few weeks. Unfortunately neither of these polymers has the viscosity producing power to be economically suitable for EOR.

Similar conclusions were reached in our laboratory when evaluating water-soluble polymers for a more hostile application in seawater at 121°C (250°F). In our tests, only polyvinylpyrrolidone survived more than a few months. A summary of some of the polymers evaluated in our study is given in Table I.

Table I. The Results of Aging Selected Water-Soluble Polymers in Synthetic Seawater at 121°C (250°F).

Water-soluble Polymer (by type)	Result
Polyacrylamide (PAm)	Precipitation & Viscosity Loss
PAm/polyvinylpyrrolidone blend	Precipitation & Viscosity Loss
Hydrolyzed PAm	Precipitation & Viscosity Loss
PAm Flocculants	Precipitation & Viscosity Loss
Poly(acrylic acid)	Insoluble
PAm Oilfield Polymer	Precipitation & Viscosity Loss
High Performance PAm	Precipitation & Viscosity Loss
Hydroxyethyl Cellulose	Viscosity Loss
Polysaccharides	Precipitation & Viscosity Loss
Stabilized Biopolymers	Precipitation & Viscosity Loss
Scleroglucan	Slow Viscosity Loss
Xanthan	Precipitation & Viscosity Loss

Table I. Continued from previous page

Water-soluble Polymer (by type)	Result
Poly(vinyl Alcohol)	Insoluble
Acrylic Acid-Acrylamide-Vinyl Pyridine Terpolymer	Precipitation & Viscosity Loss
Acrylamide-SAMPS (60:40)Copolymer	Precipitation & Viscosity Loss
Poly(methyl vinyl ether)	Viscosity Loss
Polyvinylpyrrolidone	Stable, Low Viscosity
Polyethylene oxide	Precipitation & Viscosity Loss

Sodium Poly(2-acrylamido-2-methylpropanesulfonate) (SAMPS) solutions have been shown to be relatively stable at 121°C and to hydrolyze slowly to about 4% of available acrylamide after extended aging[7]. Copolymerization of SAMPS with acrylamide does not protect the latter against thermal hydrolysis and consequentially precipitation in hard brines. Only SAMPS-Acrylamide copolymers containing less than 9% (w/w) acrylamide can tolerate a brine such seawater at 121°C. SAMPS-acrylamide copolymers have been evaluated at less hostile conditions.

An acrylamide/SAMPS 40/60 weight ratio copolymer was found to be the optimum composition for use in 93°C (200°F) applications. Since polyacrylamide is suitable for applications up to about 70-75°C (160-170°F)[2,3], inclusion of 60% of a more expensive SAMPS monomer into the copolymer to increase its usability by only 20°C does not appear attractive. It is believed that many of the so called " high temperature " polymers commercially available are SAMPS based materials containing high levels of acrylamide to reduce cost. If so these materials are not suitable for applications at temperatures significantly higher than those possible with polyacrylamide.[7]

The preparation, properties, and usefulness of copolymers of vinylpyrrolidone and acrylamide will be discussed in this paper. Although neither homopolymer appears economically viable for large scale use in hostile oil recovery environments (hard brines at temperatures in excess of 170°F), the various copolymers were found to produce solutions which resist precipitation on aging. The inclusion of vinylpyrrolidone in copolymers of acrylamide apparently protects and limits the equilibrium degree of hydrolysis. The equilibrium level is less that the total available acrylamide, and it is related to the amount of vinylpyrrolidone in the copolymer. The importance of these findings will also be discussed.

EXPERIMENTAL

Polymer solutions (25 ml) were aged in pyrex ampules (50 ml) at 121° and 150°C to determine the extent of hydrolysis and cloud points. Oxygen was removed from the ampules by alternating vacuum and nitrogen cycles. Once deoxygenated, the ampules were torch sealed under a slight vacuum (-25Kpa).

The sealed ampules were aged for the appropriate time and temperature in vented aluminum jackets in a temperature regulated oven. Although no breakage occured during the course of these experiments, the jackets were employed as a safety measure.

All solutions were placed in an oxygen free helium atmosphere glove box. The glove box was used for all sample preparation and handling. The presence of oxygen in the box was monitored using a solution of diethyl zinc. A 5% saline solution was sparged for 15-20 minutes with zero grade nitrogen. This length of sparging time was sufficent to reduce the dissolved oxygen content to less than 10ppb as measured by a Chemet dissolved oxygen kit.

The procedures for determination of the extent of hydrolysis[2/8] and cloud point[2/9] have been described at length elsewhere.

A capillary viscometer was specially constructed to determine the solution viscosity in the 40-150°C range[10/11]. This was necessary as room temperature viscosities of post-aged solutions give at best an approximation of polymer performance at reservoir temperatures. The viscometer consists of a 20 ft. preheating coil and a 40 ft. measuring coil of 0.069" ID stainless steel. Both coils are immersed in a silicone oil bath and heated using a temperature programmer at 2.0°C/min. The polymer solution is driven through the coil using nitrogen pressure. The pressure drop across the measuring coil is monitored by a differential pressure transducer. After passing throught the measuring coil the solution passes through the flow control section of the apparatus. The latter is equipped with a valve to maintain 90psi back pressure on the system. This pressure is sufficent to prevent boiling of the solution. These flow control devices maintain a flow rate of 1.2 ml/min.Control measurements of the viscosity of distilled water gave excellent agreement with literature values over the whole temperature range.

The polymers were evaluated as solutions of either distilled water or synthetic seawater. The composition of synthetic seawater is given in the following table.

Table II. The Composition Synthetic Seawater.

Compound	Quantity
$NaHCO_3$	3.69 g
Na_2SO_4	77.19 g
NaCl	429.00 g
$CaCl_2.2H_2O$	29.58 g
$MgCl_2.6H_2O$	193.92 g
water, distilled	18.0 liters

This synthetic seawater contains 10,550 ppm sodium ions, 1693 ppm combined divalent Calcium and magnesium, and 33,756 ppm total dissolved solids. Freshly prepared synthetic seawater was used in all experiments to provide consistency in pH and hardness.

N-Vinyl-2-pyrrolidone-acrylamide copolymers were prepared by aqueous homogeneous solution polymerization in either distilled water or synthetic seawater. The initial monomer level in the polymerization solutions was either 9.1% or 20% (w/w). The solutions were initiated at either room temperature or 50°C by either 0.50phm or 0.10phm commercial free radical initiators. A temperature controlling water bath was used with the 50°C polymerizations.

Two commercial free radical initiators were used in the preparation of the copolymers. They were DuPont Vazo-64 (2,2'-azobisisobutyronitrile) and DuPont Vazo-33 (2,2'-azobis(2,4-dimethyl-4-methoxyvaleronitrile)). Vazo-64 was used at 50°C and Vazo-33 at room temperature. N-Vinyl-2-pyrrolidone (Aldrich) and acrylamide (Aldrich and Mitsubishi) were used as received.

Comparative aging results, reported in Table I, were performed on many commercial and selected experimental polymers. These were in most cases used as recieved from the suppliers. A partial list of polymer suppliers includes Aldrich, Allied Colloids, American Cyanamid, Betz, CORT, Dow, DuPont, GAF, Hercules, Jetco, Kelco, Nalco, Pfizer, and Union Carbide.

RESULTS AND DISCUSSION

Poly(vinylpyrrolidone-co-acrylamide) (PVP-Am) forms readily as a high molecular weight polymer in aqueous solutions in the presence of free radical initiators. The preparation and characterization of these copolymers has been reported[12]. The copolymers prepared for this study were prepared by homogeneous polymerization as 9.1-20% (w/w) solutions. When the polymerization was complete, usually in 8-24 hours, these high solids solutions were rigid and gel-like in appearance. The linearity of the copolymer was evident on dilution, for they dissolved to form gel free, filtrable viscous aqueous solutions.

As stated earlier, an extensive laboratory review of many available commercial water-soluble polymers and selected experimental water-soluble polymers was conducted. A summary of the results is given in Table I. As shown, only polyvinylpyrrolidone survived the 121°(250°) test in synthetic seawater more than a few days. Polyvinylpyrrolidone is not however an efficent viscosifier. Combined with its high adsorption characteristics in reservoir rock (polyvinylpyrrolidone has been used in selected cases as a sacrifical agent) and high price, polyvinylpyrrolidone is regarded as unsuitable for use in EOR.

The most widely employed synthetic oil field polymer is polyacrylamide. At elevated temperatures, hydrolysis occurs on the labile amide groups. Charge repulsion in the newly formed carboxylate groups brings on an increase in viscosity. The viscosity increase can be permanent in pure water, but in brines, such as those found in oil producing reservoirs, the viscosity increase is usually followed by a dramatic loss of viscosity.

The following diagrams are examples of hydrolysis and subsequent reaction with available metal ions. As we have attempted to show, the acrylamide polymer hydrolyzes to form an acrylic acid containing polymer. The polymer containing acrylic acid groups can further react with metal ions, and if the ion is multivalent, it can eventually crosslink and precipitate.

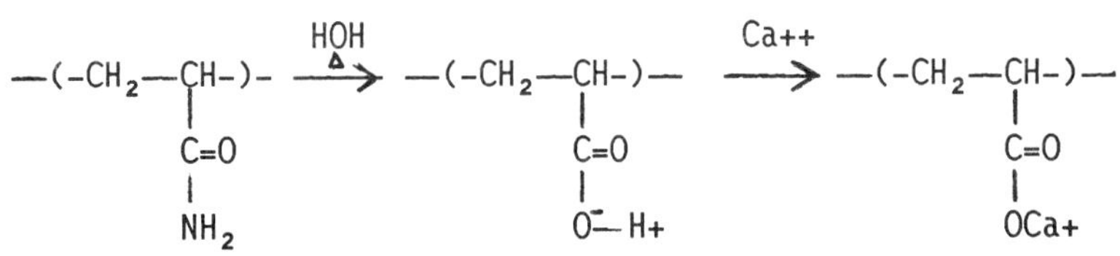

In seperate, comprehensive investigations Moradi-Araghi and Doe, and Ryles [2/3] found that the hydrolysis reaction of polyacrylamide is accelerated by heat. Given the weeks and often years required in most oil recovery processes, both reports concluded that a ceiling exists for the usefulness range of polyacrylamide. Moradi-Araghi and Doe reported the "safe limit" to be temperatures less than 75°C (67°F).

The level of hydrolysis of acrylamide and an equimolar copolymer of PVP-Am (60/40 w/w) is shown in Figure 1. Polyacrylamide, as shown, rapidly hydrolyzes. The degree of hydrolysis is nearly quantitative after about one week of aging. The polyacrylamide reported in Figure 1 became insoluble in synthetic seawater after a few days of aging.

The equimolar copolymer also hydrolyzed, but the percent hydrolysis reached an equilibrium value after a few days of aging. The percent hydrolysis was never quanitative. At 121°C (250°F), the percent of hydrolysis was about 40 percent. The value, 40%, of the available acrylamide in the copolymer is only about 16 weight % total carboxylate. PVP-Am (60/40 w/w) does not precipitate at 121°C in synthetic seawater.

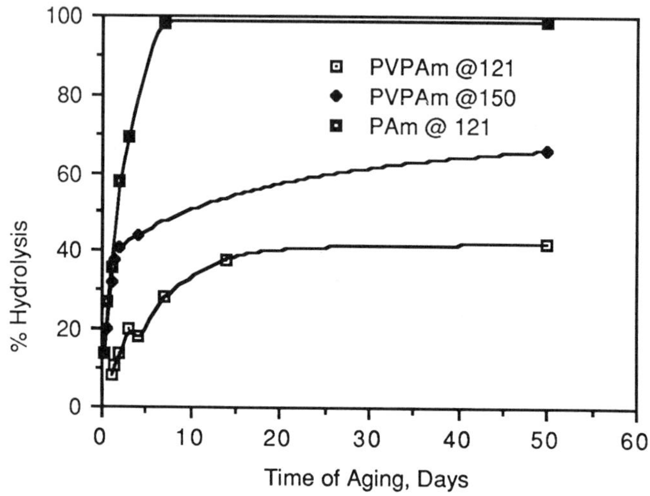

Figure 1. The Extent of Hydrolysis for PVP-Am (60/40 w/w) and a Commercial Polyacrylamide.

One sample of this copolymer has been aged in our laboratory for more than six years without precipitating in selected brines.

PVP-Am (60/40 w/w), as shown, hydrolyzes to about 60% at 150°C (300°F) after 50 days of aging. This level is close to its 150°C equilibrium value. The copolymer did not precipitate at 150°C, even after very long aging times. The equilibrium hydrolysis value of 60% of the available 40% acrylamide in the copolymer is about 24 weight percent total carboxylate.

The relationship of the degree of hydrolysis of available acrylamide and aging is shown in Figure 2. Reviewing, PVP-Am (60/40 w/w) undergoes hydrolysis to about 40%. Thus 40% of the 40% available acrylamide is about 16 weight percent total carboxylate. As shown in Figure 2, PVP-Am (70/30 w/w) equilibrium hydrolysis is about 30%. Thirty percent of 30% available acrylamide gives about 9 percent total carboxylate. The estimated total carboxylate of the copolymers shown in Figure 2 are given in Table III.

Table III. The Total Weight Percent Carboxylate in PVP-Am After Aging at 121°C (250°) in Synthetic Seawater.

Copolymer	Wt. % Am	Hydrolysis %	Total % Carboxylate
PVP-Am	80	95	76
PVP-Am	60	80	48
PVP-Am	50	65	33
PVP-Am	40	40	16
PVP-Am	30	30	9

It is evident that the available acrylamide in each PVP-Am hydrolyzes to an equilibrium level. Furthermore, the level of hydrolysis is related to the vinylpyrrolidone level in the copolymer. Copolymers with the greater amounts of vinylpyrrolidone exhibit proportionately less hydrolysis of the available acrylamide. The mechanism of protection of the acrylamide will not be addressed in this writing.

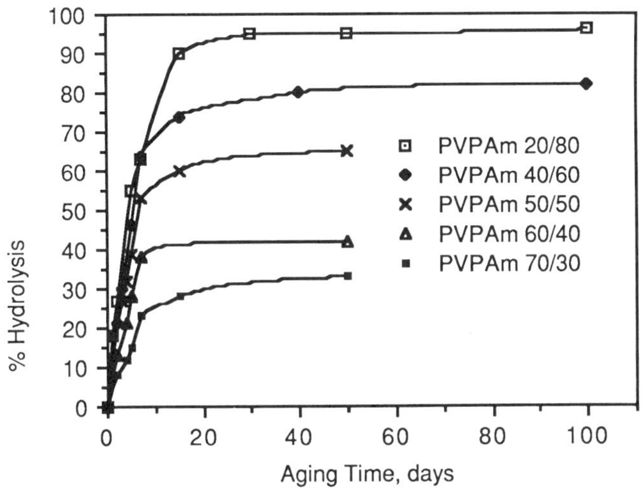

Figure 2. The Extent of Hydrolysis for Various PVP-Am Aged at 121°C (250°F).

The relationship of the level of hydrolysis, this time expressed as mole percent carboxylate, of the aged copolymer and the cloud point is shown in Figure 3. The measured cloud point is the detectable onset of precipitation[9].

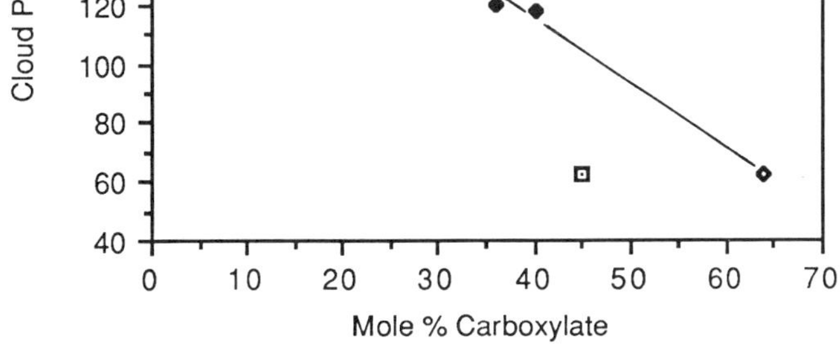

Figure 3. Effect of Carboxylate Content on the Cloud Point of PVP-Am in 5% Saline with 988 ppm each of Ca++ & Mg++.

Predictably, the cloud point decreases with increasing carboxylate content. The copolymers containing 50% (w/w) and 40% (w/w) vinylpyrrolidone did not form as much carboxylate on aging as did polymers with less vinylpyrrolidone (more acrylamide). The restriction of carboxylate level (limited acrylamide hydrolysis) by inclusion of vinylpyrrolidone is the key element to the high temperature/ high hardness stability of these materials.

To polymerflood a reservoir with seawater at 121°C (250°F), Figure 3 indicates that a copolymer must be selected so that less than 40 mole percent hydrolysis will be produced after aging. The polymer will precipitate if the level of hydrolysis exceeds this value. Figure 2 indicates that a copolymer with more than 60% (w/w) acrylamide is not suitable for this application.

The data shown in Figure 4. compares the viscosity ratio (the viscosity of synthetic seawater containing polymer to the viscosity of synthetic seawater) of several PVP-Am solutions with the mole percent acrylamide in the copolymer. Unaged solutions of the copolymers with greater amounts of acrylamide produced higher viscosity solutions. This is expected. These copolymers had the greatest viscosity loss on aging. Presumeably the acrylamide moieties in the polymers not protected by vinylpyrrolidone hydrolyze, coil, and thus have the greatest loss. Although not presented here, these same copolymers predictably have the highest viscosities in distilled water.

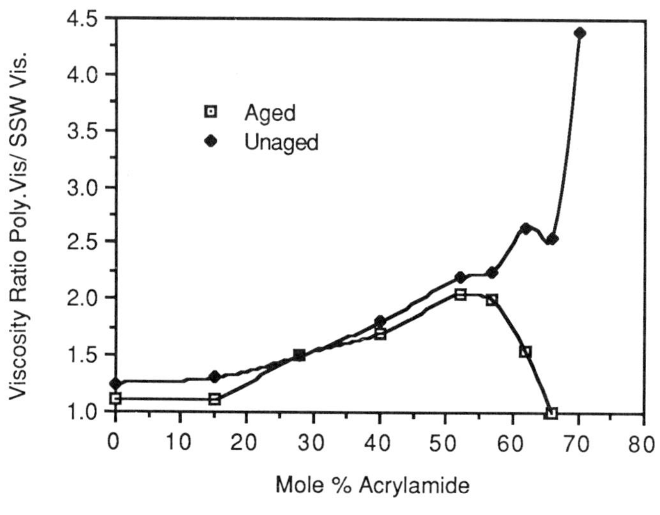

Figure 4. The Viscosity Ratio at 121°C vs. Mole Percent Acrylamide for 2500 ppm Solutions of Aged and Unaged PVP-Am in Synthetic Seawater.

PVP-Am containing equimolar amounts of each monomer retained the highest aged viscosity. These are assumed to be the most useful in EOR at 121°C. As shown a copolymer of 30 mole percent acrylamide did not lose viscosity on aging.

The retention of viscosity of PVP-Am at 121°C (250°F) is shown in Figure 5. The representative polyacrylamide hydrolyzed rapidly at 121°. It also precipitated in just a few days, a time hardly enough to be considered useful in oil recovery processes at this temperature. The data suggests that the 50/50 and 60/40 (w/w) copolymers are partially hydrolyzed. But after 30-50 days, no further hydrolysis was evident. The level of carboxylate was apparently at equilibrium. The viscosities of both copolymer solutions stabilized at about twice the viscosity of seawater at 121°C. Equimolar PVP-Am (60-40 w/w) had a slightly higher aged viscosity although the unaged viscosity was lower than the 40/60 molar copolymer (50/50 w/w).

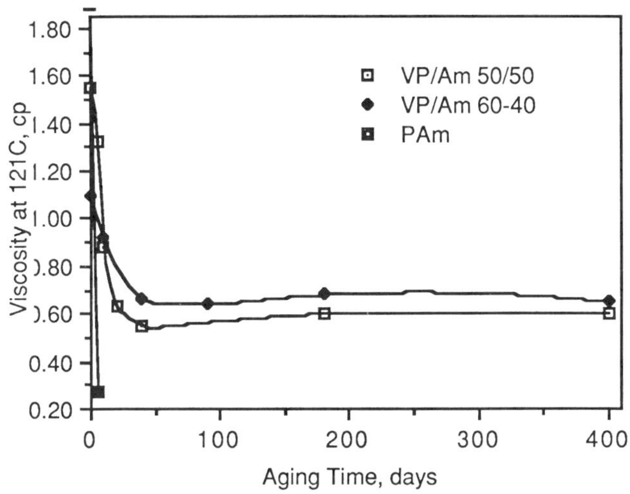

Figure 5. The Viscosities of PVP-Am and a Commercial Polyacrylamide Aged in Synthetic Seawater at 121°C.

CONCLUSIONS

1. Copolymerization of vinylpyrrolidone with acrylamide results in products with stability superior to that of polyacrylamide in the presence of divalent cations at elevated temperatures.

2 The level of hydrolysis of the available acrylamide is directly related to the level of vinylpyrrolidone in the copolymer.

3. The greatest aged viscosity at 121°C was prepared with an equimolar PVP-Am (60/40 w/w).

4. A sample of PVP-Am (50/50 w/w) has remained in synthetic seawater solution at 121°C for over six years.

5. A range of PVP-Am can tolerate hostile conditions of 121°C and synthetic seawater for an extended period of time.

6. PVP-Am should be useful for oil recovery processes such as well drilling, cementing, stimulation, and enhanced recovery.

ACKNOWLEDGEMENTS

We acknowledge the helpful discussions of H. L. Hsieh and I. J. Westerman. F. W. Blanton, M. W. Branstetter, D.B. Cunningham, and S.C. Sherrill performed many of the aging studies and viscosity measurements. We also acknowledge the polymer preparations of D.A. Westerman, R.M. Infield and Rev. J.H. Irvin. Finally, we thank the management of the Phillips Petroleum Company for their generous permission to publish this paper.

REFERENCES

1. Present address: Exxon Chemical Co., 5200 Bayway Dr. Baytown, TX 77522

2. Moradi-Araghi, A. and Doe, P.H., Soc. Pet. Engr. Reser. Engr., (May, 1987) ,189.

3. Ryles, R.G., Soc. Pet. Engr. Paper SPE 13585, (April, 1985).

4. Wellington, S.L., Soc. Pet. Engr. J., (December, 1983),901.

5. Abdo, M.K., U.S. Patent 4,141,842 (February, 1979).

6. Davison, P. and Mentzer, E., Soc. Pet. Engr. J., (June,1982),353.

7. Moradi-Araghi, A., Cleveland, D.H., and Westerman, I.J., Soc. Pet. Engr. Paper 16273, (February, 1987).

8. Scoggins, M.W. and Miller, J.W., Soc. Pet. Engr. J., (June, 1979),1151.

9. Moradi-Araghi, A., Brunning, D.D., and Doe, P.H., Rev. Sci. Instrum., 57, 2315, (1986).

10. Doe, P.H., Moradi-Araghi, A., Shaw, J.E., and Stahl, G.A., Soc. Pet. Engr., Reser. Engr.,(January, 1988), 461.

11. Cunningham, D.B., Doe, P.H., Joshi, S.D. and Moradi-Araghi, A.,Rev.Sci. Instrum., 57, 2310, (1986).

12. Stahl, G.A., U.S. Patent 4,644,020 (February, 1987).

VINYL SULFONATE/VINYL AMIDE COPOLYMERS AS TEMPERATURE- AND SALT-STABLE THICKENERS FOR EOR FLOODING PROCESSES

Walter Gulden and Sigmar-Peter von Halasz

Technical Department Oilfield Chemicals
Hoechst Aktiengesellschaft
Postfach 800320
6230 Frankfurt/Main-80 / W.Germany

In 1985 enhanced oil recovery processes represented over 11.7 % of total domestic crude oil production in Germany. Some 75 % of this supplementary production was the result of steam processes, the remaining 25 % was recovered using chemical flooding processes, in particular polymer flooding techniques.

According to a recent estimate of the German Oil and Gas Association (WEG)[1] the actual recovery of oil using presently known processes totals about 43 % of OOIP, while 57 % remains in the formation. Higher recovery rates will demand new de-oiling techniques not yet developed. There is shown a comparison (Fig. 1) with the situation in the USA where a recovery efficiency of approx. 60 % is estimated. Naturally the absolute production of crude oil in the USA is more than 100 x higher, as Germany with 4.1 mty (82,000 bbl/d) is one of the world's smaller producers.

However, the size of our reserves is no disadvantage, as the chemical industry has been very closely involved with the development of the German oilfields since the very early days of production.

At the beginning of the fifties, experiments were under way, using various surfactants and water soluble polymers in order to improve secondary recovery processes.

This included small scale field trials with both CMC thickened water and polyacrylamide derivatives. Obviously the early stages on the learning curve didn't come cheaply.

The relatively high average salinities of German oil formations limited the broader application of polyacrylamides because of their known salt sensitivity. This situation was overcome by pre-conditioning the reservoir with a fresh water preflush and/or through a two stage polymer injection. The two stage polymer injection method involves first, flooding with a less salt sensitive polymer (e.g. slightly hydrolysed PAA) to effect a displacement of the formation water, followed by an injection of the polymer solution in fresh water for the actual oil displacement.[2]

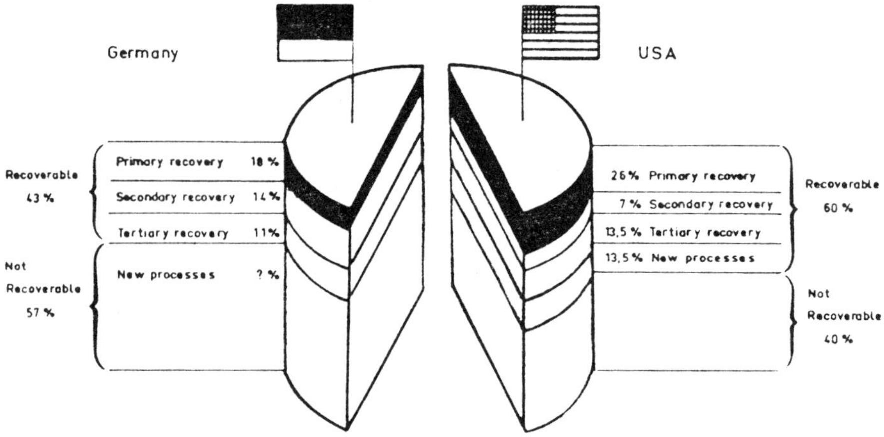

Figure 1 Crude oil production by Recovery Mechanisms. Comparison Germany / USA

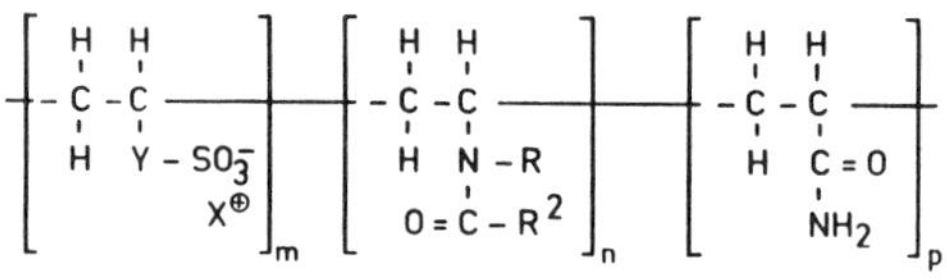

Figure 2 Chemical Composition of VS/VA/AM Copolymers.

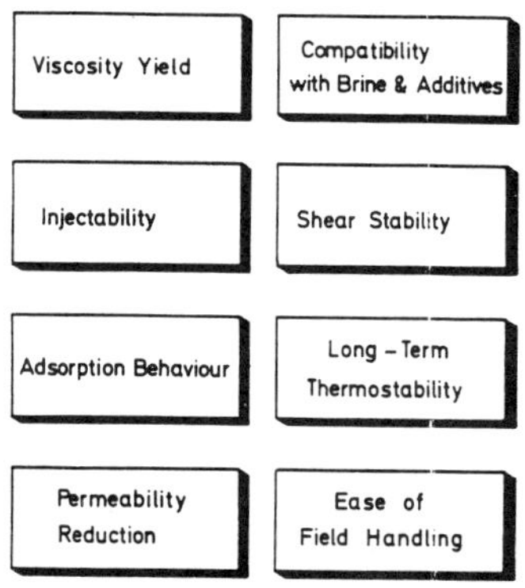

Figure 3 Requirements for EOR Polymers.

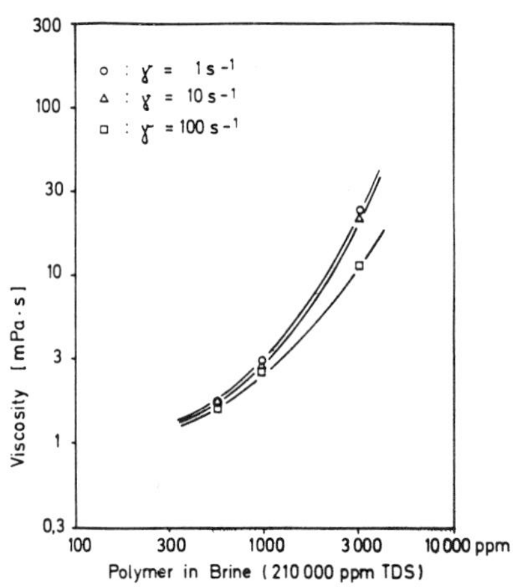

Figure 4 Viscosity at 75°C (167°F) Versus Concentration of VS/VA/AM - Copolymer.

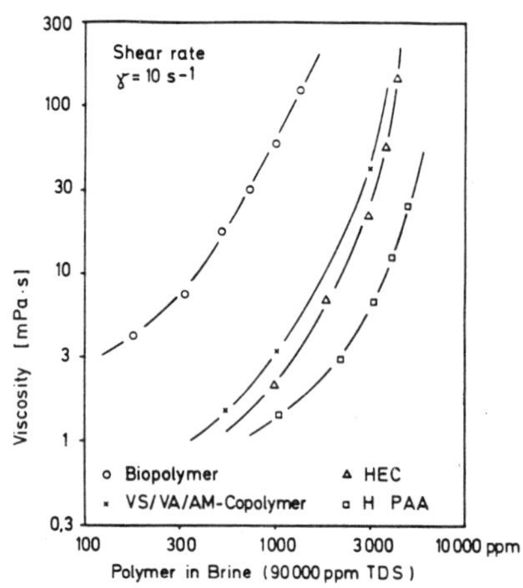

Figure 5 Viscosity at 55°C (131°F) Versus Concentration of Different EOR - Polymers.

It is noteworthy that this technology on an industrial scale achieves incremental recoveries between 10 - 20 % OOIP.[3]

This process of course involves additional costs and is more time demanding than single stage techniques. It is only suitable for moderate salinities and temperatures to 160 °F (70 °C). The method cannot be used for clay containing formations sensitive to fresh water, which are not uncommon.

The chemical industry then was confronted with the task of developing a polymer that remained stable in high salinities over a longer period, and at temperatures up to approx. 220 °F (105 °C).

Our company has long experience with temperature and salt stable copolymers, which have been used throughout the world as fluid loss additives in drilling muds under conditions of extremely high salt concentrations and temperatures to over 390 °F (200 °C).[4,5]
However, the molecular weights of these polymers at 1×10^6 (approx.) are too low to produce a reasonable viscosity/cost ratio of polymer solution in the EOR application.
The task of the research chemist therefore was to considerably increase the molecular weight of such copolymers while maintaining these desirable properties. Today we have so far succeeded, using a special gel-polymerization process, to manufacture products having an average molecular weight of $6 - 8 \times 10^6$ and accordingly showing an adequate viscosity yield. Presently a well performing copolymer is likely to contain the monomers of vinyl sulfonate, vinyl amide and acrylamide as building blocks.

Through skillful combination of derivatives of these monomers the desired properties can be targeted in the finished polymer (Fig. 2). In the course of development of these polymers it became apparent, that through variations of the relative amounts of the monomer building blocks, the characteristics of the polymer could be optimized. It should be mentioned here that during the polymerization the acrylamide component is partially hydrolized.

The oil industry makes the following special demands on the quality of EOR polymers[6] (Fig. 3):

- viscosity yield and flow properties
- compatibility with formation water and additives
- injectability
- shear stability
- adsorption behaviour
- long-term stability
- permeability reduction
- ease of field handling

The VS/VA/AM-copolymers show, in comparison to standard polyacrylamides and biopolymers, very interesting properties under severe test conditions. We would now like to discuss some of these characteristics in more detail.

Viscosity Yield and Flow Properties

A decisive criterion for the economics of a viscosifier for polymer flooding processes is its ability to perform in the existing, often saline reservoir- and flooding waters at the respective reservoir temperature. This goes just as well for the shear viscosity as for the elastic viscosity. Data on the effectiveness of the additive and the rheological behaviour of the polymer solutions are preferably obtained using a low shear rotational viscometer. In the next diagram (Fig. 4) one can see the viscosity relationship of VS/VA/AM-copolymers at 170 °F (75 °C) and at a salinity of 210 g/l (210,000 ppm). The measurements were made for various shear rates. The next illustration (Fig. 5) shows the comparison with a typical partially hydrolyzed PAA, a biopolymer and a high viscosity HEC, at 130 °F (55 °C) and a salinity of 90 g/l TDS (90,000 ppm). As may be expected the biopolymer is the most effective under these conditions (without aging!), while the VS/VA/AM-copolymer is clearly apart from the HEC and HPAA.

Injectability

In order to be able to judge the injectability of a polymer solution in a pay zone, filterability tests with surface pre-filters alone are not sufficient. As well as the influence of shear rate within the pores of the formation the interaction between the rock matrix and polymer solution should also be considered due to the effect on processes such as adsorption and retention.

For this reason we perform our injection tests on plugs of natural sandstone at temperatures of 130 - 170 °F (55 - 75 °C). By continuous injection of polymer solution (1000 ppm polymer in brine) at a constant rate over the flood section, differences in pressure that develop are recorded and are used to judge the injectability. A product is considered to be good in this respect if the pressure gradient, after the injection of 10 x pore volume of solution, increases by less than 10 % of the initial pressure gradient (after saturation!) (Fig. 6).
Experience has shown that the injection of biopolymers and HPAA at these high salinities leads to considerable problems.
From the pressure gradients determined for two flooding rates we are able to deduce the resistance factor RF, which serves as a measure of the flow resistance of the polymer solution in the pore space, as well as the residual resistance factor RRF. The factor RRF serves as a measure of the permeability reduction due to polymer material absorbed and retained in the pores. The values obtained for the VS/VA/AM-copolymers are particularly good.

Shear Stability

Aside from the viscosity dependence of shear rate with flow of structurally viscous polymer solutions within the pores, an irreversible viscosity reduction occurs under higher shear stress.
In order to measure this viscosity degradation, the polymer solution at a concentration of 3000 ppm and various temperatures is circulated through a natural sandstone plug (length 10 cm, diameter 3 cm) until an effective flow distance of 100 m has been reached.
The flow rate is about 10 m/day, which corresponds to the more intensive conditions in the vicinity of an injection well bore.
The relationship between final viscosity and original viscosity as a function of time and effective distance covered respectively serves as the criterion of the shear stability of the product under investigation.

The comparison of continuous shear stress of VS/VA/AM-copolymers with HPAA and biopolymers at 122 °F (50 °C) and a salinity of 90,000 ppm is shown in the next diagram (Fig. 7). The well known good shear stability of the biopolymers can clearly be seen, but also the large difference between standard HPAA and our own copolymer.

Figure 6 Injectability of 1000 ppm Copolymer in Brine (210000 ppm TDS). Natural Sandstone Core.

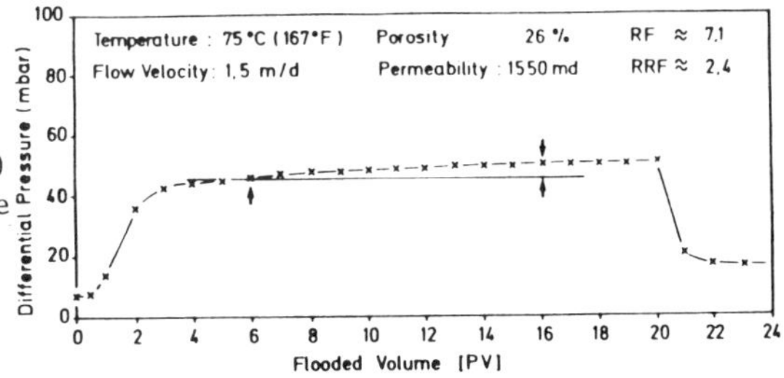

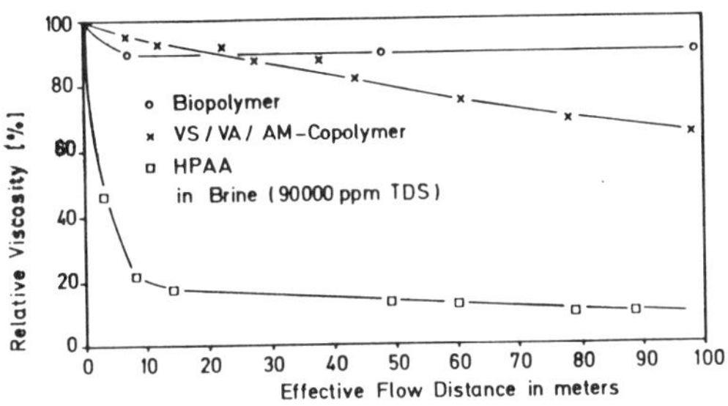

Figure 7 Long Term Shear Stress Experiment at 50°C (122°F).

Figure 8 Thermal Stability of EOR-Polymers.

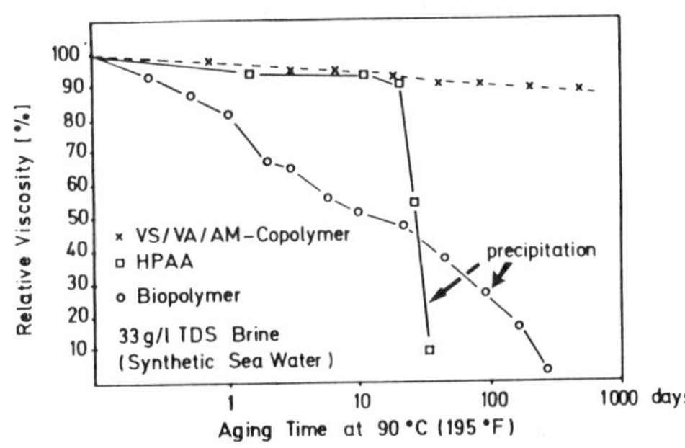

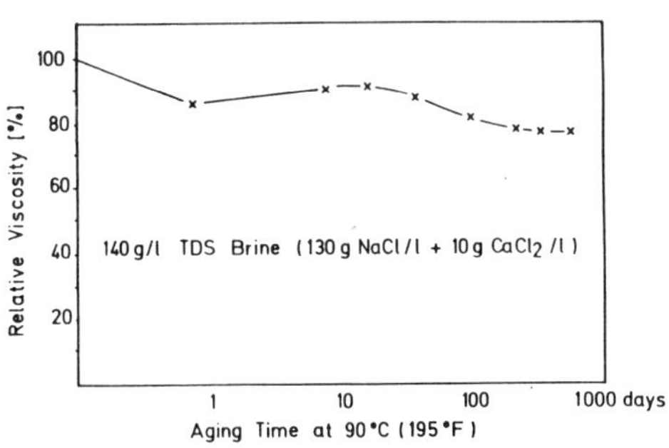

Figure 9 Thermal Stability of VS/VA/AM - Copolymer

The reaching of a plateau in the viscosity degradation curve confirms that within the first few metres of flooding, molecules or structures are destroyed that exceed a critical size in relation to the pore space, while the remaining molecules are not damaged and no further viscosity degradation is experienced. Clearly this critical size of molecular structure is more strongly influenced by the salinity than the temperature, as the shear degradation is more severe in high salinity systems (210 g/l TDS).
At the moment we don't quite know for sure if its a matter of an increased, purely mechanical shear degradation, or if colloidal chemical influences are involved. In support of the latter theory is the fact that be addition of certain additives the viscosity loss can be considerably reduced.

Thermostability

A further important criterion for assessing the suitability of intended EOR polymers is the long-term stability under thermal aging. In these tests the flooding fluid should show no substantial loss of viscosity or tendency to cloudiness and flocculation, and especially no precipitation of insoluble matter, which would lead to serious permeability loss.
The copolymers developed by us show in aqueous saline solution good thermostabilities in long-term studies in comparison to other EOR polymers that have been developed for low- or non-saline conditions and low reservoir temperatures.
The polymer solutions are stabilized with 500 ppm formaldehyde and kept practically oxygen free (<5 ppb O_2). As the next figure shows (Fig. 8), clear viscosity differences arise even at a salinity of only 33 g/l TDS (33,000 ppm) and reservoir temperatures of 195 °F (90 °C). The viscosity values as measured at room temperature are expressed as a percentage of the original values before the temperature tests.
While the biopolymer solution (Xanthan gum) continuously loses viscosity, the HPAA solution remains relatively stable in 40 day storage tests. Finally however the viscosity of the HPAA solution declines drastically. Both solutions, biopolymer and HPAA containing, show substantial clouding and precipitation that increases noticeably with the increasing test duration.
The VS/VA/AM-copolymer solution under the same test conditions and after around 400 days shows only a slight loss of viscosity to about 90 % of the original value and displays no cloudiness or precipitation.

Results (Fig. 9) are given for the storage tests of a copolymer solution at the substantially higher salinity of 140 g/l TDS (140,000 ppm), again at a temperature of 195 °F (90 °C). Under these conditions the viscosity of the copolymer solution falls to about 75 % of its original values after about 400 days.
The solutions too after more than 400 days continuous aging remain completely clear and free from precipitates.
With regard to EOR demands in higher salinity and higher temperature reservoirs, the copolymer solutions we have discussed here have shown very promising thermostability results.

Oil Recovery and Compatibility

In laboratory core-tests under various test conditions (salinity, temperature, oil quality) the excellent slug stability of VS/VA/AM-copolymers could be demonstrated, along with their ability to build up an oil bank. The pressure build up is modest and technically always manageable. As a rule in sandstone core trials an additional oil recovery of 5 - 20 % can be obtained.
EOR polymers are not just for use in polymer floods alone, but can be

used in combinations with surfactants in micellar/polymer or surfactant/polymer flood systems.
The conception of surfactant/polymer processes requires extensive testing of compatibility of both flooding mediums. Therefore at the present moment we are testing intensively the behaviour of surfactants and polymer solutions and how they interact with one another, and with particular reference to salt and temperature resistance in surfactant systems.
Although our work is not yet completely finished, the results so far obtained confirm that we are on the right track. We have already determined that we have salt and temperature resistant surfactants available that are very compatible with VS/VA/AM-copolymers.

SUMMARY

To summarize it can be stated that the VS/VA/AM-copolymers show better injectability and long-term thermostability compared to biopolymers. When compared to HPAA we find a better viscosity yield in brine, a better shear stability and lower adsorption losses.
The 'whole truth' concerning the quality and performance of these types of polymers can only be revealed in a field trial.
For a long time now, together with a German oil company, we have been preparing for such a trial. Because of its reservoir conditions (salinity, temperature, clay content), our chosen field can be treated neither with HPAA nor with biopolymers. By next year we hope to have collected the first set of field data.

REFERENCES

1. Neue Techniken der Erschließung von Erdöl und Erdgas
   W.E.G.; Hannover (Dez. 1984)
2. Neuere Entwicklungen auf dem Gebiet der Lagerstättentechnik
   M. Rieckmann, G. Pusch; Erdöl, Erdgas; Heft 6 (1984)
3. Application of Chemical Flood Processes in the Oilfield Hankensbüttel
   B. Maitin; 3nd European EOR-Conference, Rome (1985)
4. Vinyl Sulfonate/Vinyl Amide Copolymers in Drilling Fluids for Deep, High-Temperature Wells
   M. Hille; SPE 13558; Phoenix/USA (April 1985)
5. Temperature-Stable Polymeric Fluid-Loss Reducer Tolerant to High Electrolyte Contamination
   A.J. Son, T.M. Ballard and R.E. Loftin; SPE 13160; Houston (Sept. 1984)
6. Polymer Flooding, General Aspects and Case History
   Deutsche Texaco AG (Febr. 1986)

THERMALLY STABLE ACRYLIC POLYMER FOR PROFILE MODIFICATION APPLICATIONS

R.G. Ryles and R.E. Neff

American Cyanamid Company
Chemical Research Laboratories
Stamford, CT 06904

INTRODUCTION

Profile modification is a process in which flooding water is diverted from highly permeable zones into less permeable oil-bearing layers within a subterranean reservoir. In recent years this has been achieved by the use of polymeric hydrogels in which the water-soluble polymer is cross-linked by a multivalent metal ion[1]. The latter may be generated in situ from a redox couple, e.g., sodium dichromate/sodium metabisulfite or as a complexed reagent, e.g., aluminum citrate. This enables the gelation process to be retarded until the aqueous polymer solution has been placed deep into the thief zone.

The most widely used water-soluble polymer has been anionic polyacrylamide because it readily forms rigid hydrogels under these conditions and is relatively inexpensive. However, recent reports have shown that gels derived from polyacrylamide have limited thermal stability particularly when high concentrations of metal ions are present in solution[2,3]. Specifically, these systems were unsuited to reservoir temperatures above 90°C[3]. Others[9] have indicated a somewhat lower (75°C) temperature for polyacrylamide.

The primary objective of this work was, therefore, to design a thermally stable profile modification system. This paper describes some aspects of the chemistry of 2-acrylamido-2-methyl-propanesulfonic acid (AMPS®)/acrylic acid (AA) copolymers which have been shown form hydrogels with excellent thermal stability[3].

EXPERIMENTAL

Gel Copolymerization

Monomers were purified prior to use. AMPS was recrystallized from methanol and AA was distilled under reduced pressure. Free radical gel polymerizations were performed in aqueous solution at 20, 35, 50, 70, and 90 mole % AMPS at a monomer concentration of 2.48 moles per kg. Two series

AMPS® is a registered trademark of the Lubrizol Corporation.

of polymerizations were conducted, the first at pH 7.0 at >95% degree of neutralization, the second at low pH by adding acrylic acid to sodium AMPS solution. The polymerizations were short stopped at <10% conversion. The pH was then adjusted to 4.0 with either acetic acid or with sodium hydroxide. Then, the copolymer was precipitated and isolated three times in acetone, washed twice with a 80:20 acetone/water mixture and then with once with 100% acetone. The polymer was then dissolved in water and the pH adjusted to 7.0 with sodium hydroxide. After precipitation with t-butanol, the polymer was washed twice with acetone and once with ether. The polymer was then dried under vacuum to constant weight.

The composition of the isolated copolymer was determined from (a) the nitrogen elemental analysis, and (b) the sulfur elemental analysis. Both figures were corrected for the small amount (4-12%) of water associated with the polymer.

^{13}C NMR Spectroscopy

All ^{13}C spectra were recorded at 50.3 Mz using a Varian XL-200 spectrometer at ambient temperature. Gated decoupling was employed to remove all NOE. Polymers were dissolved in D_2O at 5-10% concentration. Dioxane was used as an internal standard. T1's were determined by the inversion recovery method.

Preparation of Hydrogels

A 50:50 (mole ratio) copolymer of AMPS:AA (3.75%), sodium dichromate (0.5244%) and sodium thiosulfate (2.17%) were dissolved in brine containing 5.0% sodium chloride and 0.5% calcium chloride. After pH adjustment to 6.2, the solution was sealed in a thick-walled glass jar, Gelation was afforded by heating at 120°C for 3 hours.

A sample of the gel was then mounted between 50 mm parallel plates of a Rheometrics Dynamic Mechanical analyzer. The gap between the plates was then reduced to 1 mm and the excess polymer trimmed off. Dynamic scans were then performed at room temperature (G' and G" as a function of frequency) at 50% strain.

REACTIVITY RATIO OF 2-ACRYLAMIDO-2-METHYLPROPANESULFONIC ACID WITH ACRYLIC ACID

^{13}C NMR Analysis - Spectrum and T1 Relaxation Times

A typical ^{13}C NMR spectrum of the sodium salt of an equimolar copolymer of AMPS/AA obtained at 99.9% conversion is shown in Figure 1. Peak assignments were determined from the chemical shift values and are shown in Table 1. Analysis of the ratios of the integrals shows that this spectrum overstates the amount of AMPS present in the copolymer - the ratio of carbonyls gives 54%, the ratio of side chain to backbone carbons gives 56% and the ratio of methyl to backbone carbons gave 60% AMPS. This discrepancy was found to be due to the differences in the T1 relaxation times of carbon atom nuclei (Table 1). The T1's of the carbonyl and quaternary carbons were found to be an order of magnitude longer, 1-2 secs., than the backbone and side chain methylene carbons. Furthermore, the T1 of the acrylate carbonyl was almost twice that of the amide carbonyl. Obviously in this experiment, the pulse delay time was insufficient to allow for the steady state population to be reestablished and hence the observed concentration of carbons with long T1 times was underestimated.

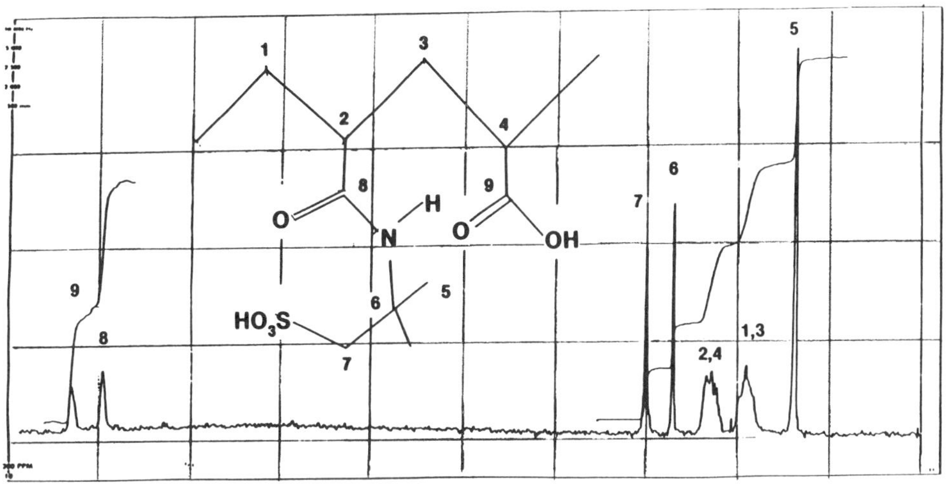

Figure 1. ^{13}C NMR Spectrum of a 50:50 AMPS®: AA Copolymer

Reactivity Ratios from Elemental Analyses

Comonomer ratio in the isolated polymer was determined by elemental analysis as described above. This method was found to be accurate to within 1% for a polymer of known composition. Also the consistency between nitrogen and sulfur data was very good for the most part as shown in Table 2.

Table 1. ^{13}C NMR T_1 Relaxation Times

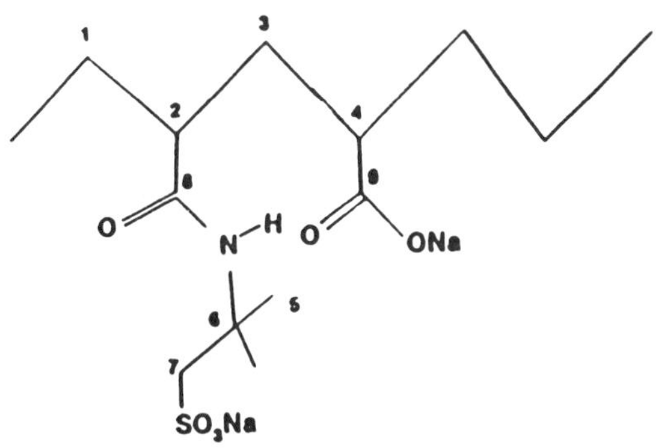

Carbon	δ (ppm)	T_1 (secs)
1 + 3	37	0.09
2 + 4	47	0.132
5	27	0.249
6	53	1.468
7	59	0.119
8	177	1.067
9	184	2.116

Table 2. Elemental Analysis of Comonomer Ratio

pH	AMPS® in Feed mole %	AMPS® in Copolymer mole % N	S
7.0	20	22.09	17.9
	35	32.23	31.53
	50	42.31	41.88
	70	52.39	53.17
	90	69.95	72.00
2.08	20	12.12	12.23
2.21	35	20.97	21.55
2.57	50	31.28	32.72
2.85	70	42.37	43.95
3.57	90	64.61	65.69

Reactivity ratios for the AMPS/AA (AMPS = monomer 1, AA = monomer 2) system were derived from Behnken's[5] non-linear regression treatment of the copolymer equation (i)

$$\text{(i)} \ldots\ldots \quad \frac{dM1}{dM2} = \frac{M1}{M2} \cdot \frac{R1\ M1 + M2}{R2\ M2 + M1}$$

in order to minimize the effects of systematic errors. These results were obtained from the combined nitrogen and sulfur analyses and are shown in Table 3. (Note: Finneman-Ross plots gave R1 = 0.194 and R2 = 0.7 at pH 7 and R1 = 0.133 and R2 = 1.60 at the lower pH's).

Table 3. Reactivity Ratio of AMPS® and AA in Aqueous Solution

Polymerization pH		Reactivity Ratio
pH 7.0	R_1	0.187 ± 0.09
	R_2	0.740 ± 0.13
pH 2-4	R_1	0.111 ± 0.03
	R_2	1.58 ± 0.15

DISCUSSION

Reactivity Ratio and Sequence Distribution

Several authors have studied the effects of degree of ionization on reactivity ratios of the acrylic acid/acrylamide system - a weak acid with a non-ionizing monomer[6-8]. From this work, it has been well established that the reactivity of an acrylate terminated macroradical with acrylate monomer decreases as the degree of ionization increases. For example, Ponratnam and Kapur showed that the reactivity ratio fell from 0.92 at a pH of 2 to only 0.33 at pH 6. This was explained in terms of electrostatic charge repulsion between the carboxylate groups at the higher pH.

From these observations, it seemed possible to be able to control the reactivity of ionizing monomers in general by proper control of solution pH. Thus, in the case of the copolymerization of AMPS, a strong acid, with AA, a weaker acid, three distinct regimes exist - the all acid system (very low pH), sodium AMPS with acrylic acid (pH 2-4), and the fully neutralized system at high pH.

Therefore, two series of experiments were performed. In the first set, polymerization was conducted at pH 7.0 in the region where both monomers were fully neutralized to the sodium salt. In the second series, the sodium salt of AMPS was copolymerized with acrylic acid.

The relative reactivities are controlled by a balance of steric and electrostatic factors. The dominant repulsive forces are summarized as follows:

i) Long range electrostatic - between negatively charged macroion and negatively charged monomer.

ii) Short range electrostatic - between the negatively charged macroion and negatively charged functionality on the approaching monomer.

iii) Steric - between macroion and monomer functionality.

(i) and (ii) will depend on the magnitude of the charge, which is pH dependent, and the distance between the charges. Steric repulsion should be greater with AMPS monomer independent of pH.

The reactivity ratio results are shown in Table 3 and discussed below.

<u>Acrylate Macroradical</u>

<u>pH = 7</u>

The rate of reaction between the acrylate terminated macroradical and monomer, (R2 = 0.74) is greater with the sodium salt of AMPS than with sodium acrylate. The rate of addition of negatively charged monomers to a negatively charged propagating radical should be depressed relative to an unionized system of lower charge (cf. these data with the acrylamide literature data above). This type (i) repulsion is approximately equal in this case since both monomers are charged. However, the charge on the AMPS monomer is located much further away from the reactive center, 7 bonds lengths for AMPS versus 3 bonds lengths away for sodium acrylate. This leads to much reduced electrostatic repulsion at close approach (type ii repulsion) resulting in greater reactivity. This factor apparently dominates over steric repulsion which would tend to favor acrylate.

<u>pH = 2-4</u>

At lower pH, the relative reactivity was found to be reversed (R2 = 1.58). Here, the carboxyl functions of the acrylate terminated macroradical are unionized and uncharged and electrostatic repulsion overall (type i) should be significantly reduced. However, the AMPS monomer and the macroradical as a whole do remain charged at this pH, since sulfonic groups are strong acids. While both type (i) and type (ii) repulsions should, therefore, be much less than at high pH, the reduction is more pronounced for acrylic acid than for sodium AMPS. Also, the larger side-chain group of AMPS would tend to inhibit the rate of addition of AMPS to the acrylate terminated radical (type iii repulsion).

AMPS Terminated Macroradical

The observed reactivity ratios both indicate a significant preference for acrylic acid and sodium acrylate over sodium AMPS.

pH = 7

Type (i) repulsion is equal for both charged monomers. Type (ii) repulsion should, however, favor AMPS since the charged sulfonate group is located further away from the reactive center. However, steric control seems to dominate and acrylic acid is more reactive (R1 = 0.187).

pH = 2-4

Again the charge on the macroion is still negative but significantly reduced. This should lead to an overall rate enhancement. As shown above, steric repulsion (type iii) favors acrylic acid. Type (i) and (ii) electrostatic repulsions also favor acrylic acid over AMPS, since only the acrylic acid is uncharged at this pH. The additive effect of all three factors results in even higher preference for acrylic acid (R1 = 0.111).

Comonomer Distribution

The effect of reactivity on monomer distribution (Figure 2) shows the integral plots of composition versus conversion in the two pH regimes for a 50:50 mole % copolymer of AMPS and AA. At pH = 7, the distribution of acrylate groups is quite even, the maximum amount ca. 60 mole %, being incorporated at low conversion. At lower pH, the distribution is broader and copolymer containing as much as 70 mole % of acrylate groups is formed, again at low conversion.

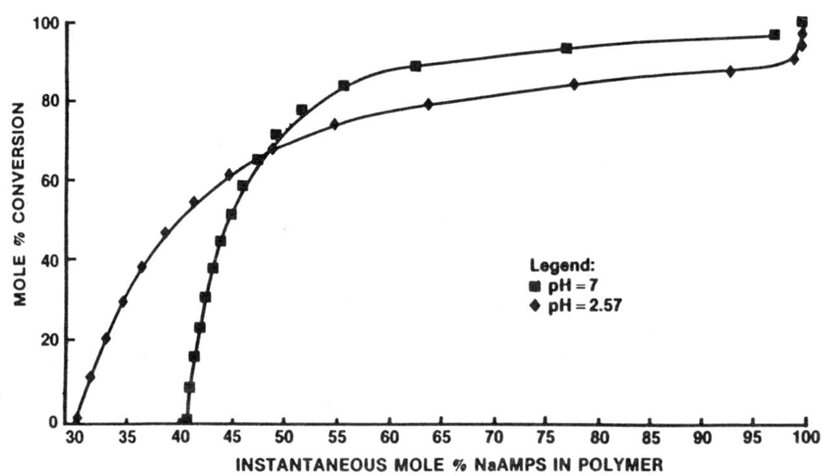

Figure 2. Integral Plot of Conversion vs. Polymer Composition (50/50 Mole Ratio)

Effect of pH on Gel Properties

A mechanism for formation of chromium ion cross-linked anionic polyacrylamide gels has been discussed by Prud'homme, et al.[9]. Theories of rubber elasticity have shown that storage modulus G' is proportional to the cross-link density v as shown below:

$$G' = g\,v\,k\,T + Gen \qquad \ldots\ldots \text{(ii)}$$

where g is a constant and Gen is a contribution due to polymer chain entanglements. Gen is small for dilute polymer solutions. Prud'homme, et al. went on to show that the rate of gelation was determined by polymer and metal ion concentration, Cp and Cm respectively as shown in equation (iii) below:

$$\frac{dv}{dt} = k'\,Cp^{2.7}\,Cm^{2} \qquad \ldots\ldots \text{(iii)}$$

Since the expression is second order in chromium, a binuclear chrome complex was postulated. Furthermore, the reaction would be expected to be second order in polymer. The higher order which was observed was explained in terms of a fraction of intramolecular cross-links which did not contribute to the storage modulus. This fraction increases with polymer dilution. Thus, equations (iii) becomes

$$\frac{dv}{dt} = k''\,Cp^{x}\,Cp^{2}\,Cm^{2} \qquad \ldots\ldots \text{(iv)}$$

where x depends on polymer concentration.

From this work, it is clear that the cross-link density and hence the magnitude of G' should be determined by the number of intermolecular cross-links. Effective cross-linking should therefore be favored by a homogeneous distribution of cross-linking monomer throughout the copolymer. Therefore, we studied how the polymerization pH affected the properties of chromium ion gels prepared from 50 mole % AMPS/AA copolymers.

Figure 3 shows the variation in G' with polymerization pH. The work described above has shown that copolymers prepared at pH 7 have a more even distribution of carboxyl groups than do copolymers prepared at lower pH and would, therefore, be expected to form more intermolecular cross-links with Cr^{3+} ions. This was indeed found to be the case. Polymers which were prepared at >50% degree of neutralization of acrylic acid gave the highest G' values, 260–330 dynes/cm^2. On lowering the polymerization pH, G' was found to fall dramatically such that the copolymer of sodium AMPS and acrylic acid gave a gel with a G' of only 60 dynes/cm^2, indicating that intramolecular cross-linking could have been more extensive under these conditions.

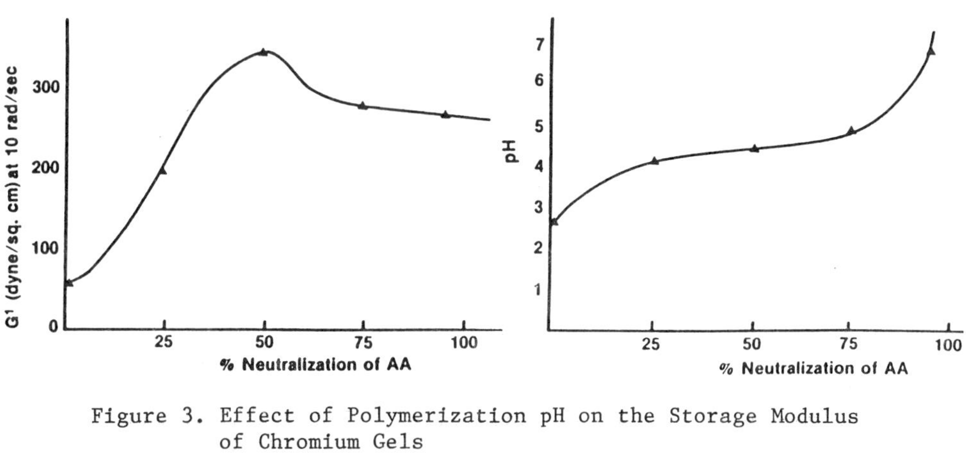

Figure 3. Effect of Polymerization pH on the Storage Modulus of Chromium Gels

CONCLUSIONS

1. Reactivity ratios for the copolymerization of AMPS with acrylic acid were determined in aqueous solution. These were found to be controlled by solution pH. This was rationalized on the basis of a combination of electrostatic and steric effects.

2. Copolymerization of the sodium salts favored a more even distribution of comonomers than when acrylic acid was copolymerized with sodium AMPS.

3. The storage modulus of chromium ion gels was found to be greater when the copolymer was prepared at neutral pH. At this pH, a more even distribution of carboxyl groups leads to optimum intermolecular cross-linking.

REFERENCES

1. J.E. Hessert and P.D. Fleming, "Gelled Polymer Technology for the Control of Water in Injection and Production Wells," presented at the 3rd Tertiary Oil Recovery Conference, Wichita, April 25-26, 1979.

2. P.M. DiGiacomo and C.M. Schramm, "Mechanism of Polyacrylamide Gel Synthesis Determined by ^{13}C NMR," SPE 11787, presented at the International Symposium on Oilfield and Geothermal Chemistry, Denver, CO, June 1-3, 1983.

3. R.G. Ryles and J.V. Cicchiello, "New Polymers for EOR Applications," SPE/DOE 14947, presented at the 5th Symposium on Enhanced Oil Recovery, Tulsa, OK, April 20-23, 1986.

4. D.W. Behnken, J. Polymer Sci., Vol. 2, 645, (1964).

5. S. Ponratnam and S.L. Kapur, Makromol. Chem., 178, 1029 (1977).

6. J. Bourdais, Bull. Soc. Chim. Fr., 485, (1955).

7. W.E. Cabaness, T. Yen Chin Lin, C. Parkanyi, J. Poly. Sci., Part A-1, 9, 2155, (1971).

8. R.K. Prud'homme, J.T. Uhl, J.P. Poinsatte and F. Halverson, Soc. Pet. Eng. J., 23(5), (1983).

9. Moradi-Araghi, A. and Doe, P.H., "Stability of Polyacrylamides in Harsh Brines at Elevated Temperatures," SPE 13033 presented at the 59th Annual Technical Conference, Sept. 16-19, 1984.

ACKNOWLEDGMENTS

The authors would like to express their gratitude to members of American Cyanamid's Analytical staff: John Lancaster performed all ^{13}C NMR experiments; Peter Achorn performed all of the rheological measurements; Peter Deng performed all of the math analyses; J.V. Cicchiello prepared all chrome gels. They would also like to thank the management of American Cyanamid for permission to publish this work.

HYDROPHOBICALLY ASSOCIATING POLYMERS

J. Bock, P. L. Valint, Jr., S.J. Pace,
D.B. Siano, D.N. Schulz and S.R. Turner*

Exxon Research and Engineering Company
Clinton Township
Route 22E
Annandale NJ 08801

INTRODUCTION

Considerable research has been conducted to identify water soluble polymers which can efficiently control the flow properties of displacement fluids for enhanced oil recovery.[1-5] Two main types of polymeric viscosifiers have emerged from this research which rely mainly on ultra high molecular weight for thickening efficiency: natural biopolymers such as Xanthan or Scleroglucan and synthetic acrylamide based polymers. Although these polymers possess many useful characteristics, the reservoir conditions in which they can provide adequate mobility control are limited. For example, the biopolymers provide excellent mechanical stability and salt tolerance, however, further improvement in high temperature stability[6] would be desirable. Viscosification with acrylamide based polymers depends not only on high molecular weight but also on chain expansion due to ionic charge repulsion or the polyelectrolyte effect. Thus solutions of these polymers are salt sensitive and exhibit poor mechanical[7] and thermal stability.[8]

An alternative approach to aqueous viscosification by ultrahigh molecular weight and chain expansion in polymers involves intermolecular association in solution to build hydrodynamic size and, hence, viscosity. The use of a small amount of polar functional groups[9] on a hydrocarbon soluble polymer has been extensively studied as a means of controlling the rheology of oil based fluids.[10,11] Furthermore, since the polymeric viscosifier must function in the presence of both brine and oil, structures analogous to surfactants[12] might be expected to provide improved performance. Reasoning by analogy to surfactants, the introduction of hydrophobic groups to a water soluble polymer

* Present address: Eastman Kodak Co., Rochester, N. Y.

should provide an associative interaction. This should result in an apparent increase in hydrodynamic size of the macromolecule in solution leading to enhanced viscosification and altered response of other solution rheological properties. In addition, if viscosification of aqueous fluids with hydrophobically associating polymers can be achieved with lower molecular weight and ionic charge, these systems should also exhibit improved mechanical stability and salt tolerance. To explore this concept, several novel classes of hydrophobically associating polymers were synthesized and their solution viscometric properties were determined.

EXPERIMENTAL

Polymer Synthesis/Systems

t-Butylstyrene-Styrene Polymers. A number of block and random polymers of t-butylstyrene and styrene were synthesized by anionic polymerization techniques.[13-15] The compositional data for the polymers are presented in Table I. The molecular weight data were obtained by gel permeation chromatography.

Table I. Copolymer Compositional Data

	Composition, wt%(mol%)			Molecular Weight	
Polymer	TBS	styrene	TBS	$M_w x10^{-5}$	$M_n x10^{-5}$
	Block				
D5-TBS-S1	5(3.3)	95(96.7)	--	1.4	1.3
D5-TBS-S2	"	"	--	2.8	2.7
D5-TBS-S3	"	"	--	5.2	4.9
D5-TBS-S4	"	"	--	7.2	6.9
D5-TBS-S5	"	"	--	10	9.4
D11-TBS-S	11(7.4)	89(92.6)	--	4.7	4.4
T5-TBS-S	2.5(1.65)	95(96.7)	2.5(1.65)	6.1	5.7
	Random				
R5-TBS-S	5(3.3)	95(96.7)	--	4.8	4.6
R10-TBS-S	10(6.7)	90(93.3)	--	5.0	4.6

t-Butylstyrene-Styrene Sulfonate Polymers. The resulting polymers were then selectively sulfonated in dichloroethane using a standard complex of triethyl phosphate and sulfur trioxide. The sulfonic acid groups on the polymer were neutralized with sodium hydroxide to provide the sodium salt form. Details of the synthesis and compositions of the precursor oil soluble block polymers and the water soluble sulfonated systems have been previously described.[16] The analytical data are presented in Table II.

Alkylacrylamide-acrylamide Polymers. Copolymers of alkylacrylamide (R) and acrylamide (AM), which we called RAM, were prepared using a micellar polymerization technique.[17]

This involved the use of a micellar surfactant solution to disperse the hydrophobic alkylacrylamide monomer into an aqueous phase containing acrylamide. The monomers were polymerized using a standard free radical initiator such as potassium persulfate or a redox initiator to yield the desired random copolymer. Further details on the synthesis and structure of these RAM polymers can be found elsewhere.[18]

Table II. Molecular Weight and Elemental Analysis of Sulfonated Polymers

Polymer	$M_w x10^{-5}$	Elemental Analysis Calc. S	Calc. Na	Found S	Found Na	Sulfonate Content, %[a]
D5-TBS-SS1	2.4	15.3	10.88	16.16	10.1	106
D5-TBS-SS2	4.7	"	"	13.78	9.54	91
D5-TBS-SS3	9.0	"	"	15.47	10.4	102
D5-TBS-SS4	12	"	"	12.58	8.19	83
D5-TBS-SS5	17	"	"	12.88	9.11	85
R5-TBS-SS	8.3	"	"	15.79	10.6	104
R10-TBS-SS	8.5	14.71	10.57	14.20	10.5	96

a. Sulfonate Content, [S(Found)/S(Calc)]x100, is a measure of sulfonate groups per styrene unit.

Surfactant-acrylamide Polymers. Copolymers of acrylamide and alkylarylpoly(ethoxy)-acrylate were prepared using standard free radical solution polymerization techniques. The alkylarylpoly(ethoxy)-acrylate monomers referred to as "Surf" monomers were water dispersible and thus additional surfactants were not needed to effect random copolymerization. These so called "PAM-SURF" polymers were prepared with a variety of Surf monomers containing different amounts of ethylene oxide and different alkylaryl functionality.[19]

Solution Rheology

Polymers were hydrated in distilled, filtered water under mild agitation until dissolution of the polymer was complete. To prepare polymer solutions containing salt, concentrated sodium chloride brine solutions were added to previously dissolved polymer in distilled water. Final polymer concentrations ranging from 1000 to 4500 ppm and salt concentration of 2.0 wt. % NaCl were prepared in this manner for viscometric study. An alternative procedure was used to evaluate the effect of salinity on solution rheology. Solid sodium chloride was slowly added to various concentrations of polymer in solution. These solutions were allowed to equilibrate for approximately 12 to 24 hours prior to obtaining viscometric measurements.

Low shear rate solution viscosity was measured on a Contraves LS 30 Rheometer using a No 1 bob and cup. The viscosity-shear profile was determined through the range of about 10^{-2} to 10^{2} s^{-1} at 25 C. Dilute solution viscosities for determining the intrinsic viscosity of the polymer systems in 2.0 wt. % NaCl were obtained in a Ubbelohde capillary viscometer using standard methods.[20,21]

RESULTS AND DISCUSSION

Hydrophobic association of polymers in solution can be inter- or intra-molecular or some combination of both. The relative amounts of these types of association and the conformation of the molecules in solution should depend strongly on polymer architecture. This was one of the first issues addressed in our study. Three structures (a block and two random architectures as shown in Figure 1.) were synthesized to investigate the relationship between polymer architecture and solution properties. To prepare these polymers, synthetic hurdles needed to be overcome.

Synthesis

Synthesis of polymers containing incompatible monomers presents difficulties. The incompatibility of the oil soluble and water soluble monomers prevents an effective concentration of one or the other of these monomeric species from being achieved at the locus of polymerization of the other comonomer. For example, simply dispersing the water insoluble monomer as fine particles in the aqueous medium containing dissolved water soluble monomers would result in low incorporation of the water insoluble monomer and would lead to a heterogeneous product of particles dispersed in a predominantly water soluble polymer. Techniques for polymerizing water soluble polymers based on the use of mutual solvents or solvent mixtures to dissolve the water and oil soluble monomers have some serious limitations. For example insolubility of the resulting polymer in the mutual solvent leads to precipitation of the copolymer before it has achieved sufficient molecular weight to provide efficient

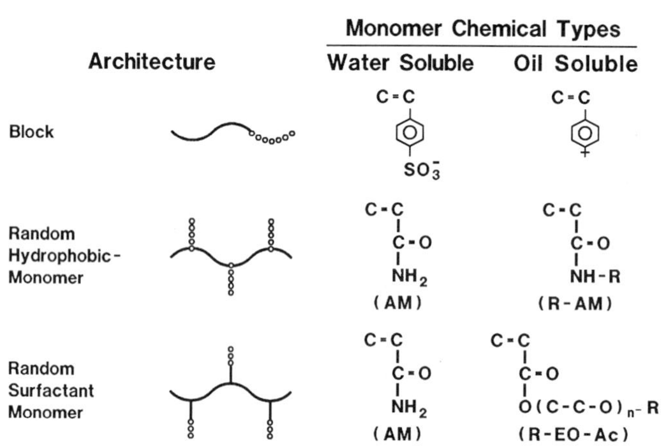

Figure 1. Synthetic approaches to associating polymers.

aqueous viscosification. The use of water miscible solvents such as alcohols, ether and acetone either alone or with water results in extremely low molecular weight (e.g. 10,000) polymers due to the high chain transfer characteristics of these solvents. Thus this approach provides polymers which are rather ineffective viscosifiers.

Three different approaches were used for preparing the hydrophobically associating copolymers shown in Figure 1. The first involved selective functionalization of a preformed polymer. Block prepolymers of styrene and t-butyl styrene were prepared by anionic polymerization and then functionalized by sulfonating the styrene units. The anionic polymerization and sulfonation techniques used to prepare these polymers and the chromatographic techniques used to characterize molecular weight and distribution were well developed. Sulfonation was clean, essentially complete and selective to the para position of the styrene monomer due to the blocking of this position by the t-butyl groups of the t-butyl styrene monomer. The synthesis[16] and characterization[22] of these polymers provided a well defined set of structures of individually controlled block size and very narrow molecular weight distribution. Homopolymers of sulfonated polystyrene and random copolymers of sulfonated styrene and t-butyl styrene were also synthesized and characterized. These systems provided a means of evaluating the importance of polymer architecture on solution viscometric properties.

Synthesis of high molecular weight random copolymers of acrylamide and alkylacrylamides required a novel aqueous surfactant micellar solution polymerization.[17] The surfactant type and concentration were chosen to provide solubilization of the hydrophobic monomer with preferably one or at most a few hydrophobic monomer groups per micelle. This reduced the tendency for blocky incorporation and consequently yielded a more random polymer. The resulting RAM polymer after isolation provided uniform and homogenous solutions for subsequent rheological study.

The third approach to hydrophobically associating polymers avoided the need for a surfactant during polymerization by employing a water dispersible surfactant like monomer such as an alkylaryl poly(etheroxy)acrylate in conjunction with acrylamide to form the PAM-SURF copolymers.[19] With this approach the relative amounts of ethylene oxide to hydrophobic alkylaryl group were used to provide a balance between water dispersibility of the hydrophobic monomer and association in solution. The incorporation of this monomer and sequencing depended on the relative amounts of ethylene oxide and hydrophobic portion of the monomer. The use of an arene group in the hydrophobic portion enabled UV spectroscopy to be used to determine monomer incorporation.

Aqueous Solution Properties

Solution rheology is a classical method for characterizing some of the major elements of polymer structure such as molecular weight and hydrodynamic size in solution. In addition, information on polymer-polymer and polymer-solvent interactions can also be obtained. Measurements in the dilute solution regime and also above the chain overlap concentration were obtained to understand the structure and properties of hydrophobically associating polymers in solution.

Block Polymers. The viscosity in distilled water of a diblock polymer with 3.3 mol% t-butylstyrene, D5-TBS-SS3, is presented in Figure 2 as a function of shear rate at concentrations ranging from 500 to 2500 ppm. As can be

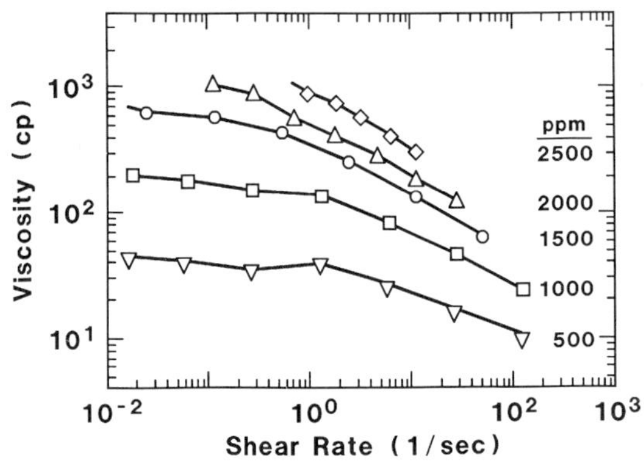

Figure 2. Solution viscosity as a function of shear rate of a diblock polymer, D5-TBS-SS3.

observed, at or above 2000 ppm, the log-log plots show a decrease in viscosity with shear rate. This pseudoplastic behavior can thus be represented by a simple power law relationship. At concentrations of 1500 ppm and below, a Newtonian or shear rate independent viscosity can be observed at low shear rates. This so called zero shear viscosity was used to construct reduced viscosity - concentration plots providing information on the hydrodynamic volume per unit mass of the polymer in solution.

The reduced viscosity - concentration curves for a 7.4 mol% t-butylstyrene diblock polymer, D11-TBS-SS, and two polystyrene sulfonate homopolymers, SS1 and SS2, are shown in Figure 3. The block polymer exhibits a reduced viscosity at least one to two orders of magnitude higher than the homopolymer, SS1, of similar molecular weight. Although block polymer, D11-TBS-SS, had a molecular weight some three times smaller than homopolymer, SS2, its viscosity was greater at concentrations above 750 ppm. Thus the presence of the hydrophobic t-butyl styrene block has resulted in a significant enhancement in viscosification efficiency. In controlling solution viscosity, this hydrophobic effect can be used to reduce polymer molecular weight while maintaining viscosity or to use less polymer at comparable molecular weight.

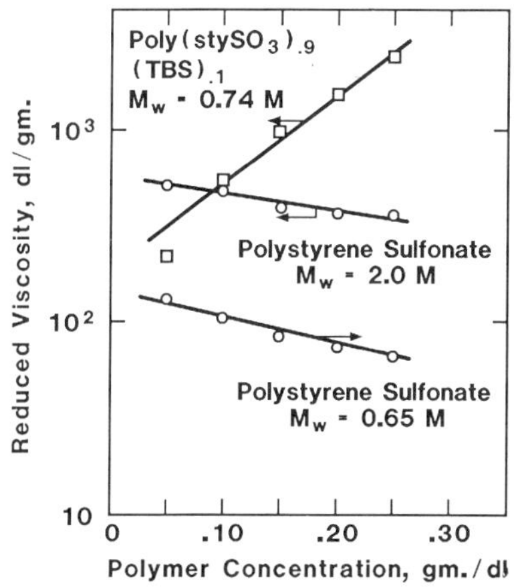

Figure 3. Reduced viscosity vs polymer concentration plots for a diblock polymer and homopolystyrene sulfonates.

The opposite slope of the reduced viscosity curve for the block polymer as compared to the homopolymers is indicative of fundamental mechanistic differences in viscosification. The decrease of reduced viscosity with increased polymer concentration for the homopolymers is a result of a decrease in effective polymer size from coiling. This is believed attributable to a reduction of charged group repulsions by shielding of the anionic charge as a result of the proximity of other polymer molecules. With the block polymer, it is postulated that the reduced viscosity increases with increasing concentration because of polymer hydrophobe association. It is envisioned that the t-butylstyrene blocks associate into aggregates due to their incompatibility with the aqueous medium. As the polymer concentration increases, the size and/or number of these aggregates increases. This results in stronger interaction and, hence, an enhancement in solution viscosity. Conversely, in dilute solution (<750 ppm) wherein polymer molecular interactions are minimized, molecular weight becomes the controlling factor. Consequently, the reduced viscosity of the block polymer is less than the homopolymer, SS2, in the dilute concentration region because of its lower molecular weight.

To define the effect of molecular weight on intrinsic viscosity, diblock polymers, D5-TBS-SS, with 3.3 mol% t-butylstyrene were prepared having weight average molecular weights from 240,000 to 1.7 million. It can be seen from the plot in Figure 4 that a power law relationship, described by the Mark-Houwink-Sakurada equation ($[\eta] = kM^a$) exists between the molecular weight and intrinsic viscosity for this series of block polymers in distilled water. The exponent, a, was determined to be 1.0 which was in good agreement with a = 0.93 reported[23] for polystyrene sulfonate in 0.005 M sodium chloride solution. The lower plot in Figure 4 represents the literature data. Therefore, the influence of molecular

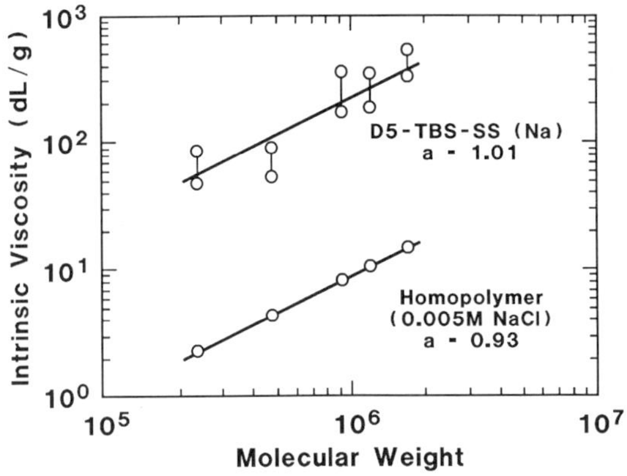

Figure 4. Effect of polymer molecular weight on solution viscosity.

weight on the hydrodynamic size of the block polymers in solution was similar to what was observed for the homopolymer analogs. The block polymers have intrinsic viscosities an order of magnitude higher than the homopolymers due to the aggregation of several polymer molecules via the association of the hydrophobic blocks. Thus while molecular weight has a significant effect on the viscosification efficiency of block polymers in solution, the presence of the hydrophobic groups has a larger effect. It thus appears that the aggregation numbers would be greater than a few to explain the level of viscosity enhancement. Fluorescence measurements would be informative to further explore this issue.

A comparison of the reduced viscosity as a function of concentration for the block polymer, D5-TBS-SS, and random polymer, R5-TBS-SS, having 3.3 mol% t-butylstyrene content and Mw's of $9.0x10^5$ and $8.3x10^5$, respectively, is shown in Figure 5. It can be seen that the random polymer is very similar in rheological behavior to that of the homopolymer and significantly different than the associating block polymer. Thus at first pass it appears that polymer architecture is a critical parameter in developing hydrophobic associations and controlling solution rheology. However, this may be due to the fact that the t-butylstyrene hydrophobic groups are similar in steric bulk to hydrophilic styrene sulfonate neighboring groups and, consequently, cannot sense the presence of like groups in order to aggregate. As we will observe with the RAM systems discussed below, the alkyl chain length of the hydrophobe has a pronounced effect on polymer association and solution rheology.

In dilute solution, the reduced viscosity of an ionically charged polymer generally increases upon dilution because of the polyelectrolyte effect.[24] As the charge centers along the backbone become deshielded by dilution, the polymer hydrodynamic size increases due to electrostatic repulsion. Surprisingly, the polyelectrolyte effect was not

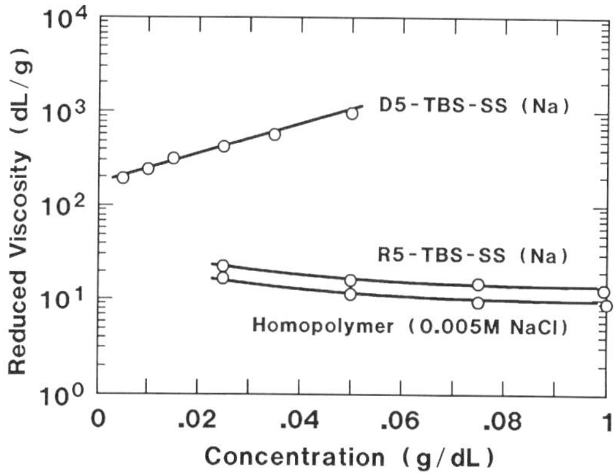

Figure 5. Effect of hydrophobic group distribution on solution viscosity.

observed with the block polymers, most likely due to the association of polymer molecules. Polymer concentrations as low as 50 ppm were used with no apparent upturn in reduced viscosity with decreasing concentration. Apparently, block polymer aggregation is present to very low concentrations. The nature of the association maintains a locally effective shielding of the sulfonate anions within the aggregate and preservation of the clustered structure. This suggests that for these block polymer structures, the energetics of maintaining hydrophobic association are more favorable than monomer dispersion due to ionic repulsion. This was further demonstrasted by the extreme salt sensitivity of these polymers to solution ionic strength. Small amounts of sodium chloride resulted in polymer precipitation and of course loss of viscosification. This precludes the use of these particular polymers for chemically enhanced oil recovery and indicates the need for nonionic functionality to provide water solubility. To further pursue this approach acrylamide based polymers were studied.

<u>RAM Polymers(Dilute Solution Viscometrics).</u> The rheology of dilute polymer solutions has been used extensively to gain insight into the structure and conformation of polymers.[24] The intrinsic viscosity, determined as the limit of reduced viscosity as polymer concentration goes to zero, provides a measure of the molecular weight of a polymer through a relationship such as the Mark-Houwink-Sakurada equation. The application of this approach to acrylamide based polymers and some of the complexity in water soluble systems can be found in the work of Klein[25] and others.[26] Another valuable parameter obtained from dilute solution viscometric measurements is the Huggins' constant. This is the slope of the reduced viscosity versus concentration plot in the linear concentration regime divided by the square of the intrinsic viscosity. It is a measure of the polymer-solvent and polymer-polymer interactions. For

random coil polymers in solution, the Huggins' constant is generally in the range of 0.3 to 0.8.

The introduction of hydrophobic associations further complicates the dilute solution viscometrics. As shown in Table III, the intrinsic viscosity tends to decrease and the Huggins' constant increases as the extent of hydrophobic groups or associations increases. For random polymers in the dilute concentration regime, intramolecular associations would lead to a contraction of the coil in solution manifested in a decrease in intrinsic viscosity. This has been demonstrated for ionomers in hydrocarbon solvents[10] and appears to be present in the hydrophobically associating polymers as well. The increase in the Huggins' constant indicates a decrease in solvent quality and an increase in associations with increasing concentration of hydrophobic groups in the polymer. Thus a Huggins' constant greater than the usual 0.3-0.8 found in conventional polymers,[27] is a good indicator of hydrophobic associations. As shown in Table III both the RAM and the PAM-SURF polymers exhibit hydrophobic associations.

Table III. Dilute Solution Viscometrics

System	Hydrophobe Type	Hydrophobe Level, mol%	[n]	Kh
PAM	--	0	7.3	0.4
RAM	n-C8	0.75	4.5	0.8
RAM	n-C8	1.0	3.4	2.5
PAM-SURF	(EO)-PhC9	0.3	2.1	5.3

<u>RAM Polymers(Semi-Dilute Viscometrics).</u> Solution viscometrics of polymer solutions above the overlap concentration indicate dramatic effects due to the associative nature of the hydrophobic groups. As shown in the viscosity - concentration profile of Figure 6, the

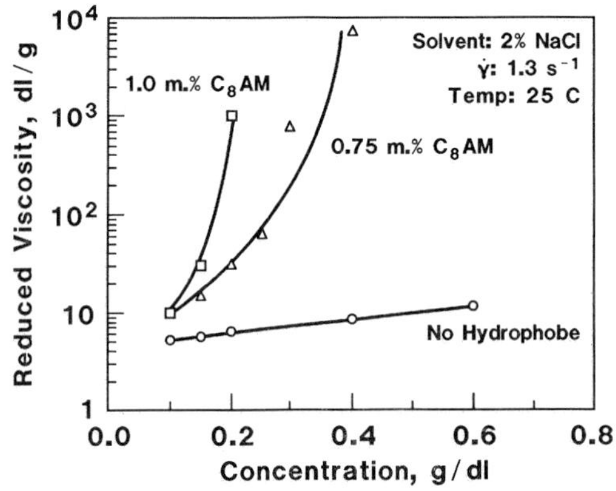

Figure 6. Effect of hydrophobe level on polymer solution viscosity.

introduction of only 0.75 mole % N-n-octylacrylamide to polyacrylamide can increase the viscosification efficiency dramatically. All of these polymers had a molecular weight of about 3×10^6 as determined by sedimentation-LALLS technique.[28] The sharp upsweep in viscosity with concentration, particularly at the higher hydrophobe levels, appears to be a general characteristic of hydrophobically associating polymers. This is probably due to an increase in the relative amount of intermolecular association, forming a network-like structure in solution. In fact at even higher polymer concentration levels, solutions of hydrophobically associating polymers exhibit gel-like behavior. However, this is not the region of interest for chemically enhanced oil recovery.

Polymer - solvent interactions can have a significant effect on polymer conformations in solution and, in turn, solution properties. This is demonstrated in Figures 7 and 8 for the effect of salinity on solution viscosity. The

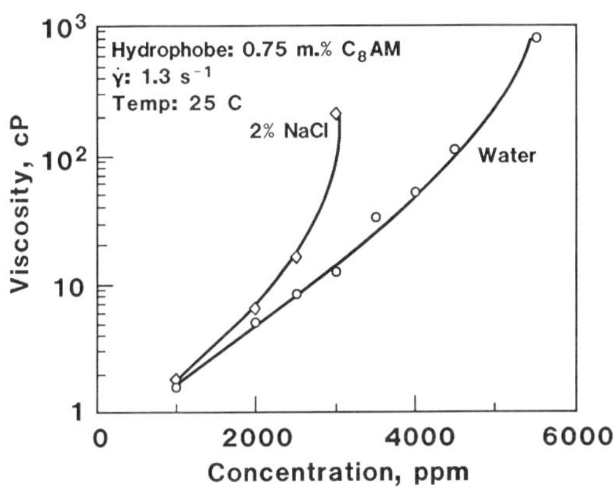

Figure 7. Solution viscosity response to salt as a function of polymer (0.75 mol% RAM) concentration.

addition of salt to a solution of a RAM polymer containing 0.75 mole % hydrophobic C_8 groups causes an increase in viscosity. The viscosity increase with salt is enhanced at higher polymer concentration. This may be explained by considering the effect of salt on surfactant molecules in solution. Salt reduces the critical micelle concentration of surfactants and enhances aggregation or micellization. The viscosity increase of the RAM polymer with salinity is thus attributed to an increase in associations. The nature of the associations in terms of the relative amount of inter- and intra-molecular assocaitions may also be important. This maybe a function of polymer concentration which could explain the sharp increase in viscosity at concentrations near the critical overlap concentration as approximated by the reciprocal of the intrinsic viscosity.

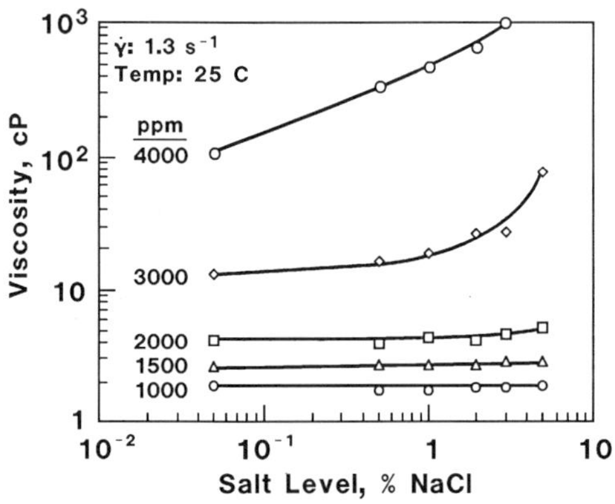

Figure 8. Solution viscosity response to polymer concentration as a function of salt.

The aggregation of these polymers has also been observed by measuring viscosity of their aqueous solutions as a function of shear rate. As shown in Figure 9, the viscosity can have rather complex responses to shear rate depending on concentration. Regions of Newtonion behavior, pseudoplasticity and shear thickening can be observed depending on the shear rate range. This is probably a result of a shifting of the relative amount of intra- and intermolecular associations with shear. However further work

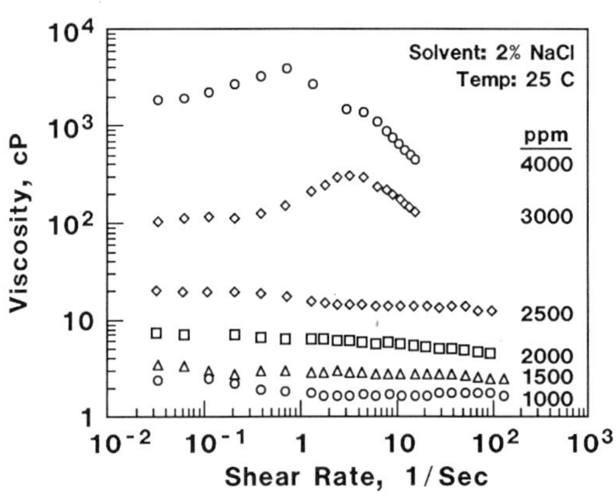

Figure 9. Effect of shear rate on viscosity of 0.75 mol% RAM as a function of polymer concentration.

would be needed to more fully understand this behavior. Nevertheless, it appears that the hydrophobically associating RAM polymer offer enhanced viscosification and improved salt tolerance; characteristics which are desireable in mobility control agents for enhanced oil recovery.

CONCLUSIONS

Several new classes of macromolecules which consist of a water soluble polymer backbone containing a relatively small amount of oil soluble or hydrophobic groups have been synthesized. Block and random structures, comprised of incompatible monomers which have significantly different solubility characteristics, required some unusual synthetic techniques for their preparation. Rheological studies of solutions of block polymers in distilled water clearly identified viscosity enhancement due to the presence of the hydrophobic t-butylstyrene endblocks. Comparison with homopolymer and random copolymers of similar composition and molecular weight demonstrated the increased viscosification with the block configuration. However the particular block polymers evaluated in this study were too salt sensitive for use as mobility control agents.

Hydrophobically associating acrylamide based polymers were explored as a means of alleviating the salt sensitivity observed in the block systems. A micellar polymerization technique was developed to enable preparation of random copolymers of acrylamide and N-n-alkylacrylamide. When these copolymers were dissolved in an aqueous solvent, the hydrophobic groups associated to minimize their exposure to water. The hydrophobic associations provided an additional dimension to polymer molecular weight and chain expansion by ionic groups for the control of aqueous fluid rheology. Above the overlap concentration, a significant enhancement in solution viscosity was observed along with an increase in viscosity with salt concentration. In addition, the solution viscosity could exhibit Newtonian, dilatant, or pseudoplastic behavior depending on the shear rate, polymer composition and concentration. While considerable further research needs to be done, it appears that the use of hydrophobic groups on water soluble polymers provides an attractive approach to improved mobility control polymers for chemically enhanced oil recovery.

ACKNOWLEDGEMENTS

We wish to acknowledge Dr. L. J. Fetters for the synthesis of the various t-butylstyrene-styrene block and random polymers done at the University of Akron. D. Griffin and S. Zushma carried out the sulfonation of the polymer precursors and W. Gallagher performed the rheological measurements. The synthesis and rheological characterization of the RAM systems were ably and carefully conducted by J. Wagensommer, E. Habeeb and P. Myer.

References

1. S. P. Gupta and S. P. Trushenski, *Soc. Pet. Eng. J.*, **1978**, *5*, 345.
2. E. Unsal, J. L. Duda and E. Klaus, "Chemistry of Oil Recovery", ACS symposium series *91*,**1979**, p. 141-170.
3. C. L. McCormick, R. D. Hester, H. H. Neidlinger and G.C. Wildman, "Surface Phenomena in Enhanced Oil Recovery", Plenum, New York, 1981, p. 741-772.

4. M. T. Szabo, *J. Pet. Tech.*, **1979**, 553.
5. P. Davison, and E. Mentzer, 55th Annual SPE Mtg., Preprint No. SPE 9300.
6. S. L. Wellington, SPE 9296.
7. J. M. Maerker, *Soc. Pet. Engr. J.*, 1975, *2*, 311-322, .
8. G. Muller, J. C. Fenyo and E. Selegry, *J. Appl. Poly. Sci.*, **1981**, *25*, 627-633 .
9. H. S. Makowski, R. D. Lundberg, L. Westerman and J. Bock, "Ions in Polymers", ACS symposium series 187, p. 3-19,**1980**.
10. R. D. Lundberg and R. R. Phillips, *J. Polym. Sci., Physics Ed.*, **1980**, *20*, 67 .
11. M. W. Kim and D. G. Peiffer, *J. Chem. Phys.*, **1985**, *83*, 4159.
12. A. Ben-Niam, "Hydrophobic Interactions," Plenum: New York, 1980.
13. M. Morton, E. E. Bostick and R. G. Clark, *J. Polymer Sci.*, **1963**, *A1*, 475.
14. M. Morton and L. J. Fetters, *Macromol. Rev.*, **1967**, *2*, 71.
15. L. J. Fetters, *J. Polymer Sci., Part C*, **1969**, *26*, 1.
16. P. L. Valint, Jr. and J. Bock, U.S. Patent 4,492,785 **1985**.
17. S. R. Turner, D. B. Siano, and J. Bock, U. S. Patent 4,528,348, **1985**.
18. S. R. Turner, D. B. Siano, and J. Bock, U. S. Patents 4,520,182 and 4,521,580, **1985**.
19. D. N. Schulz, J. J. Maurer, and J. Bock, U.S. Patent 4,463,152, **1984**.
20. R. H. Whorlow, "Rheological Techniques"; Halsted Press, John Wiley & Sons: New York, **1980**.
21. J. R. Van Wazer, J. W. Lyons, K. Y. Kim and R. E. Colwell, "Viscosity and Flow Measurement"; Interscience Publishers, John Wiley & Sons: New York, **1963**.
22. P. L. Valint, Jr., and J. Bock, presentation at 191st National ACS Meeting, Anaheim, Cal., Sept 7, **1986**.
23. J. Brandrup and E. H. Immergut, eds., "Polymer Handbook," Wiley-Interscience: New York, **1975**, p. IV-19.
24. P. J. Flory, "Principles of Polymer Chemistry," Cornell University Press: Ithaca, New York, **1953**, 635.
25. J. Klein and K. D. Conrad, *Makromol. Chem.*, **1980**, *181*, 227-240.
26. P. Molyneux, "Water-Soluble Synthetic Polymers: Properties and Behavior, Volume I and II"; CRC Press, Inc.: Boca Raton, Florida, **1985**.
27. J. Brandrup and E. H.Immergut, Eds., "Polymer Handbook", 2nd ed., John Wiley & Sons: New York, **1975**, IV-135.
28. G. M. Holzwarth, L. Soni, D. N. Schulz and J. Bock, presentation at 191st National ACS Meeting, Anaheim, Cal., Sept. 7-12, **1986**.

STRUCTURALLY TAILORED MACROMOLECULES FOR

MOBILITY CONTROL IN ENHANCED OIL RECOVERY

Charles L. McCormick and C. Brent Johnson

Department of Polymer Science
University of Southern Mississippi
Hattiesburg, MS 39406

INTRODUCTION

Copolymer structure has a marked effect on solution properties and thus behavioral characteristics of water-soluble copolymers utilized in enhanced oil recovery. Perhaps the most critical property of a candidate polymer for mobility control is its ability to maintain a large hydrodynamic volume (HDV) in the presence of mono- and divalent electrolytes. Additionally the structure and HDV of the polymer must be tailored to allow permeation through the reservoir rock without entrapment, adsorption, or shear degradation.

Hydrolyzed polyacrylamides or acrylamide/sodium acrylate copolymers have been used extensively in EOR; however, these polymers have been shown to be very susceptible to phase separation in the presence of divalent cations and undergo catastrophic viscosity losses at high electrolyte concentrations. Copolymers of acrylamide and sodium-2-acrylamido-2-methylpropane sulfonate (NaAMPS) display improved solution behavior in the presence of divalent ions. Extensive studies in our laboratories during the past few years have focused on tailored copolymer systems which maintain or increase viscosity in the presence of electrolytes including calcium ions[1-21]. In this paper, we report results for three polymer systems which appear to be especially promising for certain water-soluble applications including EOR. These are NaAMB/AM, an intramolecular calcium ion chelator; N-alkylAM/AM, an associative copolymer with hydrophobic domains; and low charge density ampholytic terpolymers. Certain copolymer compositions of each of the three possess unique solution characteristics and, thus potential commercial utility.

Copolymers of acrylamide with sodium-3-acrylamido-3-methylbutanoate, the AM/NaAMB series, are polyelectrolytes with high molecular weight. Unlike conventional acrylamide/sodium acrylate copolymers, however, viscosity loss in the presence of electrolytes is relatively low, apparently moderated by intramolecular stiffening effects. The AM/NaAMB copolymers like AM/NaAMPS show no phase separation in aqueous solutions of calcium chloride even at 100°C. The high molecular weight, viscosity maintenance, and absence of phase separation at elevated temperatures in the presence of multivalent electrolytes render these polymers particularly attractive for mobility control in EOR[22].

In some instances the benefits of utilizing ultrahigh molecular weight (> 10 million) copolymers for viscosification are lost due to associated problems of entrapment (pore and channel clogging), shear degradation, and chromatographic effects. Field problems of dissolution, mixing, and injection without formation damage or plugging are often critical. As a result our group and industrial investigators[23,24] have recently initiated research to develop low molecular weight copolymers which cooperatively associate in aqueous solutions to yield high viscosity. In theory, problems of channel and pore clogging and shear degradation would be averted since associations are reversible with increasing shear stress. "Hydrophobic" associations of this type, although widely recognized in biopolymers and associative thickeners for coatings[23-27] are poorly understood.

The acrylamide/N-alkyl acrylamide copolymers in this report contain low mole percentages of C_8, C_{10}, or C_{12} substituted monomers. The copolymerizations were conducted in an aqueous medium containing high concentrations of surfactant. This method has been called "micellar" polymerization[28,29]. Ongoing research in our laboratories leads us to beleive that micellar copolymerization may result in somewhat blocky structures. Above a critical polymer concentration the hydrophobic domains of separate polymer chains may associate. This concentration appears to be significantly lower than the critical concentration required for chain entanglement in non-associating polymers. Increasing ionic strength and increasing temperature appear to enhance domain formation while increasing shear disrupts domain association. Although the above properties are certainly intriguing, the mechanism of association must be clearly elucidated in order to tailor macromolecules of this type for EOR and related applications.

Ampholytic polymers[30-33] offer some unique characteristics which might be utilized in viscosity modification and/or permeability alteration during EOR processes. By careful molecular tailoring, charged moieties which can be responsible for large hydrodynamic volume changes with the addition of external electrolytes may be placed strategically along the macromolecular backbone. To date, few studies have been directed toward energy-related polyampholyte applications. In this paper we report initial studies on promising low charge density polyampholytes.

One additional polymer system which will not be detailed in this report is a surfactant-like polymer which in aqueous solution has little viscosity and forms independent micelles[19,20]. Upon addition of hydrocarbons to the aqueous polymer solutions, a viscosifying effect is observed. Such a system might be potentially designed to act both as a surfactant and a viscosifier for mobility control in EOR.

EXPERIMENTAL

Polymer Synthesis

NaAMB Series. 3-Acrylamido-3-methylbutanoic acid (AMBA) was synthesized via a Ritter reaction involving acrylonitrile and 3,3-dimethylacrylic acid in the presence of water and an excess of sulfuric acid. The synthesis basically followed the procedure set forth by Hoke and Robins[34]. The crude product was twice recrystallized from a mixture of methyl ethyl ketone and petroleum ether prior to use (m.p. 89-91°C). The product was analyzed by elemental analysis and FTIR. Acrylamide was twice recrystallized from acetone prior to use.

The copolymers of acrylamide with sodium-3-acrylamido-3-methylbutanoate (NaAMB), as well as the homopolymer of NaAMB, were prepared in aqueous solution at 30°C using 0.1 mole percent potassium persulfate as the initiator. The feed ratio of AM:NaAMB was varied from 95:5 to 25:75, and the total monomer concentration was held constant at 0.456 M. Following reaction, the polymers were purified by precipitation in acetone, dialysis, and lyophilization. Conversions were determined gravimetrically and compositions were determined using elemental analysis and ^{13}C NMR[2].

Associative Systems. Hydrophobic comonomers were synthesized in the following manner. N-n-octyl, decyl, and dodecylacrylamide were synthesized by the reaction of acryloyl chloride (0.55 M) with the corresponding alkyl amine (0.55 M) in THF at 10°C with triethylamine (0.60 M) present as the acid receptor. Following filtration of triethylamine hydrochloride and evaporation of THF, the crude product was twice recrystallized in acetone at -25°C. The products were characterized by elemental analysis, melting point, FTIR, ^{13}C NMR, and GC.

The N-alkylacrylamides were copolymerized with acrylamide via micellar copolymerization using sodium dodecylsulfate as the surfactant and potassium persulfate as the free-radical initiator. Monomer feed mole ratios were varied from 0.25:99.75 to 0.75:99.25 N-alkylacrylamide:acrylamide. Following reaction, the polymers were purified by precipitation in acetone, dialysis, and lyophilization. Elemental analysis confirmed the absence of surfactant in the copolymers.

Polyampholytes. Low to moderate charge density terpolymers of acrylamide (AM) with 2-acrylamido-2-methylpropane-dimethylammonium chloride (AMPDAC) and sodium 2-acrylamido-2-methylpropanesulfonate (NaAMPS) have been prepared. The synthesis was conducted via free radical initiation in water employing potassium persulfate as the free-radical initiator. A typical syntheis involved the dissolution of a specified quantity of the monomers in separate solutions, followed by the addition of an equimolar quantity of inorganic components, NaOH for NaAMPS and HCl for AMPDAC. The monomers were then mixed to form a single solution. Following adjustment of the pH to 6, the reaction mixture was purged with nitrogen for fifteen minutes then initiated with a specified quantity of potassium persulfate.

Following the collection of reaction aliquots at low and high conversion, the samples were purified by three cycles of precipitation in acetone and redissolution in water. Finally, the samples were lyophilized to remove nonsolvent acetone and the majority of water in the sample.
Terpolymer compositions were determined using elemental analysis for carbon, nitrogen, and sulfur.

Characterization

Viscosity. Solution viscosities of the NaAMB copolymers were determined using a Cannon-Ubbelohde four-bulb shear dilution viscometer (size 100). Solution viscosities of the associative polymers were determined using a Contraves low shear rotational rheometer at a shear rate of 1.28 sec^{-1}. All viscosity studies were conducted in the absence of surfactant at temperatures in the range of 25-45°C and [NaCl] ranging from zero to 0.684 M.

Light Scattering. Low angle laser light scattering measurements, used in determining weight-average molecular weights and second virial coefficients, were performed with a Chromatix KMX-6 photometer. Refractive index increments were determined with a Chromatix KMX-16 laser differential refractometer. All light scattering measurements were conducted at 25°C in 1 M NaCl aqueous solutions at a pH of 6.

^{13}C Nuclear Magnetic Resonance Spectroscopy. Carbon-13 NMR spectra were obtained at 22.5 MHz on a JEOL FX-90Q Spectrometer using a 5000 Hz window. Approximately 30,000 transients were accumulated for each sample consisting of 0.1 to 0.2 grams of copolymer in 2 mL of D_2O in 10 mm tubes.

Fluorescence. Pyrene was obtained from Aldrich Chemical Co. and recrystallized five times in methanol prior to use. All polymers were initially dissolved in deionized water. The pyrene/polymer samples were prepared in the following manner. A 10^{-4} M solution of pyrene in methanol was first prepared. From this solution, 10 mL was then pipetted into a scintillation bottle. The bottle was slowly heated to allow the methanol to evaporate. Then 10 mL of the polymer solution was placed in the bottle containing the pyrene residue. This mixture was heated at 40°C for six hours and then centrifuged for one day.

Steady state fluorescence spectra were recorded (λ_{ex} = 347 nm) on a Perkin-Elmer Model 650-10S Fluorescence Spectrometer. The absorbance of pyrene ($\sim 10^{-7}$M) in each solution was between 0.01 and 0.05 at the excitation wavelength. Fluorescence quantum yields were obtained by comparison with a quinine bisulfate standard according to the method of McClure et al.[35]. Fluorescence lifetimes were measured on a Photochemical Research Associates single-photon counting fluorescence decay apparatus. The excitation and emission wavelengths used in the fluorescence lifetime studies were 347 and 380 nm, respectively.

Elemental Analysis. Elemental Analysis to determinecarbon, hydrogen, and nitrogen content of each copolymer was conducted by M-H-W Laboratories of Phoenix, AZ.

Table 1

Molecular Weight, Second Virial Coefficient, and Zero-Shear Intrinsic Viscosity Data for Copolymers of Acrylamide (AM) and Sodium-3-Acrylamido-3-Methylbutanoate (NaAMB), and the Homopolymer of NaAMB

Sample Number	Composition (mol%)[a] AM	NaAMB	M_w $\times 10^{-6}$ g/mol	A_2 $\times 10^{4}$ mL mol/g^2	$[\eta]$ dL/g
AMB-10-2	91.6	8.4	15.6	4.28	26.0
AMB-25-2	79.9	20.1	12.9	1.52	-
AMB-60-2	50.2	49.8	14.0	3.49	11.9
AMB	0.0	100.0	3.23	4.34	3.8

[a] from elemental analysis

[b] [NaCl] = 0.514 M; T = 30°C; pH = 7.0

Table 2

Correlation between pK_a and Copolymer Sequence Distribution for Acrylamide-NaAMB Copolymers

Sample	pKa(H_2O)	ΔpKa	pKa (KCl)	Δ pKa	P_{ABA}	P_{BBB}	$P_{BBA-AAB}$
AMB-10-2	5.23±0.05	0.08	4.79±0.06	0.10	0.946	0.001	0.053
AMB-25-2	5.31±0.06	0.22	4.74±0.05	0.10	0.754	0.018	0.228
AMB-40-2	5.75±0.08	0.28	4.90±0.04	0.05	0.615	0.047	0.338
AMB-60-2	5.88±0.05	0.55	5.01±0.04	0.12	0.339	0.175	0.486
AMB-75-2	6.19±0.06	0.46	5.10±0.05	0.12	0.205	0.299	0.495
AMB	6.86±0.07	0.41	5.49±0.05	0.20			

Table 3

Acid Strength (pK_a)[a] of Acrylamide-NaAMB Copolymers

Sample	H_2O				0.36 M KCl			
	pK_a	n	pK'	r	pK_a	n	pK'	r
AMB-10-2	5.23	1.1	5.23	0.99	4.79	1.1	4.71	0.99
AMB-25-2	5.31	1.4	5.29	0.99	4.74	1.1	4.69	0.99
AMB-40-2	5.75	1.5	5.74	0.99	4.90	1.1	4.82	0.99
AMB-60-2	5.89	1.8	5.96	0.99	5.01	1.1	5.03	0.99
AMB-75-2	6.19	1.7	6.19	0.99	5.10	1.0	5.18	0.98

[a] apparent pK_a.

RESULTS AND DISCUSSION

AM-NaAMB Copolymers

Copolymers of acrylamide (AM) with sodium-3-acrylamido-3-methylbutanoate have been prepared with a range of compositions yet similar molecular weights (Table 1) in order to assess the effect of microstructure on solution properties. Particular attention has been focused on statistical monomer sequence lengths along the polymer backbone as calculated from experimental reactivity ratios ($r_1 x r_2 = 0.56$) for the monomer pair[8].

Interestingly the frequency of NaAMB units along the backbone markedly affects polymer conformation as observed by viscometry, potentiometry, ion-binding, and ^{13}C NMR relaxation times. For example, the copolymer containing 20% NaAMB units (corresponding to the maximum number of AABAA sequences) exhibits the greatest loss in viscosity as the solvent is changed from 0.514 M NaCl to deionized water[9]. The copolymer containing 8% NaAMB, however, exhibited considerably smaller viscosity loss under the same conditions indicating optimal chain stiffening and solvent interaction for this series.

Potentiometry Studies. The pKa values tend to increase with increasing NaAMB content in the copolymers (Tables 2 and 3). This behavior is expected for two reasons[36]. First, the increased electrostatic repulsion with increasing NaAMB content causes decreased stability of the ionic groups, thereby raising the pKa. Secondly, the fewer the number of NaAMB units present, the greater the number of acrylamide units available for hydrogen bonding and stabilization of the neighboring anion. This is shown in Table 2 by comparing the copolymer sequence distributions shown where A is an acrylamide group and B is a NaAMB unit. Greater amounts of ABA will give lower pKa values, while increasing BBB triads will cause greater charge-charge interaction and thus give higher pKa values. The sequences shown are calculated statistically based on reactivity ratios, run number, and copolymer composition[37].

The slope and intercept from plots of pH versus $\log(\alpha/1-\alpha)$ were made from the modified Henderson-Hasselbalch equation[38,39]. The n value, which is the slope of the line, is a measure of the average electrostatic repulsion between charges on the chain. The pK value, which is the intercept of the line, is another means of calculating the apparent pKa (Table 3). Note that in water the n values are higher, indicating that the pKa values in this medium are more affected by interactions between charges as compared to those in 0.36 M KCl. The n values increase in water with increasing carboxylate group content. This again is a result of greater charge-charge repulsion along the chains.

Turbidimetry Studies. Significantly, no phase separation was observed during turbidimetric studies for solutions of AM/NaAMB copolymers in aqueous $CaCl_2$ solutions at temperatures up to 100°C. This observation was quite surprising initially since carboxylated copolymers, particularly acrylamide/sodium acrylate, phase separate readily under such conditions and are, therefore, poor choices for viscosification.

In contrast to the NaAMB copolymers, the NaAMB homopolymer did undergo phase separation. The critical temperature was found to decrease with increasing calcium chloride concentration as shown in Figure 1. Phase separation is found at temperatures 20 to 25°C greater than those for acrylamide-sodium acrylate copolymers, suggesting that even the NaAMB homopolymer is more stable toward calcium ion binding than sodium acrylate copolymers.

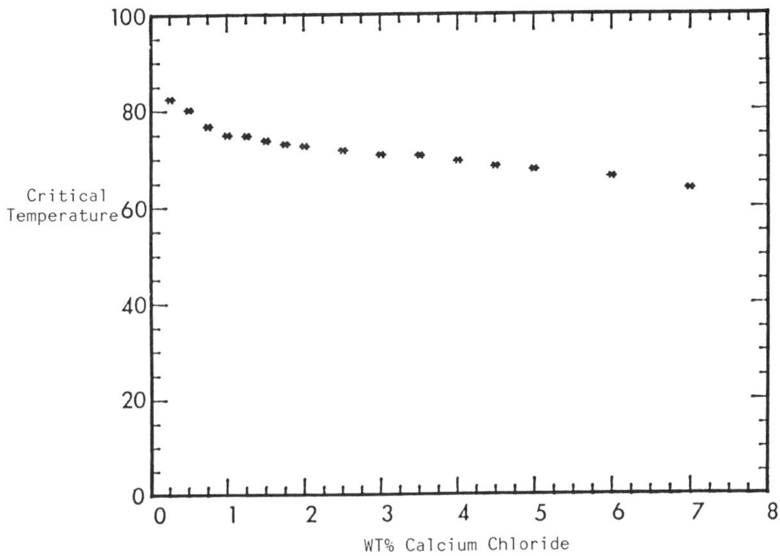

Fig. 1. Temperature of phase separation versus weight % calcium chloride for the homopolymer of sodium 3-acrylamido-3-methylbutanoate (AMB). (Polymer concentration = 1.5 grams/dL).

Copolymers of acrylamide with sodium-2-acrylamido-2-methylpropanesulfonate (NaAMPS) are known to be phase stable as a result of the strong acidity and weak ion binding nature of the sulfonate group. However, the NaAMB copolymers are thought to derive their phase stability from the geometry of the chain since the NaAMB carboxylate groups are known to be weaker acids.

Ion binding and potentiometric studies indicate strong chelation to the AM/NaAMB copolymers. We propose interactions similar to those shown in Figure 2 which allow strong intramolecular chelation of the calcium ion. This is enhanced by neighboring group effects of adjacent acrylamide units. A number of free hydrophilic acrylamide units are necessary to maintain the hydration sphere of the copolymer. This mechanism apparently extensively favors intramolecular chelation over intermolecular binding which would lead to cross-linking and eventual phase separation.

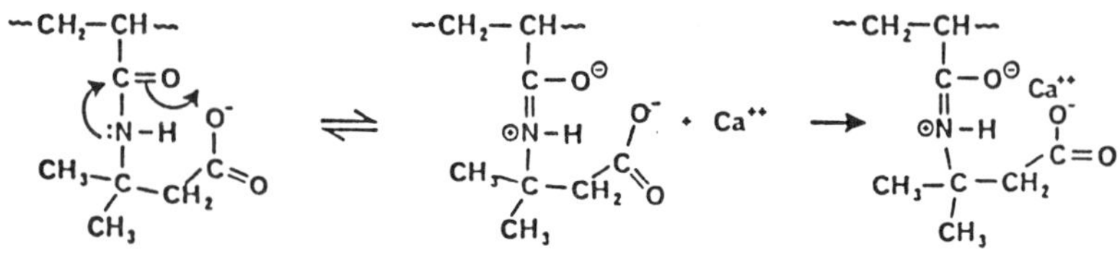

Figure 2. Proposed mechanism for the intramolecular chelation of calcium ion.

Table 4

Reaction Parameters and Solubility Properties for Copolymers of Acrylamide with N-n-Alkylacrylamides[a]

Sample Name	Monomer Feed Ratio[b]	Water Solubility
C-8-0.25	99.75/0.25	soluble
C-8-0.50	99.50/0.50	"
C-8-0.75	99.25/0.75	"
C-10-0.25	99.75/0.25	"
C-10-0.50	99.50/0.50	"
C-10-0.75	99.25/0.75	"
C-12-0.25	99.75/0.25	"
C-12-0.50	99.50/0.50	insoluble
C-12-0.75	99.25/0.75	"

[a] Total monomer concentration = 0.44 M; Sodium dodecylsulfate concentration = 0.12 M; $[K_2S_2O_8]$ = 7.86 x 10^{-5} M; T = 50°C.
[b] acrylamide / N-n-alkylacrylamide.

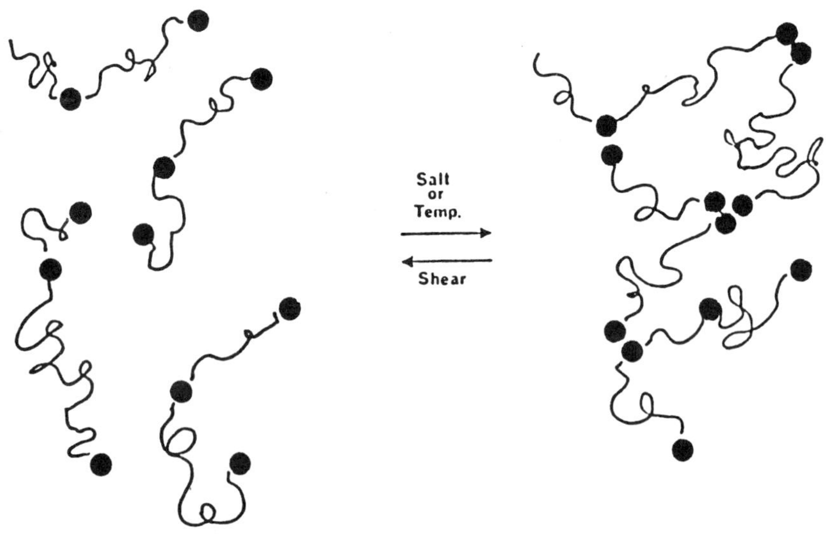

Figure 3. The extent of hydrophobic association is dependent on parameters such as salt concentration, temperature, and shear rate.

Associative Copolymers

Copolymers of acrylamide with N-alkyl acrylamides have been prepared by micellar copolymerization. Compositions of octyl, decyl and dodecylacrylamide were varied from 0.25 to 0.75 mole % in the reaction feed. As mentioned previously, all surfactant was removed by dialysis prior to polymer characterization. The reaction conditions and solubility characteristics of the copolymers can be seen in Table 4.

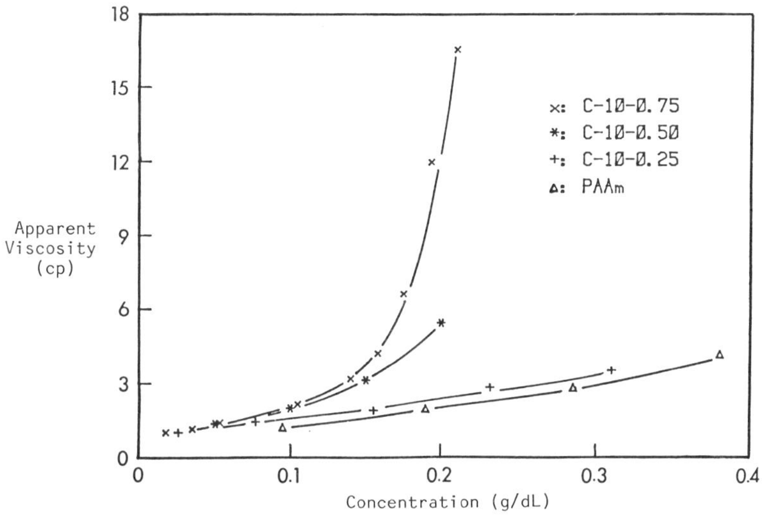

Figure 4. Effect of copolymer concentration on the apparent viscosity of copolymers of acrylamide with decylacrylamide in water at 25°C and a shear rate of 1.28 sec^{-1}.

Viscometry Studies. Viscosity behavior for the associative polymers relies on an increase in the apparent macromolecular weight which, in turn, is due to cooperative intermolecular aggregation among hydrophobic domains within the solution at some critical concentration. Research in our laboratories has shown that under high shear conditions the hydrophobic interactions can be disrupted resulting in a decrease in the apparent molecular weight hence a decrease in solution viscosity. However, under low shear conditions the hydrophobic interchain associations may again form, lending much higher viscosity fluids. Thermodynamically, associations are favored with increases in temperature and salt concentration. Figure 3 depicts the interactions which may occur among associative polymers under various conditions.

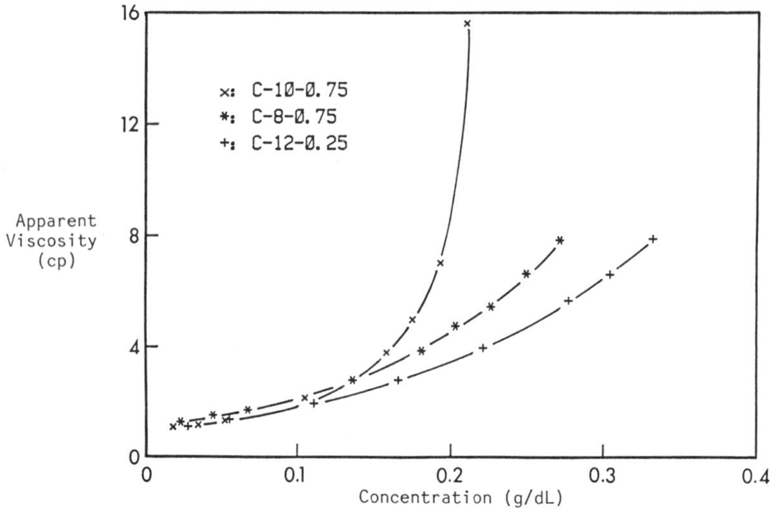

Figure 5. Effect of copolymer concentration on the apparent viscosity of copolymers of acrylamide with 0.75 and 0.25 mole % n-alkylacrylamides in 0.342 M NaCl at 25°C and a shear rate of 1.28 sec^{-1}.

The concentration dependence of the viscosity is most noticeable in the N-decylacrylamide copolymers (Figure 4). The 0.25 mole % C-10 copolymer displays a viscosity-concentration curve essentially identical to that of polyacrylamide, while the 0.75 mole % C-10 copolymer displays a dramatic increase in viscosity at approximately 0.15 g/dL. Therefore, as the concentration of hydrophobic side chains increases, the slope of the viscosity-concentration curve increases dramatically.

In Figure 5 apparent viscosity is plotted versus polymer concentration for copolymers of 0.75 mole % octyl and decyl-acrylamide and 0.25 mole % dodecylacrylamide. Copolymers of acrylamide with decylacrylamide and octylacrylamide exhibit a greater dependence of the viscosity on concentration than copolymers with dodecylacrylamide at the aforementioned compositions. These results indicate that the interchain associations increase not only with increasing content of an N-alkylacrylamide but also with the length, or hydrophobicity of the side chain.

The response of these associative polymers to changes in temperature and salt concentration over certain ranges of polymer concentration is another attractive feature. A plot of the reduced viscosity versus temperature can be seen in Figure 6 for the C-10 0.75 copolymer. As expected, the reduced viscosity of the copolymer increases with temperature while that of homopolyacrylamide essentially remains constant. The effect of NaCl and $CaCl_2$ concentration on the reduced viscosity of the C-10 0.50 copolymer can be seen in Figure 7. The viscosity of the copolymer increases with increasing NaCl concentration above about 0.09 M NaCl. In contrast, the viscosity of polyacrylamide is essentially invariant to changes in ionic strength of the solution. Such temperature and salt behavior certainly make se-hdrophobically modified polymers attractive candidates for use in EOR.

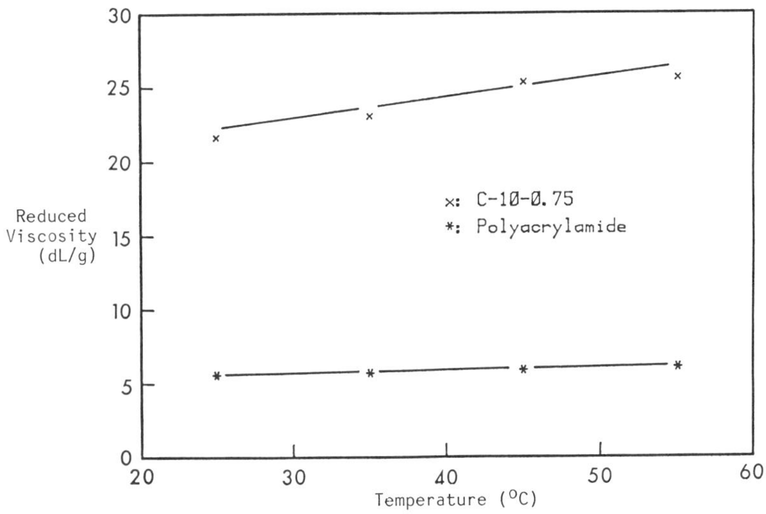

Figure 6. Effect of temperature on the reduced viscosity of polyacrylamide and a copolymer of acrylamide and with 0.75 mole % decylacrylamide and in 0.342 M NaCl at a copolymer concentration of 0.19 g/dL and a shear rate of 1.28 sec^{-1}.

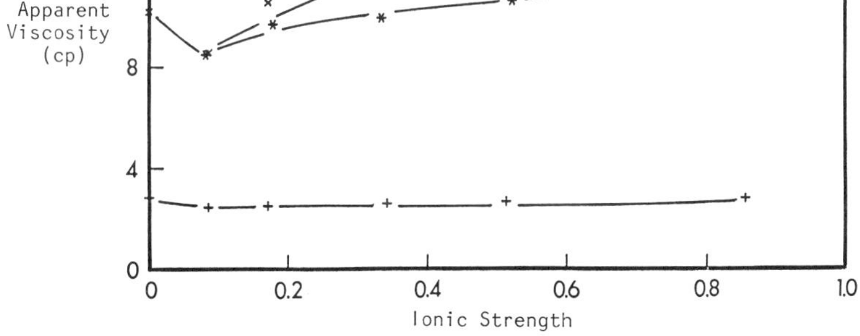

Figure 7. Effect of ionic strength of NaCl and $CaCl_2$ solutions on the apparent viscosity of a copolymer of acrylamide with 0.50 mole % decylacrylamide at 25°C, a shear rate of 1.28 sec^{-1}, and a copolymer concentration of 0.25 g/dL.

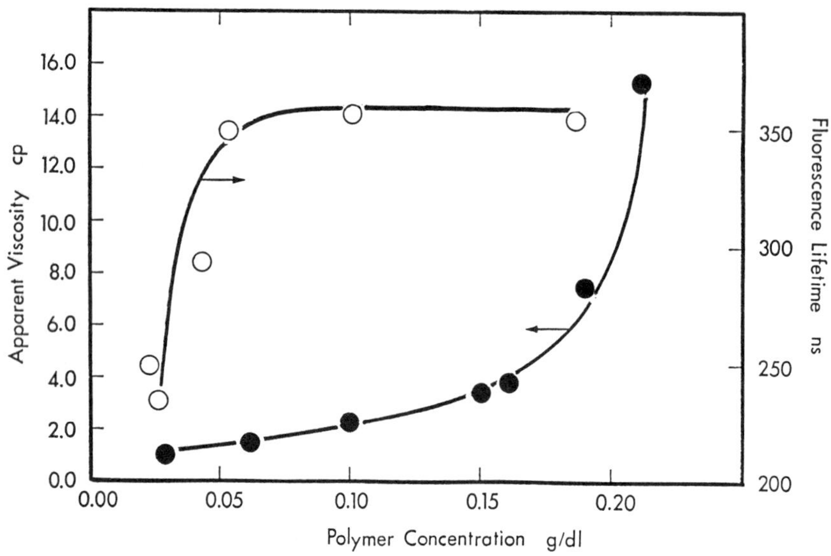

Figure 8. Fluorescence lifetime and apparent viscosity versus concentration for 0.75 mole % decylacrylamide copolymer in water at 25°C and a shear rate of 1.28 sec^{-1}.

Fluorescence Lifetime Studies. In consideration of the solubility characteristics (Table 1) and solution viscosity behavior of the copolymers, the C-10 0.75 mole % polymer exhibited the most desirable properties. Therefore, this copolymer was chosen to probe molecular associations as a function of copolymer concentration using fluorescence spectroscopy. It has been shown that an increase in the micelle aggregation number increases the fluorescence lifetime of pyrene probes incorporated into micelles. Similarly, if micelles having pyrene incorporated are disrupted, the lifetime would be expected to decrease. In Figure 8, fluorescence lifetime is plotted against polymer concentration for the C-10 0.75 copolymer. A sharp transition is seen in the polymer concentration region of 0.23 to 0.05 g/dL. This transition reflects the change from an apparent homogeneous solution of independent polymer coils to a solution where the concentration of hydrophobic groups are high enough to induce intermolecular associations among polymer molecules.

Also shown in Figure 8 is the effect of polymer concentration on the apparent viscosity of the C-10 0.75 copolymer. Interestingly, the concentration at which the asymptotic increase in viscosity is observed is significantly greater than the concentration required for an increase in fluorescence lifetime. A model which we believe could be responsible for this phenomenon is depicted in Figure 9 where three distinct domains of polymer structure are illustrated. Region I represents a polymer concentration less than 0.03 g/dL. In this region the chains act as individual macromolecules and little or no association occurs. In region II, corresponding to polymer concentrations from 0.03 to 0.20 g/dL, the formation of hydrophobic domains begins to occur. Within this concentration range, the number of hydrophobic segments is sufficient to promote intermolecular associations. However, in region II the

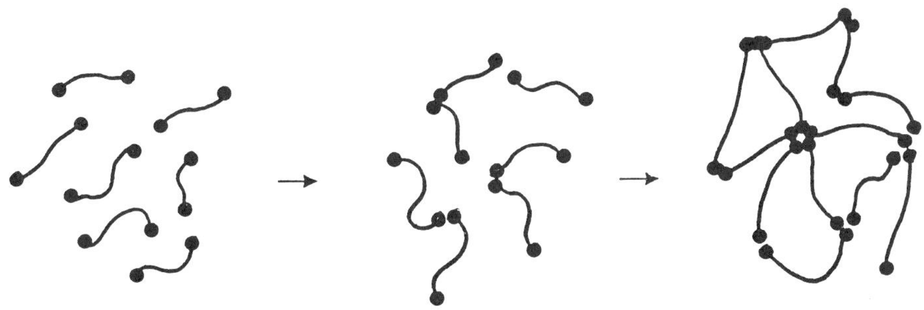

Figure 9. The three domains of polymer structure leading to large-scale hydrophobic association.

number of domains remains less than that required to result in an increase in the macroproperty of viscosity. Only at polymer concentrations greater than about 0.20 g/dL is the concentration of hydrophobic domains great enough to result in an effective intermolecular network. The formation of such a network results in the sharp increase observed in viscosity. Hence, fluorescence spectroscopy is much more sensitive than viscometry in detecting the initial occurrence of intermolecular association, while viscometry vividly displays the result of forming a large number of these interactions. Further studies are underway in our laboratories concerning the apparent usefulness of the fluorescence technique in probing macroscopic ordering and relating such ordering to rheological properties.

Ampholytic Terpolymers

Low-to-moderate charge density terpolymers of acrylamide (AM) with 2-acrylamido-2-methylpropanedimethylammonium chloride (AMPDAC) and sodium 2-acrylamido-2-methylpropanesulfonate (NaAMPS) have been synthesized with varying molar feed ratios of the three monomers. Terpolymer feed and compositional data for the AMPDAC-NaAMPS-AM, ADASAM, series are shown in Table 5. The numbers following "ADASAM" represent the mole percent of AMPDAC and NaAMPS, respectively, in the terpolymer feed. For example, ADASAM 5-10 represents a terpolymer which included 5% AMPDAC, 10% NaAMPS, and 85% AM (acrylamide) in the reaction feed.

Viscometry Studies. Apparent viscosity is plotted against polymer concentration for each of the synthesized terpolymers in pure water at 30°C and a shear rate of 1.75 sec^{-1} (Figure 10). In general, those samples containing the greatest number of charged groups display the highest viscosities, particularly in the region of chain overlap (C*). Terpolymers containing small amounts of charged groups display behavior which is identical to that of polyacrylamide. Apparently the charge density must reach a critical level in order to cause significant intermolecular ionic association. Figure 11 depicts the increase in apparent viscosity that is observed with increasing polymer concentration for ADASAM 5-10. This sample exhibits viscosities in pure water much greater than the other terpolymers, likely due to the charge imbalance in favor of NaAMPS. Copolymers of NaAMPS with AM are known to have large

Table 5

ADASAM SERIES
REACTION PARAMETERS[a] AND TERPOLYMER COMPOSITIONS

Sample Name	Feed Composition	Terpolymer Composition[b]
	AMPDAC:NaAMPS:AM	AMPDAC:NaAMPS:AM
ADASAM 0.5-0.5	0.5:0.5:99.0	1.03:1.49:97.49
ADASAM 2.5-2.5L[c]	2.5:2.5:95.0	5.20:3.36:91.44
ADASAM 2.5-2.5	2.5:2.5:95.0	5.39:3.44:91.16
ADASAM 5-5L	5:5:90	8.59:5.84:85.57
ADASAM 5-5	5:5:90	8.24:6.21:85.55
ADASAM 10-10L	10:10:80	11.14:12.56:76.30
ADASAM 10-10	10:10:80	10.69:12.57:76.74
ADASAM 15-15L	15:15:70	20.08:15.95:63.98
ADASAM 15-15	15:15:70	19.51:15.54:64.95
ADASAM 5-10L	5:10:85	8.21:10.70:81.09
ADASAM 5-10	5:10:85	8.31:10.22:81.47
ADASAM 10-5L	10:5:85	13.31:8.20:78.49
ADASAM 10-5	10:5:85	12.56:7.23:80.21

[a] pH=6.0; T = 30°C; Total Monomer Concentration = 0.45 M; $[K_2S_2O_8]$ = 4.5 x 10^{-4} M.

[b] Determined via Elemental Analysis for Carbon, Nitrogen, and Sulfur.

[c] L denotes low conversion.

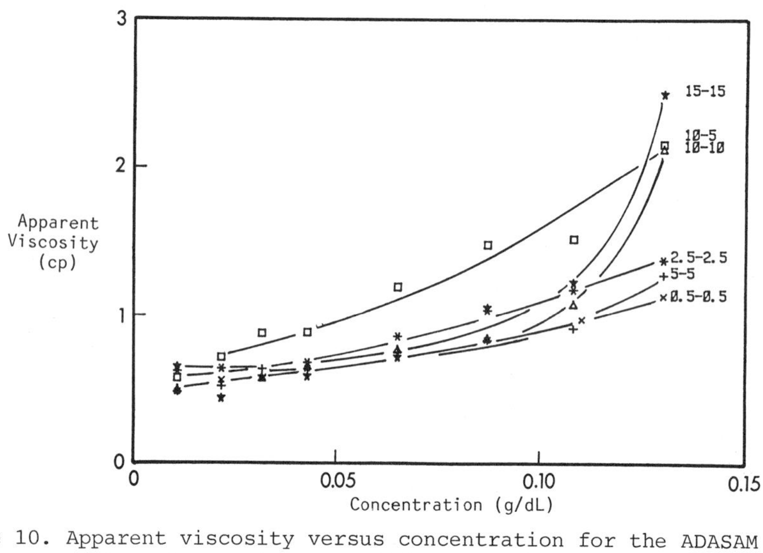

Figure 10. Apparent viscosity versus concentration for the ADASAM series of terpolymers in water 30°C and a shear rate of 1.75 sec^{-1}.

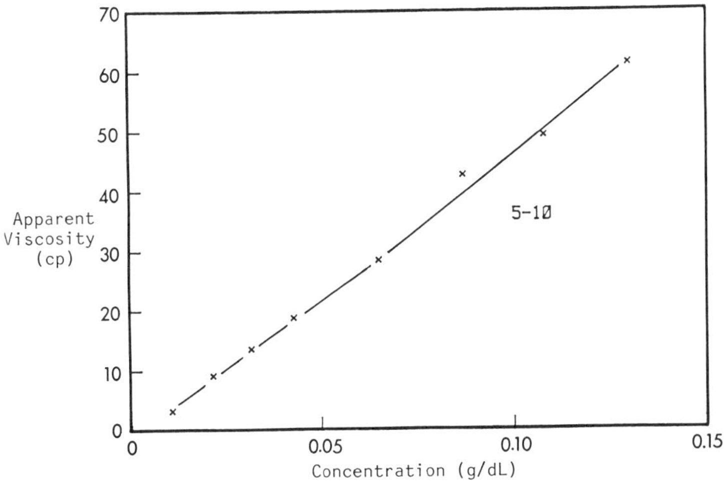

Figure 11. Apparent viscosity versus concentration for ADASAM 5-10 in water at 30^{0}C and a shear rate of 1.75 sec^{-1}.

viscosities in pure water, especially at low NaAMPS compositions, due to charge-charge repulsion.

ADASAM 10-5 exhibits viscosities among the lowest of the polymers since the molecular weight is probably reduced by the presence of a relatively large amount of AMPDAC. Molecular weight studies of a series of high charge density copolymers of AMPDAC with NaAMPS have shown that molecular weight decreases as AMPDAC incorporation increases. Therefore, incorporation of the cationic AMPDAC monomer does not seem to affect hydrodynamic volume to the same extent as incorporation of the anionic NaAMPS monomer. AMPDAC is a good chain transfer agent since it possesses abstractable hydrogen atoms on the carbons adjacent to the highly electronegative amine nitrogen.

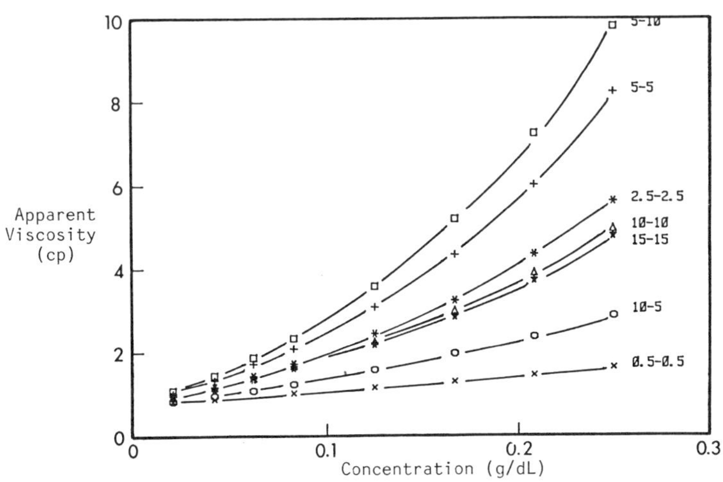

Figure 12. Apparent viscosity versus concentration for the ADASAM series of terpolymers in 0.257 M NaCl at 30^{0}C and a shear rate of 1.75 sec^{-1}.

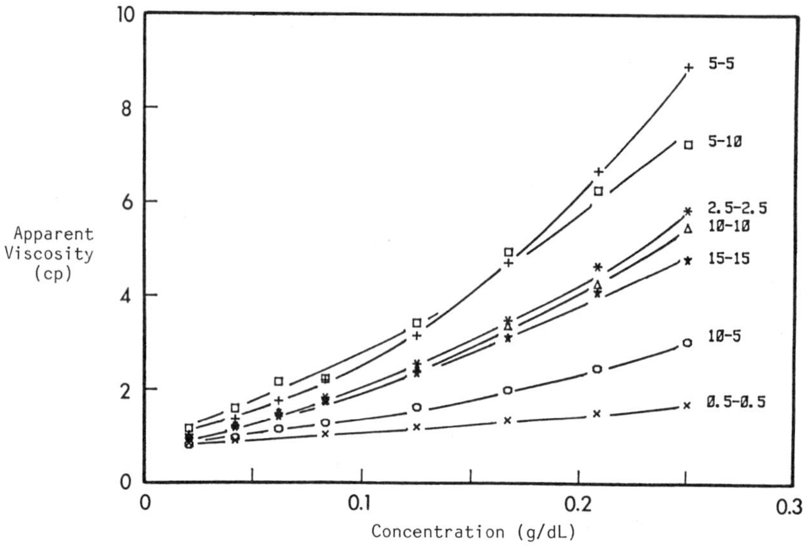

Figure 13. Apparent viscosity versus concentration for the ADASAM series of terpolymers in 0.514 M NaCl at 30°C and a shear rate of 1.75 sec^{-1}.

Apparent viscosity is plotted against polymer concentration for each of the synthesized terpolymers in 0.257 M NaCl at 30°C and a shear rate of 1.75 sec^{-1} (Figure 12). Note that the moderate salt concentration allows sufficient shielding of charges to result in the reduction of viscosity of ADASAM 5-10 within the range of the other polymers. Terpolymers containing 10 mole % or greater of AMPDAC in the feed display viscosities which are among the lowest under the described conditions.

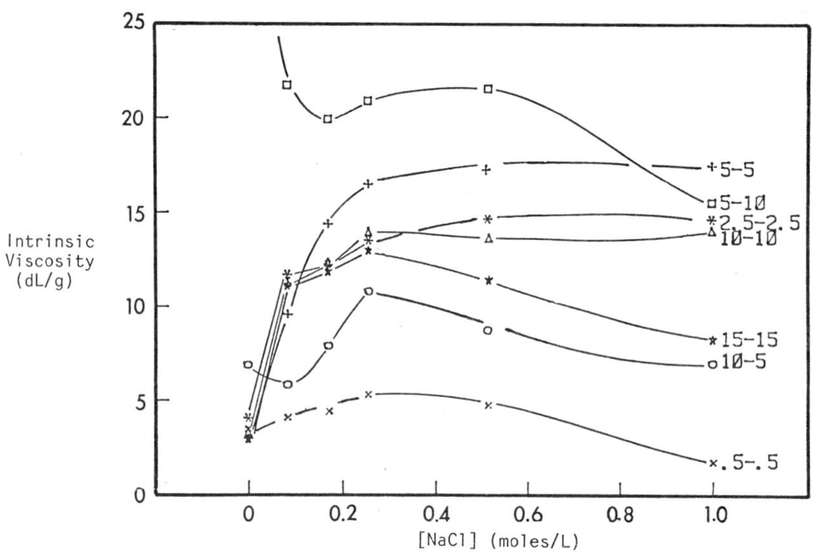

Figure 14. Intrinsic viscosity versus NaCl concentration for the ADASAM series of terpolymers at 30°C and a shear rate of 1.75 sec^{-1}.

ADASAM 0.5-0.5 apparently has too few charged groups to affect the solution properties of this polymer relative to polyacrylamide. ADASAM 2.5-2.5 and 5-5 exhibit intermediate viscosities since the charge density is large enough to warrant some intermolecular interaction but not so large that intramolecular interaction and a reduction in molecular weight adversely affect viscosity.

At a NaCl concentration of 0.514 M (Figure 13), ionic shielding results in the collapse of ADASAM 5-10 to the point that ADASAM 5-5 possesses a greater viscosity in the higher polymer concentration range. The fact that ADASAM 5-5 displays enhanced viscosities over 2.5-2.5 and 0.5-0.5 is attributed to the presence of increased amounts of NaAMPS.

In general, for each of the terpolymers with the exception of ADASAM 5-10, the reduced viscosities increased throughout the concentration range with increasing NaCl concentration. The intrinsic viscosities also increased with increasing NaCl concentration (Figure 14). This behavior is due to the fact that the majority of the synthesized terpolymers are composed of approximately an equimolar amount of oppositely charged species. Apparently, although there are a small number of ionic groups, i.e. a low charge density, intramolecular ionic interaction does serve to somewhat decrease the hydrodynamic volume of the polymer chain. Therefore as the salt concentration is increased, the charged groups along the polymer chain are shielded and the polymer chain is allowed to expand slightly.

As previously discussed, ADASAM 5-10 behaves differently from the other terpolymers because it possesses an overall polyanionic character while retaining a high molecular weight. Ordinary polyelectrolytes are known to decrease in hydrodynamic volume with increasing solvent ionic strength due to charge shielding of repulsive ionic interactions along the polymer chain.

CONCLUSIONS

Copolymers of AM-NaAMB have been shown to have high viscosities, good salt stability and surprising resistance to divalent ion binding. The copolymer containing 8 % NaAMB displayed the greatest salt tolerance, while the NaAMB homopolymer showed the least salt tolerance. All of the NaAMB polymers exhibited viscosity decreases with increasing temperature and decreasing pH; these, however are much less drastic than those observed with conventional flooding polymers. The copolymers of NaAMB exhibit pKa values which are similar to hydrolyzed polyacrylamides. These values decrease with decreasing NaAMB content, a result which may be due to hydrogen bonding with acrylamide units.

The observed viscosity differences within the NaAMB series are due to microstructural differences in the copolymers. The copolymer sequence distributions suggest that lower pKa values occur when greater quantities of acrylamide units are present, and a similar relationship may be responsible for the greater salt tolerance of the 8 mole % NaAMB copolymer. The viscosity, salt tolerance, divalent ion binding, and apparent pKa data suggest the importance of local interactions dictated by sequencing along the backbone. These include hydrogen bonding, hydrophobic, and steric interactions along the macromolecular chain. Initial core flooding experiments conducted by Sun Tech laboratories have shown the NaAMB copolymers to be quite effective for mobility control in EOR[22].

The concentration dependence of the viscosity of the associative acrylamido copolymers increases dramatically with increasing hydrophobicity. Solution viscosities of the hydrophobically modified acrylamide copolymers were increased by increases in salt concentration (NaCl and $CaCl_2$) and temperature in the range of 25 to 55°C. Fluorescence spectroscopy indicated that associations do occur in the dilute concentration regime, i.e. below the critical overlap concentration. Although intramolecular interactions no doubt occur below C^*, an increase in fluorescence with polymer concentration indicates that intermolecular associations also exist. In fact, the concentration required to increase fluorescence intensity is significantly less than that required to enhance solution viscosity. Hence, fluorescence spectroscopy is much more sensitive than viscometry in detecting the initial occurrence of intermolecular association. Viscometry vividly displays the result of forming a large number of intermolecular associations[23,24]. The large viscosity increases observed at both high and low salinity may be quite useful in EOR applications especially for tight formations. A recent patent[40] also outlines the use of such systems for formulation of fracturing fluids.

Ampholytic terpolymers containing relatively large amounts of AMPDAC are expected to have reduced molecular weights which will adversely affect solution viscosities. There is a detrimental effect brought about by the incorporation of AMPDAC units, namely, a reduction in molecular weight. Thus, there is a point at which positive effects due to increasing charge density are overcome by a decline in molecular weight of the terpolymers. ADASAM 5-5 displays viscosities which are indicative of the optimum charge density over a range of solution ionic strengths.

ADASAM 5-10 displays solution viscosity behavior which is common to ordinary polyelectrolytes since this terpolymer has an overall negative charge. However, the remainder of the terpolymers exhibit solution viscosity behavior similar to that of higher charge density polyampholytes. In other words, as the charge density of the terpolymers increases, the solution viscosities increase, particularly in the polymer concentration region above C^*. The variance of solution viscosity of the terpolymers as a function of temperature was not significant due to the fact that the charge density of the terpolymers was small. An increase in temperature did not result in the disruption of a large number of ionic interactions. In general, intramolecular ionic interactions do seem to contribute to the solution properties of these polyampholytes despite their low to moderate charge densities. To date polyampholytes have not been utilized in EOR applications. The unusual behavior of balanced low charge density systems in high brine concentrations (the opposite of that observed for conventional polyelectrolytes) may be capitalized upon in future recovery techniques.

Clearly much research remains to be done in order to properly tailor macromolecules for specific petroleum recovery operations. Intramolecular chelators like NaAMB, associative polymers, polyampholytes, and micellar polymers possess some of the most intriguing properties discovered to date. Developing these and other useful copolymers for a number of water-soluble applications will require not only creative fundamental research but also the willingness of industrial R&D management to look toward new frontiers.

REFERENCES

1. C. L. McCormick and G. S. Chen, J. Polymer Sci., Chem., 20, 817 (1982).
2. C. L. McCormick, G. S. Chen, and B. H. Hutchinson,Jr., J. Appl. Polymer Sci., 27, 3103 (1982).
3. H. H. Neidlinger, G. S. Chen, ;and C. L. McCormick, J. Appl. Polymer Sci., 29, 713 (1984).
4. C. L. McCormick and L. S. Park, J. Polymer Sci., Chem., 22, 49 (1984).
5. C. L. McCormick and G. S. Chen, J. Polymer Sci., Chem, 22, 3633 (1984).
6. C. L. McCormick and G. S. Chen, J. Polymer Sci., Chem, 22, 3649 (1984).
7. C. L. McCormick and L. S. Park, J. Appl. Polymer Sci., 30, 45 (1985).
8. C. L. McCormick and K. P. Blackmon, J. Polymer Sci. Chem., 24, 2635 (1986).
9. C. L. McCormick, K. P. Blackmon, and D. L. Elliott, J. Polymer Sci., Chem., 24, 2619 (1986).
10. C. L. McCormick and D. L. Elliott, Macromolecules, 19, 542 (1986).
11. B. H. Hutchinson and C. L. McCormick, Polymer, 27(4), 623 (1986).
12. C. L. McCormick and B. H. Hutchinson, accepted by Die Makromol. Chemie (1985).
13. C. L. McCormick and K. P. Blackmon, Macromolecules, 19, 1512 (1986).
14. C. L. McCormick, D. L. Elliott and D. L. Elliott, K. P. Blackmon, Macromolecules, 19, 1516 (1986).
15. C. L. McCormick and K. P. Blackmon, J. Macromol. Sci., Chem., A25(12), 1451 (1986).
16. C. L. McCormick, K. D. Blackmon, and D. L. Elliott, J. Macromol. Sci., Chem., A23(12), 1469 (1986).
17. C. L. McCormick and K. P. Blackmon, submitted to Polymer (1986).
18. C. L. McCormick, K. P. Blackmon, and D. L. Elliott, submitted to Polymer (1986).
19. C. L. McCormick, K. P. Blackmon, D. Angew. Makro. Chem., 144, 73 (1986).
20. C. L. McCormick, K. P. Blackmon, and D. L. Elliott, D. Angew. Makro. Chem., 144, 87 (1986).
21. C. L. McCormick and D. L. Elliott, accepted for publication in J. Polymer Sci., Chem.
22. C. L. McCormick and K. P. Blackmon, U.S. Patent 4,584,358, 1986.
23. S. R. Turner, D. B. Siano, and J. Bock, U.S. Patent 4,520,182, 1985.
24. S. Evani, U.S. Patent 4,432,881, 1984.
25. K. L. Hoy and R. C. Hoy, U.S. Patent 4,426,485, 1984.
26. W. D. Emmons and T. E. Stevens, U.S. Patent 4,395,524, 1983.
27. L. M. Landoll, U.S. Patent 4,288,277, 1980.

28. D. B. Siano, J. Bock, S. R. Turner, European Patent Application 83307999.9, Publication No. 0115703, August 15, 1984.
29. S. R. Turner, D. B. Siano, and J. Bock, U.S. Patent 4,528,348, 1985.
30. D. G. Peiffer and R. D. Lundberg, Polymer, 26, 1058 (1985).
31. J. C. Salamone, et al., from Advan. Chem. Sci., ACS (187) Ions in Polymers, 337 (1980).
32. J. C. Salamone, et al. J. Macromol. Sci., Chem., A22(5-7), 653 (1985).
33. J. C. Salamone, C. C. Tsai, and A. C. Watterson, J. Macromol. Sci., Chem, A13, 665 (1979).
34. D. Hoke, and R. Robbins, J. Polymer Sci., 10, 3311 (1972).
35. D. S. McClure, et al., J. Chem. Phys., 20, 829 (1952).
36. J. C. Fenyo, J. P. Laine, and G. Muller, J. Polymer Sci., Chem., 17, 193 (1979).
37. S. Harwood and ?? Ritchey, Polymer Lett., 2, 601 (1964).
38. J. Morcellet-Sauvage, M. Morcellet, and C. Loucheaux, Makromol. Chem., 182, 949 (1981).
39. C. Methenitis, J. Morcellet-Sauvage, and M. Morcellet, Polymer Bull, 12, 133 (1984).
38. V. G. Constien and T. King, U.S. Patent 4,541,935, 1985.

BEHAVIOR OF POLYAMPHOLYTES IN AQUEOUS SALT SOLUTION

J. C. Salamone, I. Ahmed, M. K. Raheja,
P. Elayaperumal, A. C. Watterson and A. P. Olson

Polymer Science Program
Department of Chemistry
University of Lowell
Lowell, Massachusetts 01854

INTRODUCTION

The role of ionic polymers in enhanced oil recovery is well recognized (1). Partially hydrolyzed polyacrylamide is the most widely used ionic polymer for this purpose as it produces a high viscosity in fresh water at a reasonable cost. However, anionic polyacrylamide rapidly looses its viscosity in solutions containing added electrolyte. In addition, polyacrylamides are sensitive to shear degradation.

It is known that polymers containing both positive and negative charges (i.e. polyampholytes) exhibit a higher viscosity with increasing salt concentration (2,3). During the past few years our research group has begun an investigation of the solution properties of a new class of ampholytic polymer. These polyampholytes have been derived from sulfobetaines, in which the positive and negative charges are on one pendent group (4-6), or from the homopolymerization of an ion-pair comonomer, in which the positive and negative charges either alternate or exist in a random pendent configuration (7-12). As reported previously, the ion-pair comonomers are a vinylic cation/vinylic anion salt in which no inorganic ions are present. Thus, each monomer is a gegenion to its oppositely charged counterpart.

Our detailed investigation of the aqueous solution properties of a poly(vinylimidazolium sulfobetaine) has demonstrated viscosity enhancement in the presence of increased salt concentration (6). This polyampholyte, though insoluble in water, became soluble in certain aqueous salt solutions. The solution properties of this polymer were found to be related to the type and concentration of salt employed.

In this research are reported studies of the solution behaviors of an alternating polyampholyte of 4-vinyl-N-methylpyridinium p-styrenesulfonate, 4VMP/pSS (I), and two random polyampholytes of 3-methacrylamidopropyltrimethylammonium 2-

acrylamido-2-methylpropanesulfonate, MPTMA/AMPS (II), and 2-methacryloyloxyethyltrimethylammonium 2-methacryloyloxyethane-sulfonate, METMA/MES (III).

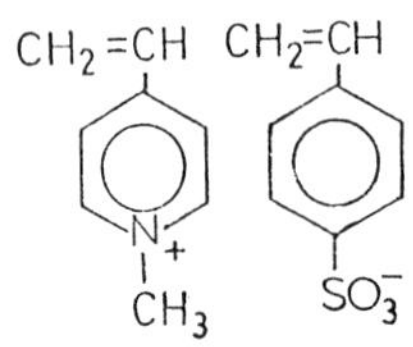

I

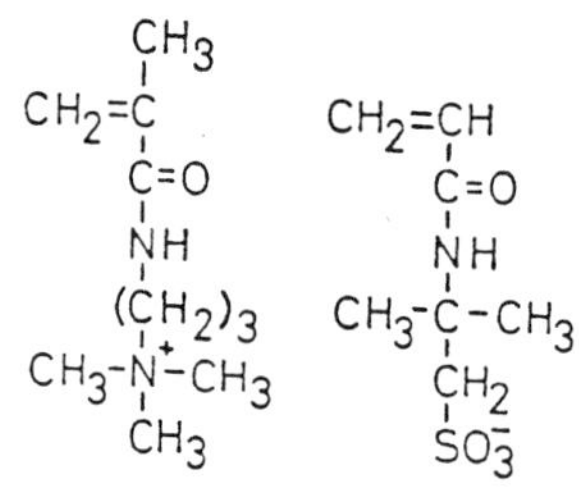

II

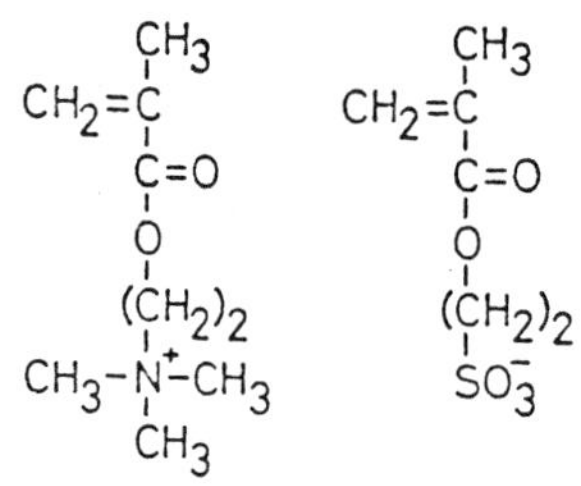

III

EXPERIMENTAL

Polymer Preparations

Poly(4VMP/pSS) was prepared by the spontaneous polymerization of a 0.6 M solution of ion-pair comonomer, 4VMP/pSS (11), in distilled water at room temperature. After an appropiate polymerization time, the precipitated polymer was dissolved in 1.5 M aqueous NaCl solution. It was then dialyzed exhaustively against salt solution followed by dialysis against distilled water to precipitate the purified polymer. The product was then filtered and dried in vacuum.

For the preparation of poly(MPTMA/AMPS) and poly(METMA/MES), the monomers (12,13) and the initiator were mixed in two polymerization ampules as follows: ampule 1: 1.60 g (4.1 mmol) of ion-pair comonomer, MPTMA/AMPS, 10.7 mg (ca. 0.93 mol%) of 4,4´-azobis-4-cyanovaleric acid, ACVA, and deionized water to make the final volume to 50 ml of solution; ampule 2: 5.50 g (15.0 mmol) of ion-pair comonomer, METMA/MES, 17.0 mg (ca. 0.40 mol%) of ACVA and deionized water to make the final volume to 50 ml of solution. Both ampules were degassed by the freeze-thaw technique, evacuated and sealed. Ampule 1 was placed in a water bath at 50°C for 24 h and ampule 2 was placed in a water bath at 60°C for 9 h. The polymer solutions were then dialyzed exhaustively against water followed by lyophilization to obtain the purified products.

Viscosity Measurements

Viscometric properties of the three polymer systems were studied under dilute (<1%) and concentrated (10%) conditions utilizing an Ubbelohde dilution viscometer for the former and a cone-plate Brookfield digital viscometer for the latter. Polymer samples were dissolved in an appropriate salt solution. Viscosities were determined at 25±0.1°C taking precautions to keep all solutions dust free.

Light Scattering Measurements

The light scattering measurements were carried out on a laser light scattering photometer, DAWN Model F (Wyatt Technology Corp., Santa Barbara, CA), which employs a He-Ne 5mW vertically polarized laser (λ = 632.8 nm) as a light-source. Measurements were made on solutions which were previously dialyzed against solvent to establish Donnan membrane equilibrium (14) and filtered through a 0.22 μm Millipore filter. Optical clarity of the solutions was assured by direct observation of the light beam within the solution coupled with constant readings on the channel monitor. The normalized scattered intensities at different angles (35° - 123°) were simultaneously measured by photo diode array and analyzed by an IBM computer using DAWNFLOW software supplied by the manufacturer. Prior to the measurements, the normalization factors for all detectors were obtained by measuring scattered intensities using a well characterized aqueous colloidal dispersion of spherical particles (15.4 nm diameter) of amorphous silica ("LUDOX" from Du Pont), which scatter light isotropically. For details of the normalization procedure, the reader is referred to the Instruction Manual of the DAWN Model F. The instrument was then calibrated using polystyrene standards of Mw ranging from 36,000 to 2,430,000 g/mol within ± 5% accuracy range.

The specific refractive index increment, dn/dc, was determined on a Brice-Phoenix differential refractometer calibrated with KCl solutions of known refractive indices. The light-source was a mercury vapor lamp fitted with a filter to isolate the 632.8 nm emission line.

Light scattering data were evaluated by the Zimm technique (15). A computer program was developed in this laboratory to generate the Zimm plot where double extrapolations were required.

RESULTS AND DISCUSSION

The two polymers poly(4VMP/pSS) and poly(METMA/MES) were insoluble in water and soluble in various aqueous salt solutions, in much the same fashion as the previously studied poly(vinylimidazolium sulfobetaine). In contrast, however, poly(MPTMA/AMPS) appeared to be soluble in water without added salt. This observation suggested that the polymers from the ion-pair comonomers 4VMP/pSS and METMA/MES consisted of either a higher degree of aggregated/ordered structure than poly(MPTMA/AMPS), or these two polymers were substantially higher in molecular weight, or both the polymers were more hydrophobic than poly(MPTMA/AMPS). Therefore, it was of interest to investigate the solution properties of the three ampholytic systems.

Viscosity Studies

Preliminary viscosity results obtained by using a variable shear Ubbelohde dilution viscometer showed that neither of the three polyampholytes were shear dependent in their dilute solution viscosity behavior, and all viscosity measurements were subsequently carried out using an ordinary Ubbelohde dilution viscometer. The results were first treated by the well known Huggins equation (Equation 1) (16):

$$\eta_{rel} - 1 = \eta_{sp} = [\eta]C + k'[\eta]^2C^2 \qquad (1)$$

The limiting viscosity number (intrinsic viscosity), $[\eta]$, was obtained from the plot of reduced viscosity (η_{sp}/C) versus concentration as the intercept on the viscosity axis. The results are given in Table 1. It can be seen from Table 1 that alternating poly(4VMP/pSS) did not follow the Huggins equation adequately. A downward curvature at lower concentrations was observed and the correlation coefficients to a straight line were poor. In case of poly(MPTMA/AMPS) a linear Huggins plot was obtained but k', the Huggins constant which represents the solute-solvent interaction, was considerably lower than the typical values for true solutions allowed from theoretical considerations (17-19). For solutions with negligible solute-solvent interactions, a modified viscosity equation was derived from the Huggins equation (9) as

$$\eta_{rel} = 1 + [\eta]C \qquad (2)$$

This equation is analogous to Einstein-Simha viscosity equation for suspended particles of variable shapes (20,21).

A comparative plot utilizing both equations (equations 1 and 2) for the three ampholytic systems in 1.5 M aqueous salt solution is given in Figure 1, and the results of the plot are given in Table 1. Treatment of the viscosity data by equation 2 produced a linear plot that fitted the data reasonably well for all three cases. However, for poly(METMA/MES) treating the data with equation 2 yielded an intercept which substantially deviated from the theoretically predicted value of 1.

The observation that the viscosity data follow only equation 2 suggests that the poly(4VMP/pSS) may be behaving as a suspension rather than a dissolved flexible coil. There could be two structural features within the polymer chain responsible for this behavior. The first concerns the ionic interaction between the charged groups which may lead to the exclusion of the solvent. A similar behavior is exhibited by polysoaps (22,23). This situation would be expected to arise from the alternately placed positive and negative charges along the polymer chain. The second feature concerns hydrophobic interactions which originate from hydrocarbon based molecules in aqueous solution (24). If the hydrocarbon portions of the chain are coalesced to minimize the interference with water structure, with aromatic rings behaving in a similar fashion, then the alternating polyampholyte would possibly be in its collapsed state existing as a finely dispersed colloidal suspension.

TABLE 1. Comparative Viscosity Data Obtained from the Usual and the Modified Huggins Equation

[NaCl] M	Huggins Equation Correl. Coeff.	Intercept [η] dl/g	k′	Modified Huggins Equation Correl. Coeff.	Slope [η] dl/g	Intercept
Poly(4VMP/pSS)						
1.0	0.99	0.52	1.05	1.00	0.64	0.99
1.5	0.91	0.84	0.61	1.00	1.00	0.99
2.0	0.89	1.09	0.59	1.00	1.23	0.99
2.5	0.95	1.18	0.59	1.00	1.49	0.98
3.0	0.96	1.10	0.51	1.00	1.24	0.98
Poly(MPTMA/AMPS)						
0.0	--	--	--	0.95	1.24	1.17
0.5	--	--	--	1.00	0.86	1.01
1.0	0.99	0.82	0.23	1.00	0.95	0.97
1.5	0.98	0.95	0.20	1.00	1.03	0.99
2.0	0.96	0.83	0.25	1.00	0.88	0.99
3.0	0.98	0.64	0.13	1.00	0.71	0.99
Poly(METMA/MES)						
0.5	1.00	1.66	0.40	0.99	2.80	0.82
1.0	1.00	1.77	0.38	0.99	3.05	0.84
1.5	1.00	1.91	0.38	0.99	3.39	0.76
2.0	1.00	1.97	0.39	0.99	3.52	0.77
3.0	1.00	2.05	0.39	0.99	3.71	0.74
5.0	1.00	2.14	0.30	1.00	3.69	0.70

In the case of poly(MPTMA/AMPS), an entirely different behavior was observed. Figure 2 represents the dilute solution viscosity behavior of the polymer in water as well as in different aqueous salt solutions. From Figure 2 it is apparent that this polymer behaves more like a polyelectrolyte at low salt concentrations and in deionized water. On the other hand, with increasing salt concentration an opposite tendency to usual polyelectrolyte behavior is found, and the polymer displays ampholytic behavior. This contrasting behavior may be visualized by the fact that even though the bulk of the polymer contains equal amounts of both charges, the individual polymer chains may not necessarily contain equal amounts of positive and negative groups, since the acrylamido portion could produce blocky character in the polymer due to higher polymerization tendency compared to the methacrylamido portion. The uneven distribution of charges should result in higher intermolecular interactions in this polymer. This could be the reason for the low k′ values observed (25,26). With diminishing intermolecular interaction at higher salt concentrations the ampholytic character of the chains is then revealed by further addition of salt. This might explain the dual character of this polymer and also the viscosity data which follow both forms of the Huggins equation. The intrinsic viscosities obtained by both equations match each other closely as shown in Table 1.

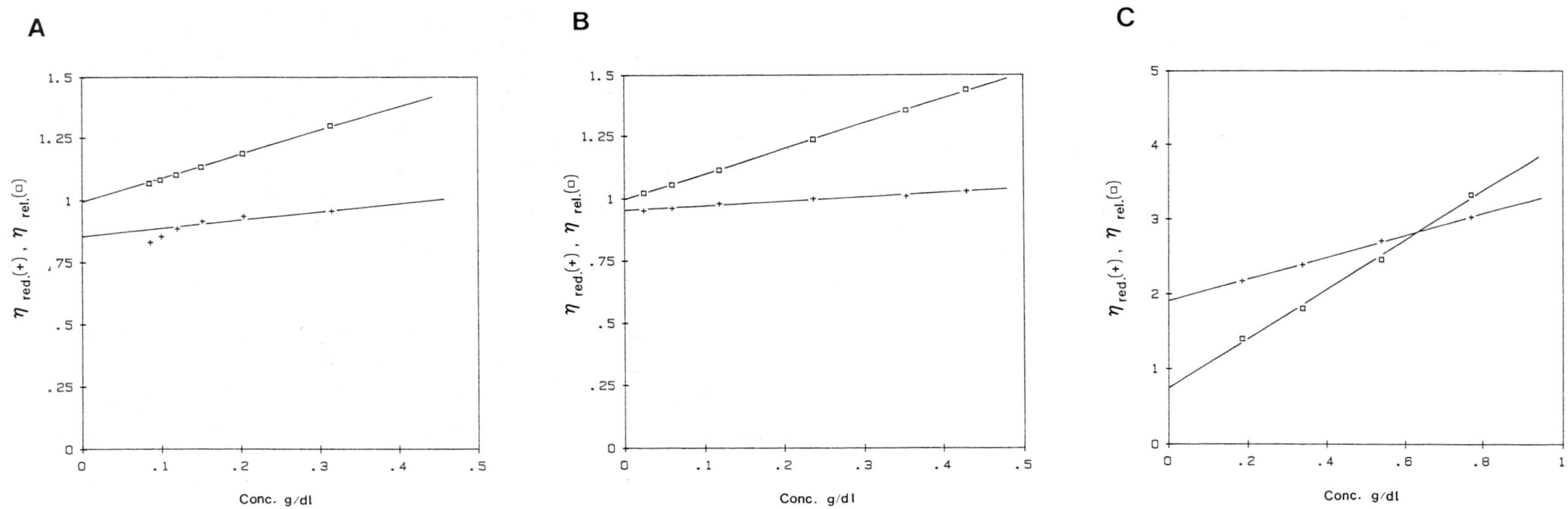

Figure 1. Comparative plots of usual(+) and modified(□) Huggins Equation for A)poly(4VMP/pSS), B)poly(MPTMA/AMPS), and C)poly(METMA/MES) in 1.5 M aqueous NaCl solution.

The third polyampholyte, poly(METMA/MES), followed the usual Huggins equation and showed typical polyampholyte behavior. This polymer should be a random copolymer of the two methacryl monomers and therefore should have a lower degree of disparity of the charges between different chains compared to poly(MPTMA/AMPS). In addition, it is intermediate in its hydrophilic character compared to the other two polymers, and is probably in an intermediate state of suspension.

Figure 3 represents the effect of added electrolyte concentration on the [η] obtained from the modified Huggins plot for poly(4VMP/pSS) and poly(MPTMA/AMPS), and the usual Huggins plot for poly(METMA/MES). The intrinsic viscosity increases with increasing salt concentration for all three ampholytic systems. Similar results are also reported for other polyampholyte-salt systems (6,13,27,28). This behavior may be rationalized on the basis of chain expansion which results in increased solute-solvent interaction. The [η] is related to the hydrodynamic volume of macromolecules in solution (29). An expansion of the chain results in the viscosity increase due to an increase in effective hydrodynamic volume of the solute in the given solvent. It is expected that the added electrolyte would disrupt the intramolecular and intermolecular interactions and allow the polymers to behave more freely. Thus, the increase in [η] may be related to extended chain conformations resulting from the increased polymer-solvent interactions.

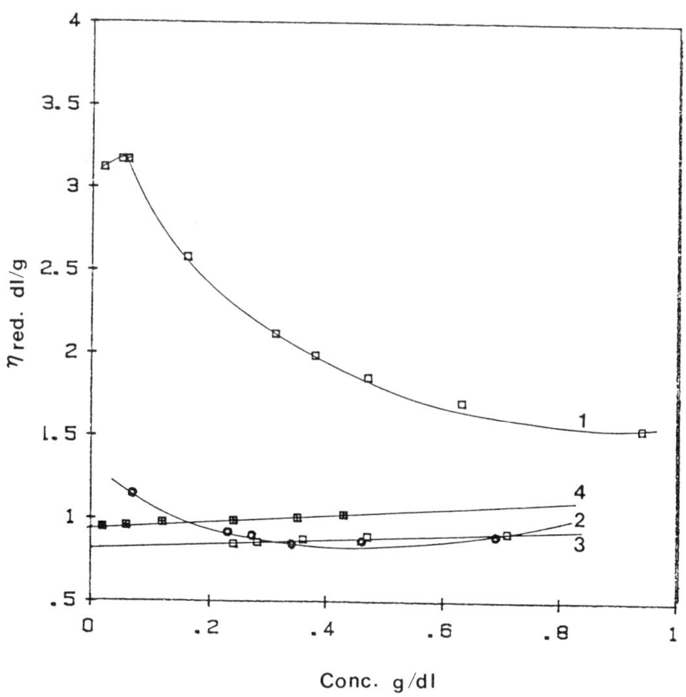

Figure 2. Reduced viscosity as a function of polymer concentration for poly(MPTMA/AMPS) in 1)deionized water, 2)0.5 M aq. NaCl, 3)1.0 M aq. NaCl, 4)1.5 M aq. NaCl.

Although the above interpretation of viscosity data agrees well for poly(METMA/MES) (Figure 3), it becomes increasingly less straightforward for other two polymers, poly(4VMP/pSS) and poly(MPTMA/AMPS), as the concentration of added electrolyte is further increased above a certain level. The interaction between polymer chain and the added ions largely depends on the site-binding ability of the mobile ions (6,30-33). Thus the specific complexing ability of the mobile ions may lead to a salting-out of the polymer above a certain salt level. As a result, the hydrodynamic volume will be reduced with a further increase of salt concentration. On the other hand, this decrease in the viscosity could result from the breaking of agglomerated structures. The situation certainly warrants further definitive measurements of molecular weight and relative parameters in various salt concentrations.

To see whether the type of salt has any effect on the viscosity behavior of the polyampholyte systems under investigation, poly(METMA/MES) was studied in the presence of different types of salt. The results are given in Table 2. The data show that the viscosity decreased in the order $Li^+ \leq Na^+ \leq K^+$ for the cations while for the anions the order was $Cl^- < Br^- < I^-$. It is also apparent from Table 2 that the effect of the anions is much more pronounced than that of the cations. This observation suggests that the viscosity behavior may be related to the charge/radius ratio of the added ions. Apparently the low charge/radius ratio favors good site-binding abilities. This ratio for the cations, for example, Na^+(1.03), is higher compared to that of the anions, e.g. Br^-(0.51). Thus, the anions are more effective in neutrali-

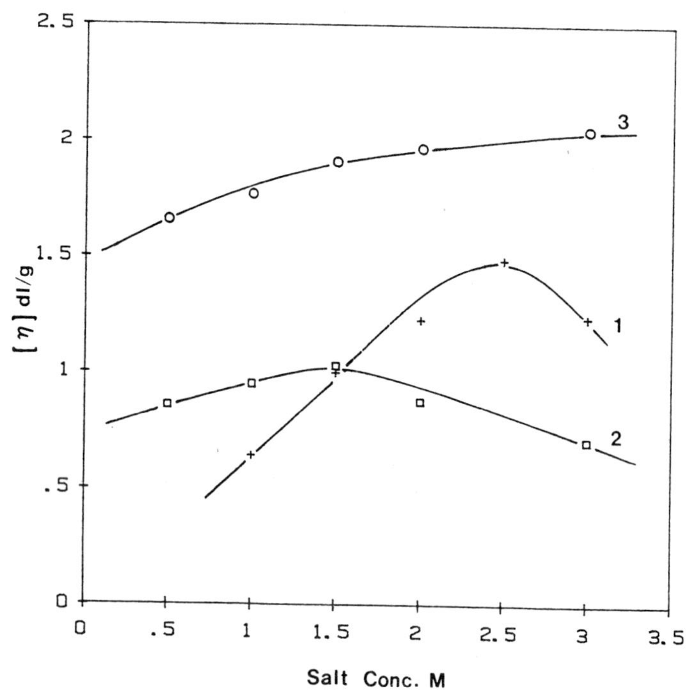

Figure 3. Intrinsic viscosity as a function of salt concentration for 1)poly(4VMP/pSS), 2)poly(MPTMA/AMPS), 3)poly(METMA/MES).

TABLE 2. Effect of Salt Type on the Intrinsic Viscosity and the Huggins Constant of Poly(METMA/MES).

Salt [1.0 M]	Charge/Radius Ratio Cation	Charge/Radius Ratio Anion	[η] dl/g	k´
LiCl	1.47	0.55	1.65	0.41
NaCl	1.03	0.55	1.77	0.38
KCl	0.75	0.55	1.80	0.39
NaCl	1.03	0.55	1.77	0.38
NaBr	1.03	0.51	2.03	0.30
NaI	1.03	0.45	2.08	0.30

zing the inter- and intramolecular ionic interactions, and hence increase the solvent power. The k´ values seem to support this view. In NaBr and NaI solutions the k´ values were 0.3, which are indicative of a thermodynamically good solvent (16,19,29). These results are in complete agreement with our previous work on poly(vinylimidazolium sulfobetaine)(6).

The Brookfield viscosity behavior was investigated to ascertain if these ampholytic systems were pseudoplastic or thixotropic. Figure 4 shows the Brookfield viscosities of the three systems as a function of shear rate. The three polymers behaved in a similar fashion. The viscosity dropped much rapidly at lower shear rates, apparently due to disentanglement of the chains and became constant (Newtonian) at higher shear rates, probably because of the ionic interactions which maintained the viscosity. With increasing salt concentration the viscosity increased due to the chain expansion as explained earlier. No thixotropic behavior was noted for any of the systems under the conditions employed.

Light Scattering Studies

It is known that classical light scattering theory can be applied to characterize a multicomponent macromolecular system specially an ion containing polymer in salt solution provided the polymer solution is dialyzed against solvent to obtain a similar chemical potential (34,36). Under this condition light scattering data can be evaluated by the Zimm technique. The Zimm method is very sensitive to such structural changes as intra- and/or inter-molecular association (36,37). Therefore, to supplement the viscosity studies it was decided to investigate the molecular characteristics of these polyampholytes by light scattering.

The Zimm plots obtained for all three polyampholytes in various salt solutions were typical, and showed only a slight scatter. Figure 5 is an illustration of a Zimm plot for poly(4VMP/pSS) in 1.5 M aqueous NaCl solution. A twisted Zimm plot was obtained for poly(MPTMA/AMPS) in deionized water. The light scattering data were inconsistent even for very dilute solutions (0.01% - 0.04%) owing presumably to the formation of aggregates. Therefore, the interpretation of data by a Zimm plot for poly(MPTMA/AMPS) in deionized water was not possible. Similar distorted Zimm plots are reported in the literature for

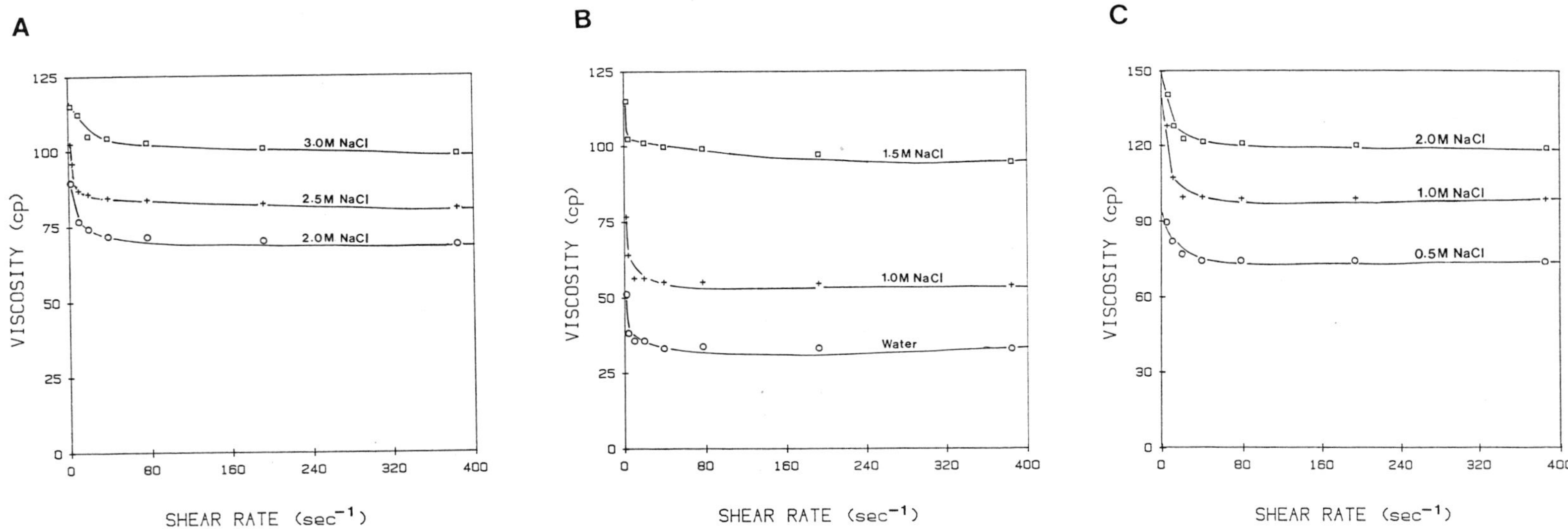

Figure 4. Effect of shear rate on apparent viscosity of 10%(w/v) solution of A)poly(4VMP/pSS), B)poly(MPTMA/AMPS), and C)poly(METMA/MES) in various salt solutions at 25°C.

polymers which show molecular association (36,37). The results obtained by the Zimm analysis of the three polyampholytes in salt solutions are given in Table 3. The light scattering data appears to be consistent with the viscosity data.

It is interesting to note that as the salt concentration is increased the molecular weight is decreased, but both the radius of gyration and the second virial coefficient are increased. The decrease in molecular weight with an increase in added electrolyte may be attributable to the breaking of agglomerated structures that are present due to the intermolecular ionic associations at low ionic strength of the medium. This also explains the drop in viscosity at very high ionic strength of the solvent since the $[\eta]$ is empirically related to $\bar{M}w$. The increase in the radius of gyration suggests that the individual polymer chains continuously expand as the ionic strength is increased, without degrading the polymer molecule itself.

Figure 6 gives a plot of the second virial coefficient, A_2, as a function of salt concentration for three ampholytic polymers (A_2 is a measure of total excluded volume effect). As reported for the other polyampholyte-salt systems (6,28), the value of A_2 for these polyampholytes also increased with increasing salt concentration, indicating an increased polymer-solvent interaction. The increase in A_2 values for poly(MPTMA/AMPS) and poly(METMA/MES) at lower ranges of the ionic strength of the medium is greater due to the breaking of agglomerated structures. As soon as the agglomerated structures are broken, the salt effect is slowed because the added electrolyte now only neutralizes intramolecular interactions.

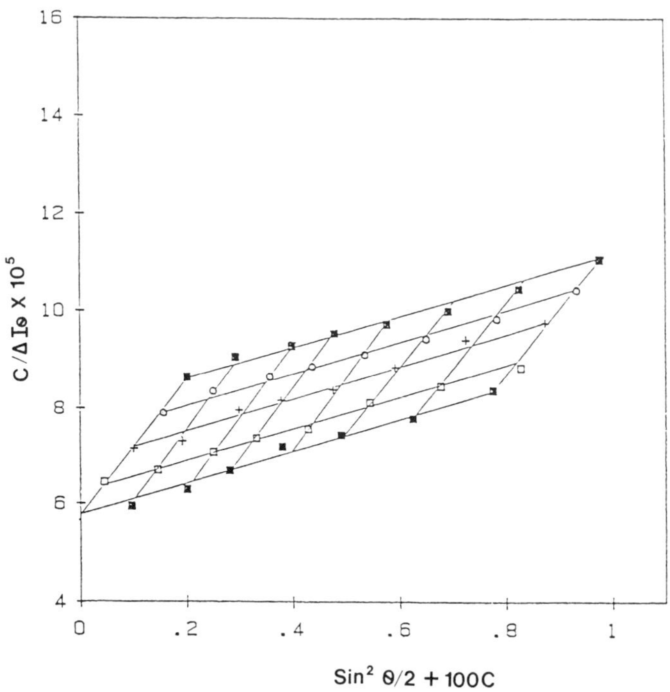

Figure 5. Zimm plot of poly(4VMP/pSS) in 1.5 M aq. NaCl solution.

TABLE 3. Molecular Characteristics of Polyampholytes in Various Salt (NaCl) Concentrations.

[NaCl] M	$\bar{M}w \times 10^{-6}$ g/mol	$A_2 \times 10^4$ ml mol/g^2	R_G Å	[η] dl/g
Poly(4VMP/pSS)				
1.5	1.78	0.7	485	1.00
2.0	1.43	1.1	483	1.23
2.5	1.59	1.5	608	1.49
3.0	1.25	1.7	587	1.24
Poly(MPTMA/AMPS)				
0.5	0.75	1.6	231	0.85
1.0	0.50	2.9	272	0.95
1.5	0.57	3.0	359	1.03
2.0	0.37	3.3	296	0.88
3.0	0.31	3.5	332	0.71
Poly(METMA/MES)				
0.5	3.11	0.1	502	1.66
1.0	2.57	1.8	721	1.77
3.0	2.20	2.4	717	2.05
5.0	2.00	3.3	812	2.14

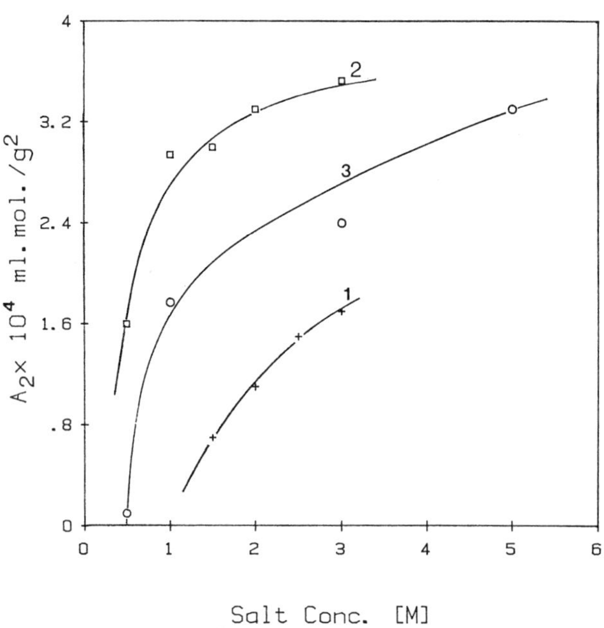

Figure 6. Second virial coefficient as a function of salt concentration for 1)poly(4VMP/pSS), 2)poly(MPTMA/AMPS), 3)poly(METMA/MES)

Additional studies for poly(METMA/MES) in various types of salts were undertaken in order to understand the site-binding abilities of the added mobile ions. The results are compiled in Table 4. The molecular parameters appear to confirm that with a salt of good site-binding ability (low charge/radius ratio), the solvent power of the medium is increased. Thus, in addition to a decrease in molecular weight in the order LiCl>NaCl>KCl>NaBr, both radius of gyration and second virial coefficient increase in the order LiCl<NaCl<KCl<NaBr. These results agree with the viscosity data.

TABLE 4. Molecular Parameters of Poly(METMA/MES) in Various Electrolytes

Electrolyte [1.0 M]	$\bar{M}w \times 10^{-6}$ g/mol	$A_2 \times 10^4$ ml mol/g^2	R_G Å
LiCl	2.70	1.3	670
NaCl	2.55	1.8	721
KCl	2.50	1.8	722
NaBr	2.27	2.2	754

CONCLUSIONS

This study confirms the concept that polyampholytes take an expanded conformation in aqueous salt solutions which is in contrast to typical polyelectrolyte behavior. Viscosity determinations in conjunction with light scattering studies has provided a general confirmation of the polyampholyte effect in the polymers derived from the ion-pair comonomer in aqueous salt solutions. This effect is related to the ion-binding capabilities of the added electrolytes.

ACKNOWLEDGMENT

This work was supported in part by the Office of Naval Research.

REFERENCES

1) For reviews, see articles presented in symposium entitled Polymers in Enhanced Oil Recovery, Polym. Prepr., 22(2), (1981).
2) G. Ehrlich and P. Doty, J. Am. Chem. Soc., 76, 3764 (1954).
3) R. Hart and D. Timmerman, J. Polym. Sci., 28, 638 (1958).
4) J. C. Salamone and S. C. Israel, Polym. Prepr., 12(2), 185 (1971).
5) J. C. Salamone, W. Volksen, S. C. Israel, and A. W. Wisniewski Appl. Polym. Symp., 26, 309 (1975).
6) J. C. Salamone, W. Volksen, A. P. Olson, and S. C. Israel, Polymer, 19, 1157 (1978).

7) J. C. Salamone, A. C. Watterson, T. D. Hsu, C.C. Tsai, and M. U. Mahmud, J. Polym. Sci., Polym. Lett. Ed., 15, 487 (1977).
8) J. C. Salamone, C. C. Tsai, A. P. Olson, and A. C. Watterson, J. Polym. Sci., Polym. Chem. Ed., 18, 2983 (1980).
9) J. C. Salamone, C. C. Tsai, A. P. Olson, and A. C. Watterson, "Ions in Polymers," A. Eisenberg, ed., Adv. Chem. Ser., 187, Chap.22 (1980).
10) J. C. Salamone, C. C. Tsai, A. C. Watterson, and A. P. Olson, "Polymeric Amines and Ammonium Salts," E. Goethals, ed., Pergamon Press, New York, 1980, pp 105-112.
11) J. C. Salamone, M. K. Raheja, Q. Anwaruddin, and A. C. Watterson, J. Polym. Sci., Polym. Lett. Ed., 23, 12 (1985).
12) J. C. Salamone, N. A. Mahmud, M. U. Mahmud, T. Nagabhushanam, and A. C. Watterson, Polymer, 23, 843 (1982).
13) J. C. Salamone, L. Quach, A. C. Watterson, S. Krauser, and M. U. Mahmud, J. Macromol. Sci.-Chem., A22, 653 (1985).
14) M. B. Huglin, "Light Scattering from Polymer Solutions," Academic Press, London, 1972, Chaps. 15&16.
15) B. H. Zimm, J. Chem. Phys., 16, 1099 (1948).
16) M. L. Huggins, J. Am. Chem. Soc., 64, 2716 (1942).
17) M. Bohdanecky and J. Kovar, "Viscosity of Polymer Solutions," A. D. Jenkins, ed., Elsevier Sci. Publ. Co., New York, 1982, Chap. 3.
18) J. M. Peterson and M. Fixman, J. Chem. Phy., 39, 2516 (1963).
19) T. Sakai, J. Polm. Sci., A2(6), 1535 (1968).
20) A. Einstein, Am. Phys. Leipzig, 19, 289 (1906).
21) R. Simha, J. Phys. Chem., 44, 25 (1940).
22) U. P. Strauss, and N. L. Gershfeld, J. Phys. Chem., 58, 747 (1954).
23) U. P. Strauss, N. L. Gershfeld, and E. H. Crook, J. Phys. Chem., 60, (1956).
24) C. Tanford, "Physical Chemistry of Macromolecules," J. Willey & Sons, Inc., New York, 1961, p.130.
25) C. Wolff, A. Silberberg, Z. Priel, and M. N. Layec-Raphalen, Polymer, 20, 281 (1979).
26) J. G. Watterson and H. G. Elias, Makromol. Chem., 157, 237 (1972).
27) V. M. M. Soto and J. C. Galin, Polymer, 25, 254 (1984).
28) D. N. Schulz, D. G. Peiffer, P. K. Agarwal, J. Larabee, J. J. Kaladas, L. Soni, B. Handwerker, and R. T. Garner, Polymer, 27, 1734 (1986).
29) H. Morawetz, "Macromolecules in Solution," 2nd ed., J. Wiley & Sons, Inc., New York, 1975, Chap. VI.
30) D. G. Peiffer and R. D. Lundberg, Polymer, 26(7), 1085 (1985).
31) H. Morawetz, "Macromolecules in Solution," 2nd ed., J. Wiley & Sons, Inc., New York, 1975, Chap. II.
32) U. P. Strauss and Y. P. Leung, J. Am. Chem. Soc., 87, 1476 (1965).
33) U. P. Strauss, C. Helfgott, and H. Pink, J. Phys. Chem., 71, 2555 (1967).
34) B. E. Boyd and K. Bunzl, J. Am. Chem. Soc., 96, 2054 (1974).
35) T. Ooi, J. Polym. Sci., 28, 459 (1958).
36) L. H. Tung, J. Polym. Sci., 2A, 4875 (1964).
37) Van R. Wijk and A. J. Staverman, J. Polym. Sci., A2, 1011 (1966).

THE AQUEOUS CONFORMATION AND SOLUBILITY OF POLYVINYLPYRROLIDONE IN RELATION TO THE USE OF POLYMERS IN OIL RECOVERY

Philip Molyneux

Macrophile Associates
53 Crestway, Roehampton
London SW15 5DB, U.K.

1. INTRODUCTION

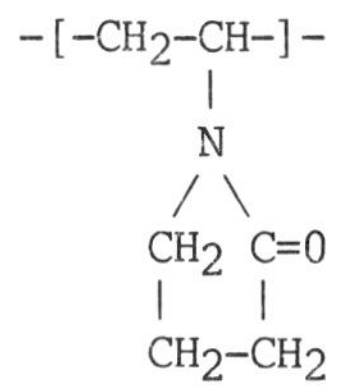

[I] PVP monomer unit

Polyvinylpyrrolidone (PVP) (monomer-unit structure [I]) is a water-soluble synthetic polymer which is stable physically and chemically in aqueous solution (1-3), and which is commercially available in a diversity of molecular weight grades (4). From these features alone, PVP has potential for use in oil-recovery and related applications; similarly, its monomer unit has potential as a component of copolymers tailored for these applications (5). In any such applications, it is important not only that the polymer remains in solution under all conditions of temperature, pressure and salinity that may be encountered, but that its solutions also retain the desired rheological behavior. Ensuring that this happens, especially with novel copolymers, requires that we have a good understanding of the molecular interactions which occur in these aqueous polymer systems.

The present paper reviews the effects of small-molecule solutes (i.e., cosolutes) on the aqueous solubility and conformation of PVP, both to demonstrate some of their practical consequences in the applications of this particular polymer, and also to illustrate some general features of aqueous polymer systems from the molecular interaction viewpoint.

2. POLYMER SOLUTION INTERACTIONS AND SOLUBILITY

Before considering the behavior of PVP specifically, it is necessary to make some general comments on molecular interactions in polymer solutions, to put the subject in its broader context. In the simplest analysis, the behavior of the system will be controlled at the molecular level by the variously named "exchange" or "dilution" or mixing process [1]:

$$M{\sim}M + S{\sim}S = M{\sim}S + M{\sim}S \quad [1]$$

where M is the monomer-unit of the polymer, S is the solvent molecule, and the symbol "$\sim$" indicates the particular combination of noncovalent interactions involved for each pair of species. More exactly, S should represent not just one solvent molecule, but the average number of solvent molecules which interact in this way with the monomer unit. The species $M{\sim}M$ then represents the so-called "long-range" interaction, that is, between two distantly connected parts of the same chain which are in contact at that moment, and equivalently between parts of two different polymer molecules which are in contact at that moment. The precise nature of the $M{\sim}M$ interaction is unclear even for relatively simple and well-studied polymers in non-aqueous systems. It may in fact involve a number of monomer units, which would enhance the overall interaction energy because of the summation of the forces involved, although at the same time there would be a penalty in entropy terms from the loss of conformational freedom of the chain units. We may view the process [1] as a quasi-chemical equilibrium, with an equilibrium constant which, because it is determined by the characteristics of all three underlying interactions, must therefore be expected to depend upon the natures of the polymer and the solvent as well as on temperature, pressure and the natures and concentrations of any other solutes in the system.

The process [1] goes to the right in the dissolution of the pure polymer, in the dilution of a polymer solution from above the critical concentration where the molecules are in contact, and in the expansion of an essentially isolated polymer molecule in dilute solution. Conversely, the contraction of an essentially isolated polymer molecule, or the precipitation of the polymer from dilute solution (where the "precipitate" is a more concentrated solution of the polymer) involves the process [1] going to the left. Thus anything that favors an increase in the number of M M contacts causes molecular contraction and then (possibly) precipitation. Because of this parallel between solubility and conformation, measurements of molecular dimensions, or equivalently (and more simply) of the intrinsic viscosity $[\eta]$, can be used to distinguish those effects which may eventually lead to the precipitation of the polymer from those which enhance the solubility of the polymer. Such conformational studies are particularly useful because solubility for polymers is very much an "all-or-none" phenomenon, so that superficially one cannot distinguish a system which is on the verge of precipitation from one well removed from this condition.

3. SOME GENERAL FEATURES OF PVP

As pointed out previously (1,3), the behavior of PVP is closely related to its amphiphilic character, since each monomer unit [1] contains nonpolar, hydrophobic groups (four methylene and one methine) together with a highly dipolar and hydrophilic amide group; dipole moment (1) and molecular dimension data (6) for the monomer unit analog N-methylpyrrolidone (NMP: structure [II], below) indicate that this dipole corresponds to around one-third of a unit negative charge on the oxygen atom and one-third of a positive charge on the nitrogenatom. One important aspect of the solution behavior of PVP is the hydrogen-bond accepting properties of the amide group. It is likely that this is confined to the oxygen atom, for the fractional positive charge on the

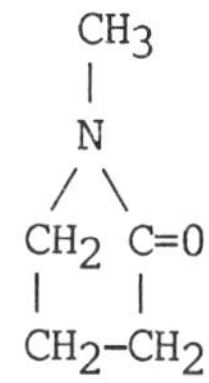

[II] N-Methylpyrrolidone (NMP)

nitrogen atom together with steric hindrance from the surrounding groups must inhibit the hydrogen-bond accepting properties of this atom. This is supported by the results of recent work (6) on the crystal structure of the 1:1 complex between NMP and hydroquinone (HOPhOH), where only the oxygen atom of the amide group is hydrogen-bonded to the hydroquinone hydroxyls, with the remaining hydroxyls preferring to bond to each other in a chain structure.

Although in the present review we are concerned specifically with the behavior of PVP in aqueous solution, it is significant that it is also soluble in a diversity of organic liquids, especially alcohols and partially halogenated alkanes (1). On the other hand, it is not soluble in aliphatic or alicyclic hydrocarbons, or in their fully halogenated derivatives such as carbon tetrachloride. However, PVP is soluble in the aromatic hydrocarbon benzene, so long as the system is absolutely anhydrous (7). This is evidently the result of strong dipole/induced-dipole forces between the polymer amide groups and the polarisable conjugated system of the solvent, which must also be an important factor in the marked binding of aromatic cosolutes by PVP in aqueous solution (Section 5.4).

We now turn to the behavior of PVP in aqueous solution, both alone (Section 4) and in the presence of cosolutes (Section 5).

4. PVP IN AQUEOUS SOLUTION

For PVP in aqueous solution at normal temperatures, the exponent in the Mark-Houwink-Sakurada equation (relating intrinsic viscosity to molecular weight) has a value of near to 0.65; recollecting that this exponent can range maximally (for flexible-chain nonionic polymers) from a lower limit of 0.5 for the theta solvent, to an upper limit of 0.8 for the best possible solvent, we see that water is a "medium" solvent for the polymer (1). Its solvent power decreases with increasing temperature, as is shown by the fall in the intrinsic viscosity, $[\eta]$, reflecting the contraction of the polymer coil through the formation of more M$\sim$M contacts. It follows that for this sytem, the exchange process [1] (Section 2) is exothermic, that is, the "hydration" of the monomer unit is stronger than the average of the monomer-monomer and the water-water interactions. Correspondingly, the concentration of a salt such as ammonium sulfate which is required to precipitate the polymer (Section 5.1) decreases with increasing temperature, showing that the cloud points are T_{p+} values (where the subscript positive sign denotes precipitation on increase of temperature). Extrapolation to zero salt concentration gives a theta-temperature (θ_+) of around 145 ^{o}C for PVP in water (1).

5. EFFECTS OF COSOLUTES ON PVP IN AQUEOUS SOLUTION

We follow our previous twofold classification of these phenomena (2), that is, into [a] "generic effects", which require cosolute concentrations of at least about 100 millimolar, and which are attributable to the general influence of the cosolute on the molecular interactions in the system, and [b] "specific effects", which appear at concentrations of less than about 100 millimolar, and which are attributable to binding (that is, complexing) of the cosolute molecule A by the polymer chain:-

$$A + M = A\sim M \qquad [2]$$

It should be noted that the generic effects discussed below for PVP with inorganic salts (Section 5.1) are seen also with many other water-soluble polymers, so that the principles outlined are applicable to these polymers in common. By contrast, the specific effects with surfactants (Section 5.3) and phenols (Section 5.4) are often much more pronounced with PVP than with other water-soluble polymers, as a result of the exceptional binding ability of PVP towards cosolutes as a whole.

The cosolutes will be considered under four main headings, as listed below.

5.1 Inorganic Salts

Many oil-drilling fluid formulations contain inorganic salts (necessarily so when sea-water is used as the make-up solvent), while there is generally a high saline content in the water formations of oil-bearing strata.

Simple uni-univalent salts such as NaCl have little effect on PVP even at the 1 molar level. Salts with polyvalent anions such as ammonium sulfate or trisodium phosphate are much more effective, contracting the polymer coil and ultimately precipitating the polymer at the 0.1 to 1 molar level. This behavior, which is seen with other nonionic water-soluble polymers, is closely analogous to the "salting-out" of small-molecule nonionic compounds (8). Since the precipitation generally occurs only at relatively high cosolute concentrations, i.e., above 0.1 molar, it seems to be a "generic efect" from the ions indirectly influencing the molecular interactions in the system. It is common to find these types of effect discussed in terms of the influence of the cosolute ions on the "water structure", i.e., "structure-breaking" or "structure-forming"; however, this only relates to the $S\sim S$ interactions, whereas it is essential to consider the effects on all three types of interaction ($M\sim M$, $M\sim S$ and $S\sim S$), since it is the balance between them that determines the equilibrium constant for the controlling process [1] (Section 2).

5.2 Nonionic Aliphatic Compounds

The commonest cosolutes of this type are aliphatic liquids which are miscible with water or at least appreciably soluble in it. Many of these, such as the aliphatic alcohols, are quite good solvents for the polymer, so that their effects on conformation or solubility are generally quite minor. Some other water-miscible organic liquids, notably acetone, can precipitate the polymer from aqueous solution; even acetone, however, is a solvent for the shorter chains, so that it is not a very efficient precipitant.

5.3 Ionic Surfactants and Other Organic Salts

Ionics surfactants (which are mostly long-chain aliphatic salts) may be used in conjunction with polymers in oil-recovery procedures, and may be present along with water-soluble polymers in oil-drilling fluids and similar formulations.

Anionic surfactants as cosolutes have a marked effect on PVP in solution, causing large increases in $[\eta]$ which are equivalent to an increase in solubility. Similar effects are seen with simple aromatic anions (benzoate, naphthoate, etc.). These effects are attributed to the binding of the anions, as shown by such experimental techniques as equilibrium dialysis. This converts the chains into polyanions, with the consequent repulsions between the bound ions causing the chain expansions observed.

Addition of a nonbound salt (such as NaCl) to the anionic surfactant system reduces the molecular expansion of the polymer because the added counterions (i.e., Na^+) screen the repulsions between the bound surfactant anions.

Cationic surfactants, by contrast, show much smaller effects than anionic ones, which reflects the much weaker binding of organic cations. This difference in the behavior of the two charge-types seems to be related to the different environments of the two ends of the amide dipole (Section 3). The oxygen atom, although it is exposed, is surrounded by hydrating water molecules which apparently shield its fractional negative charge from cationic cosolutes. On the other hand, the nitrogen atom, although surrounded by nonpolar groups, is thereby put into an environment of low dielectric constant so that its fractional positive charge can still attract anionic cosolutes.

5.4 Phenols and Related Nonionic Aromatic Compounds

These cosolutes are important in the present context since phenolic compounds are commonly included as antimicrobial agents or as antioxidants in aqueous formations which may also contain a water-soluble polymer such as PVP.

Phenols are well known for their ability to precipitate water-soluble polymers of all kinds - natural, modified-natural (semi-synthetic), and purely synthetic. This is shown most markedly by polyhydric phenols such as tannin (tannic acid). As would be expected from the preceding discussion (Section 2), precipitation is foreshadowed at lower cosolute concentrations by contraction of the polymer coil, i.e., a reduction in $[\eta]$.

In the case of PVP, we have shown (9,10) that these effects may be interpreted as a two-step process, that is, the binding of the phenol by the polymer, followed by the reversible cross-linking of the polymer chain by the bound cosolute molecules. which leads effectively to an increase in the number of M$\sim$M contacts in the system. The distinction between the two steps is made clear by the behavior of the corresponding hydroxyethyl derivatives. For example, 2-phenoxyethanol is bound about as strongly as phenol itself, but the bound cosolute molecule shows only a very weak cross-linking ability; the consequence is that the hydroxyethyl compound is unable to precipitate the polymer even at its saturation concentration of 200 millimolar, at which it causes only a 10% reduction in $[\eta]$, showing that the system is still far from precipitation. By contrast, with phenol precipitation occurs at about 80 millimolar, where $[\eta]$ is reduced by approximately 90%. The fact that phenoxyethanol cross-links so weakly, even although like phenol it still carries a hydroxyl group able to act as a hydrogen-bond donor, shows the importance of the delicate balance between the various noncovalent interactions. Evidently, the higher acidity of the hydroxyl group in phenol (pK_a 10.0) compared with that in phenoxyethanol (estimated pK_a 14.7) is sufficient to make the phenolic hydrogen a marginally stronger hydrogen-bond donor. This is confirmed by the behavior of 4-nitrophenol, which is even more acidic (pK_a 7.15), and which shows an even lower value (17 millimolar) for its critical precipitation concentration with PVP.

It is possible that these conformational and solubility effects could be put to good use for rheology control. For example, adding a phenol to a system containing the polymer and an anionic surfactant would be expected to reduce, neutralise or even reverse the viscosity-enhancing effect of the bound anions. However, care would have to be taken not to add so much phenol that it overwhelms the surfactant and precipitates the polymer.

REFERENCES

1. P. Molyneux, "Water-Soluble Synthetic Polymers: Properties and Behavior", vol I, CRC Press, Boca Raton, FL (1983), chapter 4.
2. P. Molyneux, "Water-Soluble Synthetic Polymers: Properties and Behavior", vol II, CRC Press, Boca Raton, FL (1984), chapter 2.
3. P. Molyneux, The physical chemistry and pharmaceutical applications of polyvinylpyrrolidone, in: "Proceedings of the International Symposium on Povidone", G. A. Digenis and J. Ansell, eds., University of Kentucky, Lexington, KY (1983), p. 1.
4. P. Molyneux, Polyvinylpyrrolidone - some problems and pitfalls, Brit. Polym. J. 18:32 (1986).
5. D. H. Lorenz, E. P. Williams and H. S. Schultz, 'Increasing Molecular Weight of Vinylpyrrolidone Polymer', US Patent 4,190,718, February 26, 1980 (assigned to GAF Corp).
6. P. Behmel and G. Weber, Untersuchungen zur Struktur des Adduktes aus Hydrochinon und Methylpyrrolidon-(2), J. Mol. Struct., 116:271 (1984).
7. M. J. Lyndrup and R. F. Quadrel, A viscometric and light scattering determination of the solution properties of poly (N-vinylpyrrolidone), Polym. Prepr. Amer. Chem. Soc., Div. Polym. Chem., 16:675 (1975).
8. F. A. Long and W. F. McDevit, Activity coefficients of non-electrolytes in aqueous salt solutions, Chem. Rev., 52:119, (1952).
9. P. Molyneux and S. Vekavakayanondha, The interaction of aromatic compounds with polyvinylpyrrolidone in aqueous solution. Part 5. Binding isotherms for phenols and O-substituted phenols, J. Chem. Soc. Faraday Trans. 1, 82:291 (1986).
10. P. Molyneux and S. Vekavakayanondha, The interaction of aromatic compounds with polyvinylpyrrolidone in aqueous solution. Part 6. Polymer precipitation and viscosity studies with phenols and O-substituted phenols, J. Chem. Soc. Faraday Trans. 1. 82:635 (1986).

SIZE CHARACTERIZATION OF EOR POLYMERS IN SOLUTION

R. D. Hester and A. D. Puckett

Department of Polymer Science
University of Southern Mississippi
Hattiesburg, MS 39406

INTRODUCTION

To be successful in the flooding processes used in enhanced oil recovery, polymer molecules must have large hydrodynamic size in aqueous solution. Large hydrodynamic size permits high aqueous solution viscosity at low polymer concentrations. When high polymer concentrations are required to enhance solution viscosity reservoir flooding is not economically attractive. Typical flooding processes use polymer solutions having concentrations of one-half to one gram per liter (500 to 1000 ppm). At these concentrations, solution viscosities from 5 to 10 times that of water must be obtained for sufficient mobility control during flooding.

Economic restraints and performance requirements force the use of inexpensive water soluble polymers having average hydrodynamic diameters ranging from 1500 to 2000 Angstroms. EOR polymers must have molecular weights in the millions to reach this size. Macromolecules this large present special characterization problems, particularly, determining hydrodynamic size in solution and/or molecular weight. It is the purpose of this paper to briefly review several analytical methods used for EOR polymer characterization and to emphasize some of the problems associated with each method. The methods discussed include solution viscometry, size exclusion chromatography, and dynamic light scattering.

Before we can characterize EOR polymers we must prepare an aqueous solution. This apparently simple task can sometimes be difficult. The rate at which very large polymers go into solution can be so low as to require several days of gentle agitation to obtain true solution equilibrium conditions. Harsh agitation can physically break macromolecules and over long time periods some polymers can be degraded by microorganisms. If polymer degradation is not present during solution preparation the approach to solution equilibrium can be monitored by observing the change in solution viscosity with time. If true equilibrium solution conditions do not exist then the polymer aggregates remaining will falsify all characterization results made with these solutions.

SIZE CHARACTERIZATION BY SOLUTION VISCOMETRY

Viscometry is inexpensive, simple and the most widely used method to determine polymer hydrodynamic size. However special problems are present when characterizing EOR polymers. These problems will become apparent as we outline the relationships that quantify this method.

For spheres in dilute solution which are non interacting and do not perturb the fluid flow field, Einstein showed that the solution viscosity, η, is related only to the solvent viscosity, η_o, and the volume fraction spheres in solution, ϕ[1].

$$\eta = \eta_o [1 + 2.5\phi] \tag{1}$$

If we consider polymer molecules in solution to be coils which are approximately spherical, then the volume fraction can be calculated as

$$\phi = \frac{CvN_A}{M} \tag{2}$$

where C is the polymer concentration, M is the polymer molecular weight, N_A is Avogadro's number and v is the volume occupied by a single polymer molecule and the solvent within the coil.

If we combine the above two equations, and rearrange terms we can show that

$$[(\eta/\eta_o)-1]/C = 2.5 \text{ v } N_A/M \tag{3}$$

The left term of equation (3), called the reduced viscosity, can be experimentally determined by measuring solution viscosities at very low solute concentrations which minimize polymer coil interaction. In addition low shear rate conditions must be maintained during the viscosity measurements or the polymer coils will be perturbed by the flow field. At the limits of "zero" concentration and "zero" shear rate, the reduced viscosity is defined as the intrinsic viscosity $[\eta]$.

$$[\eta] \equiv 2.5N_A v/M \quad \text{as} \quad C \rightarrow 0 \tag{4}$$

Therefore, under dilute solution conditions, equation (1) can be expressed as:

$$(\eta/\eta_o) - 1 = [\eta]C \tag{5}$$

Equation (5) shows that the left term, which is called the specific viscosity (ηsp), is directly proportional to the product of intrinsic viscosity and concentration. This product is referred to as the "coil overlap parameter" $\Phi = C[\eta]$. Polymer molecules with high ratios of hydrodynamic volume to molecular weight (large intrinsic viscosities) will have high solution viscosities. Thus, if we can expand the hydrodynamic coil size without increasing polymer molecular weight we will increase solution viscosity without increasing polymer concentration. This has been accomplished by using polymers made up of monomers having charges (polyelectrolytes) which repel one another and thus expand the hydrodynamic coil[2]. Also, coil size has been increased by producing polymers having monomer structures which sterically interfere with one another and thus force coil expansion[3].

Equation (5) applies only to very dilute solutions in which the coil overlap parameter is less than about 0.1. At higher overlap parameters, interaction between polymer coils becomes significant and equation (5) must be expanded to account for coil-coil interactions.

$$\eta_{sp} \cong \Phi + \tfrac{1}{2}\,\Phi^2 + \ldots \qquad (6)$$

Figure 1 shows the usual relationship between specific viscosity and overlap parameter. This figure shows that for EOR polymer solutions, which must have specific viscosity between 5 and 10 for mobility control during flooding, the overlap parameter must be about 3. Also, economics dictates that EOR polymer flooding solution concentrations must be 1000 ppm or less. Therefore a polymer must have an intrinsic viscosity of about 30 dl/g to be an effective and economical EOR candidate.

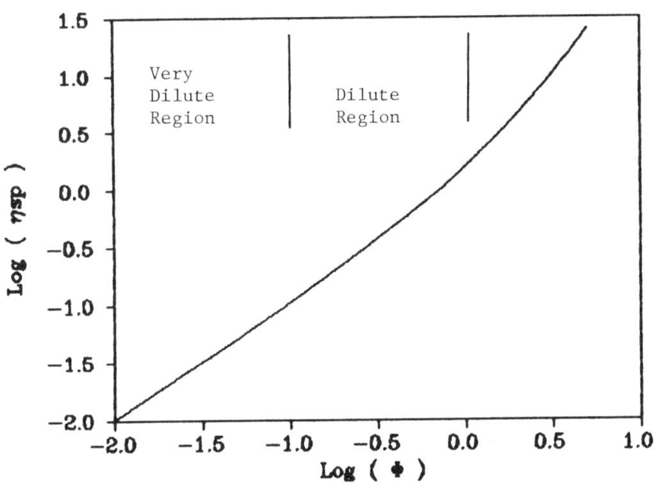

Figure 1. Polymer Solution Specific Viscosity (η sp) Versus Overlap Parameter (Φ).

Measurement of intrinsic viscosities which are less than 5 dl/g are not difficult. Solution and solvent viscosities can be determined with simple and inexpensive capillary viscometers. These viscometers usually make measurements at fluid shear rates greater than 50 sec^{-1}. These shear rates do not greatly perturb small polymer coils. However, large polymer coils are perturbed by large-fluid flow shear fields and equation (1) would not be valid. For EOR polymers with larger intrinsic viscosities, more sophisticated rheometers must be used which can make viscosity measurement in the low shear rate region which is $< 1 \text{ sec}^{-1}$. When measurements of EOR polymer solution viscosity are made by simple capillary viscometers, the shear rate conditions are too high. If corrections are not made, the apparent intrinsic viscosities determined by simple capillary viscometers will be less than the true intrinsic viscosities. Fortunately, an inexpensive four bulb shear dilution capillary viscometer which measures apparent solution viscosities at four shear rates can be used to make these corrections. The viscosity data at four shear rates are extrapolated to "zero" shear rate. The extrapolation is non-linear but can be accomplished by fitting the shear rate viscosity data to a rheological equation describing the viscosity dependence on shear rate[4]. The extrapolated "zero shear" solution viscosities can be then used in the standard concentration versus reduced viscosity plot to estimate the true intrinsic viscosity which is the reduced viscosity at zero concentration.

A second major problem that causes error in EOR polymer intrinsic viscosity measurements concerns the uncertainty that dilute solution conditions exist during viscosity measurements. As shown by Figure 1, dilute solution conditions exist only when the overlap parameter is less than one. Extrapolation to "zero" concentration reduced viscosity is linear only if viscosity data is taken in the dilute region. However, a polymer solution having an intrinsic viscosity of 30 dl/g is dilute only if the concentration is less than 1/30 g/dl (33 ppm). At concentrations higher than this, excessive coil overlap may exist. Solution reduced viscosities determined above the dilute solution region can not be easily extrapolated to zero concentration. When linear extrapolation is done, the apparent intrinsic viscosities which are obtained are larger than the true intrinsic viscosity.

In conclusion, polymer hydrodynamic size estimations from intrinsic viscosity measurements can be in error because shear rate and/or concentration conditions are too high. If intrinsic viscosities are correctly measured, then a viscous average molecular weigh can be determined from the Mark-Houwink equation provided that "K" and "a" values are known.

$$M_v = (K/[\eta])^{1/a} \tag{7}$$

However, for most EOR polymers, these values are unknown because the required high molecular weight polymer standards of known molecular weight do not exist and therefore the "K" and "a" values cannot be experimentally determined.

POLYMER CHARACTERIZATION BY SIZE EXCLUSION CHROMATOGRAPHY

A very wide distribution of molecular weights or hydrodynamic coil sizes are usually found within each EOR polymer solution. Because of this polydispersity, the molecular weights or hydrodynamic sizes determined by viscosity measurements are only averages. No information on macromolecular polydispersity is obtained from viscosity measurements. However, characterization of polydispersity is important because EOR polymers which have molecular weight or coil size distributions containing a large proportion of smaller macromolecules are not economical in a flooding process. Smaller polymer coils within the distribution are less effective in enhancing solution viscosity than larger coils.

For example, if we have two dilute polymer solutions at the same polymer concentration, but solution one contains polymer molecules having a molecular weight one-half that of the polymer in solution two, then the ratio of specific viscosity for the two solutions can be determined from equation (5).

$$\eta_{sp2}/\eta_{sp1} = [\eta]_2/[\eta]_1 \tag{8}$$

Both intrinsic viscosities can be related to molecular weight by using the Mark-Houwink equation.

$$\eta_{sp2}/\eta_{sp1} = (K_2 M_2^a)/(K_1 M_1^a) \tag{9}$$

If the polymer used in both solutions differ only by their molecular weight, then $K_1 = K_2$ and $M_2 = 2M_1$ and thus $\eta_{sp2} = 2^a \eta_{sp1}$. Values for "a" range from 0.5 to 1.0. Therefore the specific viscosity of the higher molecular weight polymer solution would be $\sqrt{2}$ to 2.0 times larger than the specific viscosity of the lower molecular weight solution. Thus, about 30% more mass of low molecular weight polymer would have to be used to make a solution having the same viscosity as the high molecular weight polymer solution. Hence, an economic incentive exists to produce EOR polymers with narrow polydispersities made up of mostly larger macromolecules.

Although larger molecular sizes are required to give the solution viscosity increase needed for mobility control, when the polymer coils are too large, they can be physically trapped in the pores at the face of the injection well[5]. The resulting face pluggage increases the injection pressure required to continue forcing polymer solution into the reservoir. An excessive pressure increase may force discontinuation of reservoir flooding.

Thus a control and knowledge of a polymer's hydrodynamic size distribution is necessary to achieve optimum flooding conditions. One method to determine the distribution of hydrodynamic sizes is size exclusion chromatography (SEC).

In SEC analysis, molecules are separated according to their hydrodynamic size. This separation is accomplished by using a solvent to force a polymer sample in solution through columns packed with porous particles. Smaller size polymer molecules are retained for a longer time within the columns because a greater fraction of the fluid volume located within the packing particles is accessible to smaller molecules. Some type of detector is used to sense the existence of polymer molecules in the eluent stream as it leaves the columns. A plot of detector signal versus the total mass or volume of eluent that has passed through the column since the injection of the polymer sample is called an SEC chromatogram. From the chromatogram shape and time of elution, a molecular weight and/or size distribution of the polymer sample can be determined.

To assure efficient SEC separation, the number of pores within the packing must be large when compared to the number of polymer molecules. Therefore, the total mass of polymer sample injected must be small. In addition, the concentration of the injected polymer should be low to assure dilute solution conditions. If dilute conditions exist, each polymer coil will occupy an independent volume domain. As a result, SEC separation of individual coils occurs and not separation of coil aggregates.

Typically, the polymer solution injected should not be greater than one-half the reciprocal of its intrinsic viscosity. Under this condition, polymer molecules do not interact and dilute solution conditions exist. It becomes obvious that, as the molecular weight or intrinsic viscosity of the polymer increases, the sample concentration injected must decrease. For example, if a polymer sample has an intrinsic viscosity of 30 dl/g, the maximum polymer concentration that should be injected is 0.017 g/dl (170 ppm).

Because of dispersion and size separation, a polymer sample may be diluted by a factor of 50 during its flow through the columns. Therefore, the polymer concentration in the eluting stream at the detectors is only 3 or 4 ppm. Usually in organic SEC fluorescence or ultraviolet detectors are used at these low polymer concentrations. However, due to the lack of active polymer chromophores, these detectors are not usable in aqueous SEC and a refractive index detector (RI) must be employed.

An RI detector measures the eluent refractive index. Hence, both temperature and solvent compositional changes in the eluent, as well as the presence of polymer solute, will be detected without discrimination. Unfortunately, small changes in solvent composition or temperature may result in a large RI response which may completely hide or alter the modest signal due to the small presence of polymer. Thus, many chromatograms obtained by using RI detectors are not usable due to the presence of phantom signals and/or highly unstable detector baselines. As an alterative to RI detectors in aqueous SEC, a continuous capillary pressure (CP) detector has been developed to measure eluent viscosity[6].

A CP detector senses the pressure drop developed when a fluid flows through a long, small diameter capillary. When polymer is present in the fluid, the solution pressure drop across the capillary, P, is greater than the pressure drop of pure solvent, P_o. The larger the volume fraction of polymer in the fluid, the larger the pressure drop. For laminar flow conditions where the pressure drop is proportional to the fluid viscosity, this pressure increase can be used to determine the specific viscosity of the eluent, $\eta_{sp} = (P-P_o)/P_o$. The specific viscosity is directly proportional to the coil overlap parameter which is in turn equal to the volume fraction polymer in the eluent stream. A plot of eluent specific viscosity versus elution volume produces a specific viscosity chromatogram. The apparent intrinsic viscosity of a polymer sample is the area within its specific viscosity chromatogram divided by the total mass of polymer sample injected. True intrinsic viscosities can be found by making shear rate and concentration corrections to the apparent viscosities.

Larger molecules, which have larger intrinsic viscosities, will have larger specific viscosity chromatograms. This detection characteristic gives an advantage to CP detectors. Since an RI detector's signal is proportional to the mass of sample injected and because the concentration of polymer must be decreased as the polymer molecular weight increases (to insure dilute solution conditions), an RI signal will decrease as the molecular weight increases. However, because the specific viscosity increases with molecular weight, the CP detector signal will increase with increasing molecular weight. This helps to offset the decrease in CP signal due to the reduction of polymer concentration.

In addition, the CP detector is relatively insensitive to small temperature fluctuations and trace quantities of low molecular weight impurities within the eluent which cause RI detector signal instabilities. Thus a CP detector has inherent advantages over a RI detector and has the potential to become the detector of choice for aqueous SEC analysis of EOR polymers.

Besides detection difficulties in aqueous SEC characterization of EOR polymers, problems concerning the absence of efficient, large pore-size packing materials limit the upper polymer coil sizes that can be separated. For adequate SEC separation, the packing must have many pores which are several times larger than the hydrodynamic diameter of the largest polymer coils. Thus pores as large as 1,000 nm in diameter are required. Only a few packing materials can even approach these requirements. Sephacryl S1000 packing, a crosslinked gel, has been used with some success and apparently can be used for EOR polymers which have hydrodynamic diameters less than 250 nm[7]. Other packing materials made from porous glass or silica gel have less success in separating EOR polymers. Better packing materials with greater separation resolution are needed for more effective SEC characterization of EOR polymers.

Calibration of a packed column also presents difficulties when using SEC to characterize EOR polymers. No large molecular weight, monodispersed water-soluble polymer standards are available to relate elution volume to molecular weight. Consequently, the usual SEC calibration methods are not applicable and special calibration techniques must be applied.

Recently, the SEC elution volume V_e penetrated by a polymer molecule having a hydrodynamic diameter d was described by[8]

$$V_e = V_o + V_p/\mathrm{EXP}[d/D] \qquad (10)$$

The three parameters in equation (10) are the interstitial volume V_o, the total packing micropore volume V_p, and the most dominant pore entry size D. The volumes V_o and V_p can be experimentally obtained. Polymer hydrodynamic diameters can be determined by dynamic light scattering which will be discussed in the next section of this paper. As equation (10) shows, if $\ln[(V_e-V_o)/Vp]$ values are plotted versus d values, a linear SEC calibration curve should be obtained. The calibration curve will have a zero intercept and a slope equal to $-1/D$. See Figure 2. The dominant pore diameter, D, is approximately equal to the ratio of packing micropore volume to surface area[9]. Hence an SEC calibration can be accomplished by using dynamic light scattering to measure polymer hydrodynamic size.

Dynamic light scattering (DLS) (sometimes referred to as intensity-fluctuation spectroscopy, self beating spectroscopy, or quasielastic light scattering) depends upon the analysis of the spectrum of light scattered from a dilute polymer solution. In a typical experiment, scattered light is frequency (Doppler) shifted from the incident laser light frequency due to rotational, translational, and internal motion of the macromolecules. The scattered light spectrum depends upon the time scales characterizing the motions of the macromolecule. Homodyne digital autocorrelation uses scattered light spectral intensity fluctuations to compute a time-based autocorrelation function.

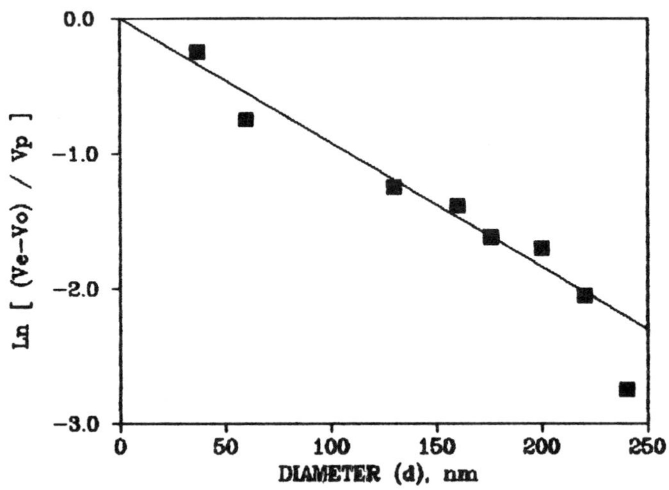

Figure 2. Calibration Curve os Sephacyl S1000 Packing. See Reference 7.

The reciprocal of scattering field vector, defined by equation (11), characterizes scattering conditions.

$$S = \lambda/[4\pi \sin(\theta/2)] \tag{11}$$

In equation (11), θ is the scattering angle, and λ is the radiation wavelength in the polymer solution. The scattered light spectrum produced by a polymer solution depends upon the mass, polarizability and number of polymer coils in the solution, and the magnitude of the scattering field vector.

If the solution is too concentrated, macromolecular entanglements will scatter radiation at different time scales than single polymer coils. To prevent this, DLS is done on dilute solutions and results are extrapolated to zero concentration conditions.

When the reciprocal of the scattering vector , S, is large in comparison to the size of the polymer coil, the DLS fluctuation spectrum is due to macromolecular coil center of mass motions. These motions are generated by translation (Brownian) diffusion and/or any bulk fluid flow of the solution. Unwanted scattering can be produced by slight thermal differences or vibration in the solution which can cause bulk fluid flow.

If the scattering vector is increased then other macromolecular motions such as chain rotation will enter into the DLS spectrum. These motions, which are much faster than motions due to translational diffusion, can cause error if present in the spectrum.

Autocorrelation is used to analyze the scattered light intensity spectrum. An autocorrelation function is developed from the spectrum by an autocorrelator, a specialized high speed computer. The autocorrelation function provides a means of determining the pattern or periodic features existing in the spectrum.

Mathematically, the autocorrelation function $G(\tau)$ at delay time τ is defined as:

$$G(\tau) = \lim_{T \to \infty} \frac{1}{2T} \int_{-\infty}^{+\infty} I(t)\, I(t + \tau)dt \tag{12}$$

In equation (12), $I(t)$ is the scatter light intensity signal at time t and $I(t + \tau)$ is the signal delayed in time, τ , from signal $I(t)$. See Figure 3.

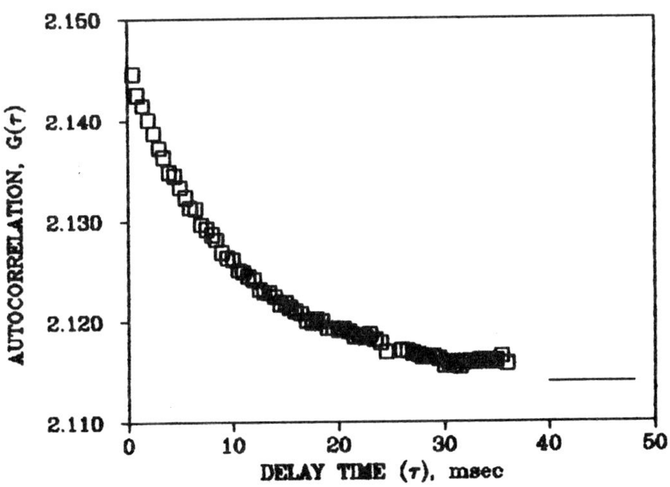

Figure 3. Typical Dynamic Light Scattering Autocorrelation Plot.

For a solution of homodispersed polymer molecules, the autocorrelation function generated will be expressed as a single exponential

$$G(\tau) = AB[EXP(-\Gamma\tau)]^2 + B \tag{13}$$

where A is a machine constant, B is the noise or base line correlation contribution and Γ the decay constant. The Γ value can be extracted from a plot of

$$\ln[(G(\tau) - B)/(AB)]^{\frac{1}{2}} \quad \text{versus } \tau$$

The value $\sqrt{[G(\tau)-B]/[AB]}$ is defined as the normalized autocorrelation function $g(\tau)$. The translational diffusion coefficient can be calculated from Γ and the scattering vector.

$$D = \Gamma S^2 \tag{14}$$

In turn, the translational diffusion coefficient for homodispersed polymer molecules in a solvent at infinite dilution is related to an equivalent hydrodynamic diameter d through the Stokes-Einstein equation

$$d = kT/(3\pi\eta_o D_o) \tag{15}$$

In the above equation, k is the Boltzman constant, T is the absolute temperature, η_o is the pure solvent viscosity and D_o is the "zero" concentration macromolecular translational diffusion coefficient i.e., the diffusion coefficient at infinite dilution.

For a system of homodisperse molecules, the value of a single decay constant can be determined from equation (13), a single exponential. However, scattering from a solution of polydisperse molecules results in a distribution of exponentials; consequently, analysis of DLS data must be made using a probability function which accounts for the distribution of molecular sizes.

For a system of polydispersed macromolecules undergoing only translational diffusion the normalized autocorrelation function, $g(\tau)$ is related to a sum of exponents which is expressed by a Fredholm integral equation

$$g(\tau) = \int_0^\infty F(\Gamma)\mathrm{EXP}[-\Gamma\tau]d\Gamma \tag{16}$$

where F (Γ) is the normalized probability distribution of decay constants such that

$$\int_0^\infty F(\Gamma)d\Gamma = 1 \tag{17}$$

We must solve the integral equation for $F(\Gamma)$. A solution is not trivial because the equation is "ill conditioned." For even small levels of noise in the scattering data, there exists a large set of possible $F(\Gamma)$ solutions. Several numerical methods have been developed to find the "best" $F(\Gamma)$ solution. Some of the methods include constrained regularization[10], subdistribution,[11,12] Fourier transform[13] and function fit techniques[14]. These methods are complicated and difficult to apply but can produce acceptable $F(\Gamma)$ solutions.

If the distribution $F(\Gamma)$ can be determined, conversion to a molecular weight distribution is usually desired. For a homologous series of polymers, that is polymers of different molecular weights but consisting of the same monomers,

$$D_o = \alpha M^{-\beta} \tag{18}$$

where α and β are constants whose value depends on temperature and the solvent-polymer system. These values can be determined experimentally; however, they can be estimated from Mark Houwink "K" and "a" values ($\beta = [a + 1]/3$).

An uncomplicated solution technique that can be applied to equation (16) is the method of cumulants. In 1972 Koppel showed that the logarithm of the normalized autocorrelation function was identical to the cumulant generating function for the distribution of decay constants[15]. The coefficients of the cumulant expansion can be related to the moments of the $F(\Gamma)$ distribution. The Koppel equation can be expressed by

$$\ln[g(\tau)\sqrt{AB}] = \ln[\sqrt{AB}] - \bar{\Gamma}\tau + \tfrac{1}{2}\sigma^2\tau^2 + \cdots \tag{19}$$

where $\bar{\Gamma}$ is the mean decay constant and σ^2 is the variance of $F(\Gamma)$.

From the complete set of moments it is theoretically possible to regenerate the exact distribution function. In practice however, even small amounts of noise in the data can cause large error in all but the first two or three moments.

The mean decay constant $\bar{\Gamma}$ is related to the "z" average translational diffusion coefficient, D_z

$$D_z = \bar{\Gamma} S^2 \tag{20}$$

Under the conditions of no interactions between molecular (infinite dilution), no intramolecular interference (small scattering vector) and no dependence of polarizability per unit mass of molecule on molecular weight then:

$$D_z = [\Sigma DiNiMi^2]/[\Sigma NiMi^2] \tag{21}$$

where Ni and Di give the number and translational diffusion coefficient of all molecules having molecular weight Mi.

The infinite dilution "z average" translational diffusion coefficient may be used in equation (15) to calculate an effective coil diameter. This diameter which represents the effective spherical dimension of the polymer molecules undergoing translation has been used to calibrate SEC systems[16].

An average molecular weight can be determined by combining intrinsic viscosity and translational diffusion information in the Mandelkern-Flory relationship

$$M = \left[\frac{kT(\Phi^{1/3}/P)}{D_o \eta_o [\eta]^{1/3}}\right]^3 \tag{22}$$

In addition to values for D_o and $[\eta]$, accurate values for the hydrodynamic coefficients Φ and P are required to use equation (22). The individual values of Φ and P vary unpredictably for solvent-polymer solution combinations. However several investigators have observed that the quantity ($\Phi^{1/3}/P$) is invariant for many polymer-solvent systems. After an extensive literature review, Tsvetkov, et al[17] have suggested that ($\Phi^{1/3}/P$) be assigned a value of 2.3 X 10^6 mole$^{-1/3}$ for flexible polymer chains and a value of 2.68 X 10^6 mole$^{-1/3}$ for rigid chain polymers. The average molecular weight determined by equation (22) is approximately equal to geometric mean of the weight average and z average molecular weights, i.e. $M = \sqrt{M_w M_z}$[18].

There are several experimental problems encountered when DLS is applied to EOR polymers. Dust particles, always present in the aqueous polymer solution, make it difficult to obtain accurate and/or reproducible scattering results. The large hydrodynamic size of EOR polymers hampers efforts to remove only dust by filtration and/or centrifugation techniques. Another problem associated with a large polymer size is that light scattering is very angle dependent. Therefore it is necessary to make scattering measurements at very low angles or to extrapolate measurements made at high angles.

Although DLS appears to be promising for EOR polymer characterization it has not been extensively tested. Only a few high molecular weight polymer-water systems have been investigated by DLS. The few systems studied include polymers of xantham polysaccharides[19], hydrolyzed polyacrylamide[20] and acrylamide copolymers[21,22].

CONCLUSION

There are inherent theoretical and experimental problems associated with all the available methods for size characterization of EOR polymers. Of the methods discussed in this paper, dynamic light scattering appears to have the greatest potential for characterization of an average molecular size. Theoretically, the method is applicable to any polymer-solvent combination and is not bounded by a upper polymer size limit. DLS measurements of diffusion coefficients can be combined with intrinsic viscosity measurements to obtain a molecular weight using the Mandelkern-Flory relationship. Currently DLS is used primarily to obtain an average size for polydispersed polymers. However DLS has the potential to extract information about the distribution of macromolecular sizes in a EOR polymer. This information will allow an understanding of the effects that EOR polymer size distributions and macromolecular compositions have on polymer solution properties.

References

1. A. Einstein, Ann. Physik., 19, (4), 289 (1906).
2. C. L. McCormick, K. P. Blackmon, J. Macromol. Sci., A23, (12), 1451 (1986).
3. C. L. McCormick, B. H. Hutchinson, Polymer, 27, (4), 623 (1986).
4. P. J. Carreau, D. DeKee, M. Daroux, Can. J. Chem. Eng., 57, 135 (1979).
5. K. D. Dreher, W. B. Gogarty, J. Rheology, 23 (2), 209 (1979).
6. C. E. Lundy, R. D. Hester, J. Liq. Chrom., 7 (10), 1911 (1984).
7. C. E. Lundy, R. D. Hester, J. Poly. Sci. A, 24, 1829 (1986).
8. R. D. Hester, P. H. Mitchell, J. Poly. Sci., A., 18, 1727 (1980).

9. R. D. Hester, P. H. Mitchell, J. Liq. Chrom., 7, (8) 1511 (1984).
10. S. W. Provencher, Makromol. Chem., 180, 201 (1979).
11. E. Gulari, Y. Tsunashima, B. Chu, J. Chem. Phys., 70, 3965 (1979).
12. C. Y. Cha, K. W. Min, J. Poly. Sci. B., 21, 807 (1983).
13. J. G. McWhirter, E. R. Pike, J. Phys. A, 11 1729 (1978).
14. R. A. Vaidya, M. J. Mettile, R. D. Hester, American Chemical.Soc. Symposium Series 332, Chapter 4, (1987).
15. D. E. Koppel, J. Chem. Phys., 57, 4814 (1972).
16. C. L. McCormick, R. D. Hester, Improved Polymers for Enhanced Oil Recovery--Synthesis and Rheology, DOE/BC/10321 (DE85000141), U. S. Dept. of Energy, 107 (1985).
17. V. N. Tsvetkov, P. N. Lavrenko, S. V. Bushin, Russian Chemical Reviews, 51, 975 (1982).
18. M. E. McDonnell, A. M. Jamieson, Biopolymers, 15, 1283 (1976).
19. A. M. Jamieson, J. G. Southwick, J. Blackwell, J. Poly. Sci. B, 20, 1513 (1982).
20. G. Muller, Polymer Bulletin, 11, 391 (1984).
21. A. Puckett, Ph.D. Dissertation, "A New Look at the Mandelkern-Flory Relationship: Determination of the Molecular Weight of Water Soluble Polymers," University of Southern Mississippi (1986).
22. J. S. Linder, Ph.D. Dissertation, "Solution Properties of Water Soluble Macromolecules for Enhanced Oil Recovery Applications," Mississippi State University (1985).

ABSOLUTE MWD'S OF POLYACRYLAMIDES BY SEDIMENTATION AND LIGHT SCATTERING

G. Holzwarth*, L. Soni,† D. N. Schulz,† and J. Bock†

*Department of Physics
Wake Forest University
Winston-Salem, NC 27109
and
†Exxon Research and Engineering Co.
Annandale, NJ 08801

INTRODUCTION

There is at present no rapid, reliable commercial method for determining the molecular weight distribution of polyacrylamides, especially for molecular weights exceeding 5×10^6. Size exclusion chromatography, or SEC, which might be expected to be the method of choice, has molecular weight, polarity, and shear degradation limitations for very high MW polymers.[1-5] Recently, we reported a new absolute MWD method for high molecular weight water-soluble polymers based upon the combination of band sedimentation and low-angle laser light scattering, termed S/LALLS.[6] The advantages of this new method include applicability for polymers with M = 10^6 to 10^8, automatic sample clarification, minimal adsorption problems, no shear degradation, and the use of commercial components. Two other new methods, hydrodynamic chromatography and field-flow-fractionation, have also been developed recently to meet the need for new separation techniques.[7,8]

This paper describes the principles and operation of S/LALLS as applied to polyacrylamide and to copolymers of acrylamide and acrylic acid (HPAM). Where appropriate, the results of this new method are compared with SEC/LALLS and supplier's data. The method is also applied to the study of the shear degradation of HPAM polymers in their flow through porous media.

PRINCIPLES OF THE METHOD[6]

Band sedimentation is a technique for separating molecules according to their sedimentation velocity. It was first used in 1951 by Brakke[9] to purify viruses and has since been extensively developed for biochemical applications.[10] Curiously, it has been largely neglected as a separation technique by the polymer science community, even though all the apparatus is commercially available. It is especially well suited to high-molecular-weight, water soluble polymers. The centrifugation can be carried out in either an analytical, or, more commonly, in a preparative centrifuge.

In our use of band sedimentation, the sample was applied as a dilute solution in a thin band onto the top of a centrifuge tube previously filled with a solvent. The tube was then placed in a swinging-bucket roter and was spun for 2-5h at 20,000-40,000 rpm in a preparative ultracentrifuge. During centfrigation the polymer molecules moved slowly from the starting band toward the bottom of the tube with velocity $dr/dt = s\omega^2 r$, where s is the sedimentation coefficient, ω is the angular velocity of the rotor, and r is the radial distance of the molecule from rotor center. The sedimentation coefficient s reflects the net effects of three forces acting on the molecule: centrifugal force $M\omega^2 r$, buoyant force $M\bar{v}_2\rho\omega^2 r$, and frictional force $f dr/dt$; here $\bar{v}_2$ is the partial specific volume of the polymer, ρ is the solvent density, and f is the friction factor. The Svedberg equation combines the effect of these three forces:

$$s = M\,(1-\bar{v}_2\rho)/N_A f \qquad (1)$$

and relates s to M. Now f, of course, also depends upon M and upon the shape of the molecule. For a homologous polymer series, $f \propto M^{0.5} - M^{0.7}$, so that $s = HM^a$ where $0.3 \leqq a \leqq 0.5$ and H is a constant. Thus a homologous polymer series will have a range of values of s which can be used to achieve a separation. In practice, the values of ω and t were chosen so that the largest molecules almost reached the bottom of the tube. In our apparatus, this required a radial movement of about 5 cm from original band location near the meniscus. The centrifuge was then gently stopped, and the tubes were analyzed. Because of the density gradient, which prevented convection, the separated polymers were stable in the tube for many hours, allowing ample time for analysis of the separated distribution.

Displacement of the separated polymers from the tube through the detectors was accomplished by injecting a dense fluorocarbon chase solution into the bottom of the tube by means of a syringe pump. The chase solution displaced the tube contents upward out of the top of the tube through a filter (Nuclepore or Millipore), a flow-through UV detector, and finally, the LALLS detector.

Concentration determination was by UV absorbance at 214 nm, where PAM and HPAM show substantial absorbance. Fig. 1 shows the UV spectra between 200 and 300 nm for solutions of a PAM (Polysciences 2806) and an HPAM (Dow 700E) after purification. The observed absorbance arises from amide $n \rightarrow \pi^*$ and amide or carboxylate $\pi \rightarrow \pi^*$ transitions, which have peaks between 190 and 200 nm. The absorbance of carboxyl and carboxylate differ greatly,[11] so pH was set at 7.0 with buffer.

Light scattering. The excess total intensity of light i_s scattered from a polymer solution of concentration c_p at angle θ is given by[12]

$$i_s = I_o K c_p M P(\theta)/(1 + 2A_2 c_p M) \qquad (2)$$

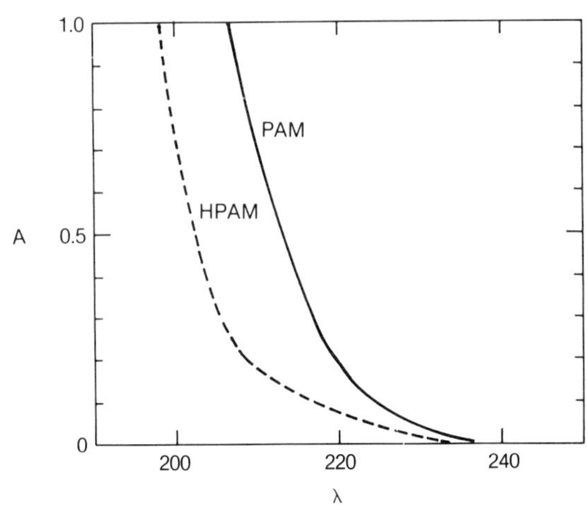

Fig 1. Ultraviolet absorption spectra of PAM 2806 and HPAM 700E, both at 0.1 mg/ml, in 0.01 M NaH_2PO_4-Na_2HPO_4 buffer, pH 7.0. Cell pathlength 1.00 cm.

where I_o is the incident intensity of polarized light, K is a constant depending on instrument and solution but not on polymer M, P(θ) expresses interference between light scattered from different regions of the same molecule, and A_2 is the second virial coefficient expressing interference between light scattered from different molecules. When θ becomes small, $P(\theta) \rightarrow 1$; this condition is met here, and eliminates the need to measure i_s at several angles, then extrapolate to zero angle [see Discussion]. Thus light scattering data provide a direct measure of M through equation 2 if I_o, K, c_p, and A_2 are known: $M = i_s/(Kc - 2A_2c_pi_s)$.

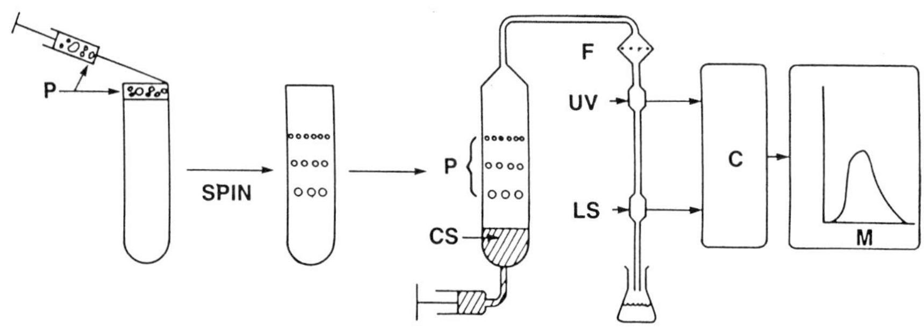

Fig 2. Overall scheme of S/LALLS experiment. P = polymer molecules; CS = chase solution; F = membrane filter; UV = ultraviolet absorbance detector; LS = low-angle light scattering detector; C = computer. (Reprinted with permission from Macromolecules 19, 422-426. Copyright 1986 American Chemical Society)

Overall scheme. The overall scheme of the analysis is shown in Fig. 2. The polydisperse polymer sample was applied to the centrifuge tube as a narrow band and was separated by centrifugation. The separated distribution was then forced slowly through a post-filter, the UV detector, and the light-scattering detector. The UV and LALLS electrical signals were sent to a computer, which computed M from Eq. 2 at each of about 80 points in the distribution and plotted or tabulated the results.

MATERIALS AND METHODS

Density gradients. Linear gradients of 5 to 10% Na_2SO_4 containing 50mM Na_2HPO_4/NaH_2PO_4 buffer at pH 7.0 were placed into plastic centrifuge tubes by an Isco Model 570 gradient maker. The resultant density gradient ranged from 1.046 g/cm^3 to 1.087g/cm^3. The salt Na_2SO_4 was chosen because it is transparent at 214 nm and relatively benign toward the stainless steel encountered in a parallel SEC/LALLS study.

Polymer samples were as follows:

1. Polysciences 2806, nonionic polyacryamide, nominal $M = 5\text{-}6 \times 10^6$;
2. Polysciences 8249, nonionic polyacrylamide, nominal $M = 0.5 \times 10^6$;
3. Dow 700E, poly(acrylamide-co-acrylic acid), nominal $M = 7 \times 10^6$;
4. Nitto 1012B, poly(acrylamide-co-acrylic acid)

Samples were purified before measuring UV spectra by dissolving in water, precipitating with ethanol in the presence of Na_2SO_4, redissolving in water, dialyzing thoroughly against distilled water, and then lyophilizing.

Centrifrigation was carried out in a Beckman Model L8-70 preparative ultracentrifuge with SW 40 swinging bucket rotor.

Displacement of tube contents was carried out with an Isco Model 185 density gradient analyzer. Each tube contained 14 mℓ of solution and was analyzed in about 20 minutes by injecting ISCO's dense and water immiscible "Fluorinert" displacement fluid into the bottom of the tube at a flow rate of 0.75 mℓ/min.

Filtration was through 0.6 to 1.2 micron Millipore or Nuclepore filters mounted in low-volume filter holders. The filter was located between the centrifuge tube and the UV detector.

UV detection of concentration was carried out with an Isco UA-5 photometer with Type 9 head. This provides monochromatic light at 214 nm. A flow-through cell with 1.0 cm light path was placed into the photometer. A full-scale sensitivity of 0.2 OD was generally employed, corresponding to 38 µg/mℓ PAM and 116 µg/mℓ HPAM full scale. Actual peak concentrations were about 1/3 these values. Solvent baselines were fixed by linear interpolation between two points judged free of polymer at the top and bottom of the tube.

LALLS detection was carried out with a Chromatix KMX-6 (Milton Roy Co) at 6-8°. A special Teflon flow-through cell with 2 cm optical path and polycarbonate insert, having inlet at the bottom and outlet at the top, was employed. The sample compartment was modified slightly to allow sample to enter through the bottom so as to prevent density inversions. Baseline or solvent scattering was subtracted from the raw light-scattering signal by linear interpolation between polymer-free points at the top and bottom of the tube.

Plumbing (tubing) between tube filter, UV detector, and LALLS detector was kept as short as possible to minimize hold-up and band-spreading. The components were arranged so that fluid flowed upward at all times to preclude mixing because of density inversion.

Data acquisition was through two A/D units, acting upon signals from the UV and LALLS detectors. The digital data were fed to an HP 9836 computer. Measurements of UV and LALLS were taken at about 80 points from top to bottom of each tube: each of the 80 LALLS readings was averaged from 200 LALLS measurements.

The LALLS averaging employed a simple algorithm to identify and reject individual data points corresponding to dust spikes in the LALLS signal; because of the extremely small scattering volume of the KMX-6, dust gave noise which was not white but occurred instead as occasional brief spikes imposed upon a steady, largely noise-free signal level. The UV and LALLS data were then saved for further analysis at a later time.

Determination of dn/dc was carried out with a Chromatix KMX-16 differential refractometer operating at 633 nm. For our previous report,[6] dn/dc was measured at only one salt concentration appropriate to the middle of the centrifuge tube (~ 6% Na_2SO_4), and this one value was used for all calculations of M for a particular type of polymer. Because dn/dc enters into the calculations of M as $(dn/dc)^2$, we decided to examine more carefully the effect of the gradient in Na_2SO_4 on dn/dc and thus on M. Five different Na_2SO_4 concentrations between 2.5 and 10% were used. For each salt concentration and polymer type(PAM or HPAM), 4 polymer solutions having concentrations between 500 and 2000 μg/mℓ were dialyzed against the buffer. The refractive index increment dn was determined with the KMX-16 and c_p was determined photometrically at 214 nm, for each of the dialyzed samples. The value of dn/dc was obtained from the slope of a plot of dn vs. c_p. Fig. 3 shows the value of dn/dc plotted against salt concentration for PAM (Polysciences 2806) and HPAM (Dow 700E). In both cases dn/dc decreased by about 0.018 from the top (2.5% Na_2SO_4) to the bottom (10% Na_2SO_4)) of the tube. If one compares the value of M obtained from the correct dn/dc to that obtained from a mean dn/dc (6% Na_2SO_4) one finds that $M_{correct}$ = $1.05M_{mean}$ at 10% Na_2SO_4 and $M_{correct} = 0.95M_{mean}$ at 2.5% Na_2SO_4: neglecting the gradient in dn/dc systematically narrows the distribution by a small amount.

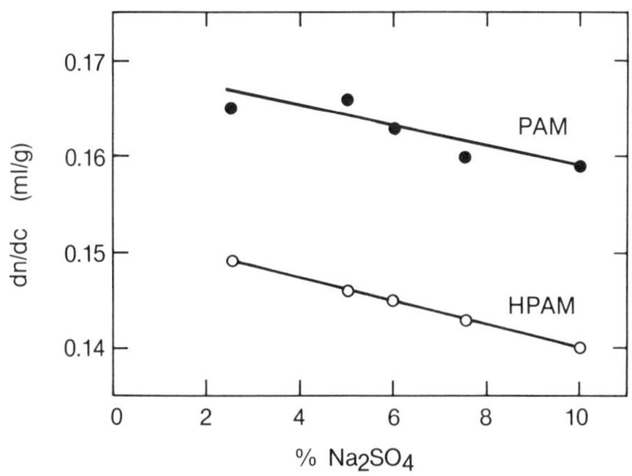

Fig 3. Refractive index increment dn/dc for PAM 2806 and HPAM 700E in solvents corresponding to different positions in the centrifuge tube. Solvent included 50 mM phosphate buffer, in addition to varying amounts of Na_2SO_4, to ensure pH 7.0.

Smoothing of the calculated values of M was carried out as described previously,[6] using the measured values of s and an L_1 fit of the M points to the relation $s = HM^a$. An L_1 fit is a "least-first-power" linear fit of log s to log M; it is more robust, that is, less sensitive to a few inaccurate data points than the traditional L_2 or "least-squares" fit.

Static LALLS measurements of M_w, as distinct from the above flow-through LALLS experiments, were carried out on unfractionated polymer which was stationary in the sample cell of the KMX-6 spectrometer. Measurements were made for a series of concentrations and the excess scattering was extrapolated to zero concentration.

RESULTS

Fig. 4 shows raw, unsmoothed data output of an S/LALLS experiment as obtained during the analysis of a single tube bearing the HPAM-700E polymer of nominal $M = 7 \times 10^6$. Signals from both the light scattering (LS) and UV detectors are shown. The LS trace is notably free of the "dust" spikes which normally plague aqueous light-scattering experiments, because sedimentation not only separated the polymer MWD but also sent "dust" to the bottom of the tube. Because of the density gradient, this "dust" did not remix with the clarified polymer solution.

Fig. 5 shows the MWD curve computed from the distribution. Note that the distribution is quite broad: there is a substantial amount of polymer with $M > 12 \times 10^6$; SEC could not resolve polymer in this size range. The values of M_w and M_n were obtained from this distribution using their standard definitions; $M_w = 6.6 \times 10^6$ and $M_n = 2.2 \times 10^6$. The value of M_n is less certain than that of M_w because it depends strongly on a small number of data points near the original band. Because this material had moved only slightly from the original starting band, it was more difficult to quantify. Specifically, the accurate determination of UV and LS baselines was difficult near the perturbations of the starting zone.

Data similar to those in Fig. 4 were obtained for the two PAM's, and each of these samples was also examined by SEC/LALLS and static LALLS on the unfractionated polymer. The SEC/LALLS setup included 3 60-cm columns [TSK 6000 PW, TSK 6000 PW, and TSK 5000 PW], a Chromatix CMX-100 LALLS, an Altex 156 DRI detector, and Chromatix data-acquisition and software systems. Table I summarizes the results obtained by SEC/LALLS and S/LALLS for the three polymers, expressed in terms of M_w and M_n.

Table I

COMPARISON OF SEC/LALLS WITH S/LALLS RESULTS

Sample	Source	$M \times 10^6$ Supplier	Static LALLS	SEC/LALLS		S/LALLS		M_w/M_n SEC/LALLS	S/LALLS
			M_w	M_w	M_n	M_w	M_n		
PAM-2806	Poly-science	5-6	5.0	3.6	2.3	5.6	2.1	1.6	2.5
PAM-8249	Poly-science	0.5	0.6	0.90	0.42	1.0	0.64	2.1	1.7
HPAM-700E	Dow	7	-	2.8	CD*	6.6	2.2	CD*	2.5

*CD: could not determine because of adsorption problems

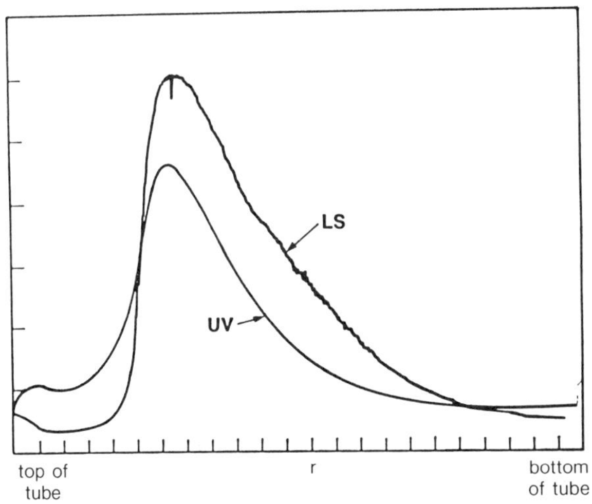

Fig 4. Raw data output of a S/LALLS experiment on an HPAM polymer(DOW 700E). Signals from both the light-scattering(LS) and UV detectors are shown. Small molecules are the first to appear, on the left side of the figure; large molecules move further down the tube during sedimentation and appear at the right end of the figure, in contrast to SEC analyses. The LS trace is notably free of the "dust" spikes which normally plague aqueous light scattering experiments and aqueous SEC/LALLS.

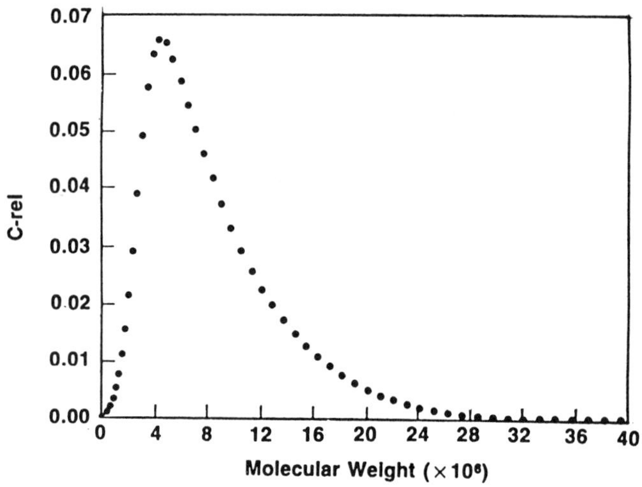

Fig 5. Processed MWD data obtained from the raw data of Fig 4. Notice that this HPAM sample shows a high M tail extending to 40 million, well beyond the exclusion limit of any aqueous SEC column. C_{rel} is the relative concentration of polymer of a given M, normalized so that the sum of C_{rel} over all fractions is 1.00.

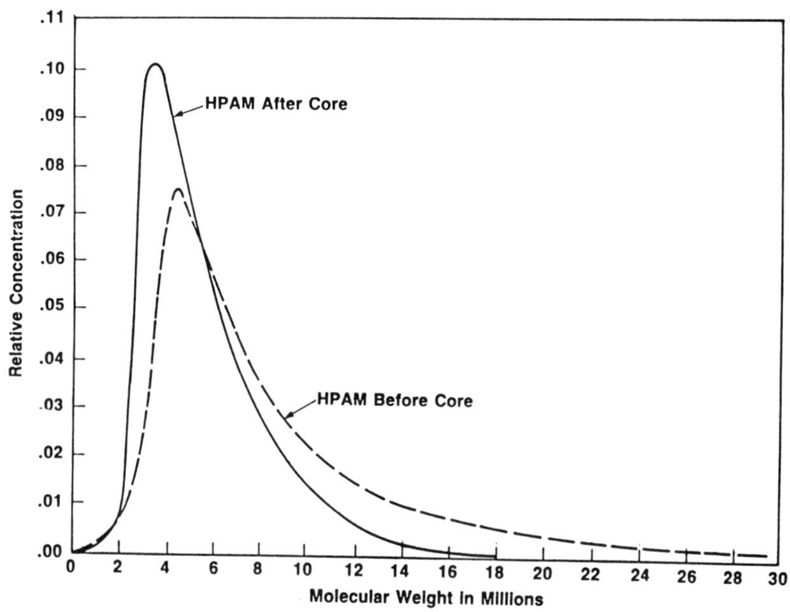

Fig 6. MWD of HPAM polymer before and after passage through a core. Note the nearly complete loss of the high M tail of the distribution upon passage through the core.

The S/LALLS apparatus can of course also be used to study the effect of various treatments of a polymer on its MWD. An example of this is the irreversible drop in viscosity commonly seen in HPAM's upon flow through the porous rock of oilfield reservoirs; this drop is a major problem when HPAM polymers are used as drive fluids in chemically-enhanced oil recovery. A similar sharp decrease in viscosity was seen in a laboratory experiment designed to simulate this effect. In this experiment a viscous HPAM solution was forced at a frontal advance rate of 600 ft/day through an 0.2" thick core of Berea sandstone(570 mD permeability). The viscosity decreased from 25 cP to 5 cP at 1.3 s^{-1} shear rate. It is believed that shear degradation is responsible for such a reduction in mobility control. The MWD's of this polymer before and after passage through the core, as measured by S/LALLS, are shown in Fig. 6. A substantial fraction of the polymer used, Nitto 1012B, had $M > 15 \times 10^6$ before passage through the core. During passage through the core, this high M material was preferentially degraded. Such detailed information on the MWD can be valuable in predicting reservoir performance of a polymer solution.

DISCUSSION

The S/LALLS method for determining MWD bears many similarities to SEC/LALLS[13-15] except that the separation was achieved by sedimentation velocity rather than by differences in partitioning between a mobile and stationary phase in a column. The absence of a column in S/LALLS brought three immediate benefits: 1)adsorption was eliminated; 2) shear degradation, a major problem in SEC of high M polymers, was precluded; 3) the highest M which could be analyzed was increased to at least 4×10^7.

These benefits are apparent from comparison of the SEC/LALLS and S/LALLS results for the two PAM's and the HPAM in Table 1. For PAM 8249, with supplier's $M = 0.5 \times 10^6$, both techniques gave very similar values for M_w, 0.90×10^6 and 1.0×10^6. The corresponding values for M_n also agreed fairly well. By contrast, for the high molecular weight PAM 2806, with nominal M of $5-6 \times 10^6$, SEC/LALLS gave $M_w = 3.6 \times 10^6$ whereas S/LALLS gave 5.6×10^6. The S/LALLS figure was in good accord with the static LALLS value of unfractionated material, 5.0×10^6. It is almost certain that the low value obtained by SEC originates in a failure to separate the high M tail of the distribution.

The last two columns of Table 1 give the values of M_w/M_n obtained by the two mthods: this ratio is a measure of the breadth of the MWD distribution. The value of M_w/M_n from SEC, 2.1, is reasonable for the sample with $M_w \sim 1 \times 10^6$ but is an unrealistically low 1.6 for $M_w \sim 5 \times 10^6$. The failure of SEC to separate the MWD for $M > 5 \times 10^6$ is the most likely reason. Our new technique, S/LALLS, gives a rather low M_w/M_n, 1.7 for the M $\sim 5 \times 10^5$ PAM; probably M_n is too low for the reasons already noted. For PAM 2806, $M_w \sim 5 \times 10^6$, the observed M_w/M_n, 2.5 more likely correct.

Similar problems of failure to resolve material of high M were even more apparent for HPAM 700E, whose nominal M is 7×10^6(Table 1). The SEC/LALLS value for M_w was only 2.8×10^6, whereas S/LALLS yielded a more realistic 6.6×10^6. A further problem with SEC/LALLS of HPAM 700E, but not of PAM, under our conditions, was adsorption tailing of the chromatograms. This made determination of M_n impossible by SEC. By contrast, S/LALLS was able to yield M_w, M_n, and a reasonable M_w/M_n, 2.5.

Another major benefit of S/LALLS over SEC/LALLS was the automatic sample clarification which occurred during sedimentation. This clarification occurred also for samples bearing significant fractions of gel (data not shown). The gel traveled rapidly to the bottom of the tube and remained there. In SEC, such samples tended to clog the column unless they were precleaned.

It should be noted that S/LALLS did not require any instrument construction, although some computer programming was necessary. It utilized proven commercial components in a new combination.

There are limitations to S/LALLS. The auxiliary parameters A_2 and dn/dc must be determined for each chemically distinct polymer family, and the polymer must exhibit significant absorbance at a wavelength at which the solvent is transparent. If the polymer absorbance is very low and M is very high, it may prove difficult to carry out the sedimentation at low enough concentrations(well below c^*) such that the polymer molecules move individually. Fortunately, PAM and HPAM have enough absorbance at 214 nm. For polymers with $M < 0.5 \times 10^6$, S/LALLS requires long centrifugation times and offers few advantages over SEC. Indeed, it is difficult to measure M at the low end of most distributions because these molecules move only slightly from the disturbances of the starting zone. It is then difficult to set the baseline accurately; this leads to uncertainties in M_n and thus also in M_w/M_n. It would be preferable to characterize an MWD by the peak M, the half-width, and the skewness of the distribution rather than the historically sanctioned M_w/M_n.

The upper M limits of S/LALLS in its present form arise from 3 factors. First, it is necessary to stay below c^* to get a good separation. As M increases, c^* necessarily decreases and it becomes difficult to measure c_p with a UV detector. Second, as M increases, intramolecular interference begins to affect the LALLS data even at 6-7° scattering angle. The effect is readily calculated if the radius of gyration of the polymer is known. For PAM, the root mean square value of $R_G = 7.49 \times 10^{-3} M^{0.64}$ nm in 0.1 M NaCl.[16] This is reasonablly close to our conditions (0.35 - 0.70 M Na_2SO_4). It is then straightforward to estimate P(θ) for a given M and θ:

$$P(\theta) = 1 - 16\pi^2 R_G{}^2 \sin^2[\theta/2]/3\lambda^2 \qquad (3)$$

For example, if $M = 10 \times 10^6$ and $\theta = 6.5^o$, $P(\theta) = 0.98$. Setting P(θ) equal to 1.00 introduces only a very small error. However, at the extreme end of the PAM and HPAM MWD's, $M = 40 \times 10^6$. At this moleular weight, for $\theta = 6.5^o$, $P(\theta) = 0.89$, which begins to be significant. To improve the accuracy in M near 40×10^6, angles less than 6.5^o could be utilized. The KMX-6 photometer is capable of measuring LALLS at angles of only 2 - 3°, in which case P(θ) would be 0.98.

A third limitation is the need for an A_2 correction in computing M from Eq 2. Three complications must be recognized. First, as M increases, the correction term $2A_2c_pM$ becomes larger with respect to unity(eq. 2). Second, one can expect A_2 to depend somewhat on molecular weight.[17] Third, one expects A_2 to vary with ionic strength for a polyelectrolyte, hence to vary at different positions in the density gradient. Thus, as M increases, the accuracy of A_2 can become a limiting parameter to the accuracy of M. For example, at the extreme ends of the MWD analyzed for PAM and HPAM in this study, $M = 4 \times 10^7$ and $c_p = 10 \times 10^{-6}$ g/ml. Assuming $A_2 = 5 \times 10^{-4}$, which is typical of polyacrylamides,[16] a 20% uncertainty in A_2 leads to an 8% uncertainty in M. The problem worsens as M increases.

One factor which affects the accuracy of M regardless of c_p or θ is the dependence of dn/dc on salt concentration and thereby on position within a tube. The data in Fig 3 are sufficient to account for the effect with PAM and HPAM in the data processing program. Since the correction to M at the top and bottom of the tube was only 5% and -5% respectively, and since most of the polymer is concentrated fairly near to the center of the tube, the effects on M_w and the MWD are actually quite small.

It should be noted that the variation of dn/dc with solvent composition can be estimated quite accurately from very simple physical-chemical considerations[18,19] if the polymer and solvent do not exhibit highly specific binding interactions. The contribution of each molecular species to the bulk n is then simply additive, and

$$(dn/dc)_b = (dn/dc)_a - \bar{v}_2(n_b - n_a) \tag{4}$$

Here $(dn/dc)_b$ is the value of dn/dc in solvent b, $(dn/dc)_a$ is the value of dn/dc in solvent a, and $\bar{v}_2$ is the partial specific volume of the polymer. Using $\bar{v}_2$ = 0.70 ml/g for polyacrylamide,[16] this formula predicts a variation of dn/dc with Na_2SO_4 concentration which is in excellent agreement with the data in Fig 3. This suggests that the correction for the effect of the salt gradient on dn/dc can be adequately handled through equation 4 rather than by direct measurement of dn/dc in a series of solvents.

Acknowledgments: We thank Dr. Wolf Schulz for many valuable discussions and for obtaining the SEC/LALLS data reported in Table 1, and Sal Pace for assistance in preparing the polymer samples degraded by flow through porous rock. GH also thanks Research Corporation for financial support.

REFERENCES

1. E. L. Slagowsky, L. J. Fetters, and D. McIntyre, Macromolecules 8:691 (1975).
2. L. Soltes, D. Berek, and S. Mikulsova, J. Colloid Polym. Sci. 258:702 (1980).
3. A. MacArthur and D. McIntyre, Polym. Prepr. (Am. Chem. Soc., Div. Polym. Chem.) 24(2):102 (1983).
4. W. Siebourg, R. D. Lundberg and R. W. Lenz, Macromolecules 13:1013 (1980).
5. H. G. Barth, J. Chromatogr. Sci. 18:409 (1980).
6. G. Holzwarth, L. Soni and D. N. Schulz, Macromolecules 19:422 (1986).
7. D. A. Hoagland, K. A. Larson, and R. K. Prud'homme, in "Modern Methods of Particle Size Analysis": H. G. Barth, Ed., Wiley, New York, 1984, Chapt. 9.

8. L. E. Schallinger, W. W. Yau, and J. J. Kirkland, Science 225:434 (1984).
9. M. K. Brakke, J. Am. Chem. Soc. 73:1847 (1951).
10. C. A. Price, "Centrifugation in Density Gradients", Academic Press: New York, 1982.
11. R. McDiarmid, and P. Doty, J. Phys. Chem. 70:2620 (1966).
12. W. Kaye and A. J. Havlik, Applied Optics 12:541 (1973).
13. A. C. Ouano and W. Kaye, J. Polym. Sci. 12:1151 (1974).
14. A. C. Ouano, J. Chromotogr. 118:303 (1976).
15. R. C. Jordan, J. Liq. Chromatogr. 3:439 (1980).
16. W.-M. Kulicke, R. Kniewske and J. Klein, Prog. Polym. Sci. 8:373 (1982).
17. P. J. Flory, "Principles of Polymer Chemistry"; Cornell University Press, Ithaca, NY, 1953.
18. M. B. Huglin, "Light Scattering from Polymer Solutions", Academic Press, New York, 1972, p. 184.
19. H. Eisenberg, "Biological Macromolecules and Polyelectrolytes in Solution", Clarendon Press, Oxford, 1976, p. 62.

A COMPARATIVE STUDY OF XANTHANS BY LIGHT SCATTERING

E. A. Lange

Exxon Production Research Company
P. O. Box 2189
Houston, Texas 77252-2189

ABSTRACT

Xanthan samples from four commercial manufacturers were studied in synthetic reservoir brines and a dilute NaCl solution. The samples included broth and powdered xanthan products. The weight-average molecular weight, M_w, for each sample was measured in the different solvents by low angle light scattering. M_w was not affected by the brine composition, and no evidence for increased aggregation in high-salinity, high-hardness brines was found. M_w varied from 4×10^6 g/mol to 12×10^6 g/mol. Variations in molecular weight were also observed among samples from one manufacturer. The translational diffusion coefficients of the xanthans could not be measured by a dynamic light scattering technique. Other modes of motion in addition to translational diffusion may have been detected at the 40° scattering angle as a direct result of the high polymer molecular weights.

INTRODUCTION

The polysaccharide xanthan is an attractive candidate for chemical flooding applications in low-temperature, high-salinity reservoirs. The salt content of reservoir brines can exceed 20% TDS, and these brines often contain high levels of divalent cations. As part of an ongoing study of xanthan behavior in brines, several commercial xanthans have been examined using light scattering techniques. The first objective of this work was to measure xanthan physical properties in high-salinity, high-hardness environments. Although the properties of xanthan have been reported extensively in the past, most previous studies have been conducted in dilute NaCl solutions. A second objective of the work was to investigate the cause of variations in some solution properties, such as viscosity, among samples. Product variations are a concern for large-scale applications of xanthan in chemical enhanced oil recovery projects. This study has focused on broth-form xanthan products.

The polysaccharide xanthan is produced in a bacterial fermentation. The chemical structure of this molecule has been determined by Jannson and coworkers.[1] The polymer consists of a β-D-glucose backbone with a side chain of three sugar units extending off every other glucose. The side chain bears ionic groups, making xanthan a polyelectrolyte. Both single-stranded and double-stranded helical structures have been proposed as the native conformation of this molecule.[2,3,4]

EXPERIMENTAL METHODS

Materials

The xanthans were obtained from four commercial manufacturers. The products are denoted Xanthan A, B, C and D. Xanthans A, B and C were supplied as the gelatinous crude fermentation broths containing 3-8 wt% xanthan. Xanthan D was supplied as a powder. Six different production lots of Xanthan A were examined, and these polymers are labeled as A-1,...A-6. The compositions of the two test brines are presented in Table 1. Brine 1 was a moderate-salinity brine containing only NaCl and $CaCl_2$. Brine 2 was the synthetic equivalent of the resident brine in one of Exxon's reservoirs. This brine was a high-salinity and high-hardness solution containing several divalent cations. Finally, a 0.1 M NaCl solution was used in some experiments to enable direct comparison of the results to previous work.

Solution Preparation

The broth xanthans were dispersed directly into the brines as dilute solutions (1.0 g/L) by brief mixing in a Waring blender at medium speed. The powdered xanthan was resuspended in a two step procedure.[5,6] The dry xanthan was first dispersed into distilled water as a concentrated solution (5.0 g/L) using a blender at high speed. This solution was then diluted and salts were added to obtain the desired composition. Remnant bacterial cells were present in all solutions. These bacterial cells were seen easily by phase-contrast or dark-field microscopy.

The xanthan solutions were purified without precipitation or centrifugation. Most xanthan solutions could be clarified by one pass through a glass fiber filter (Millipore, Type AP15). The clarification was most effective at high ionic strength conditions. Less than 3% of the xanthan was lost from a 1.0 g/L solution by this procedure. The clarified polymer solutions were then dialyzed against the solvent to remove low molecular weight sugars.

An additional filtration step was required for some xanthan solutions to remove other particulates, such as undissolved clumps of the powdered xanthan. A 0.22 μm or 0.45 μm filter usually removed this debris. The xanthan concentration was measured by the phenol-sulfuric acid colorimetric assay for carbohydrates.[7] Solution viscosities were determined using a coni-cylindrical (cup-and-bob) rotational viscometer.

Instruments

A low angle light scattering photometer (LDC/Milton Roy, KMX-6) was used in the molecular weight determinations. This instrument contained a HeNe laser light source (λ=633 nm), a flow-through scattering cell, and a series of annular apertures that allowed measurement of the time-average scattered light intensity at very low forward angles. A 6-7° scattering angle was chosen for the present study. A syringe pump was used to push the

Table 1. Test Brine Compositions

Brine 1	Brine 2
30.0 g/L NaCl	92.0 g/L NaCl
3.0 g/L $CaCl_2$	8.0 g/L $CaCl_2$
	5.0 g/L $MgCl_2$
	0.1 g/L $BaCl_2$
	0.2 g/L $NaHCO_3$

polymer solution through a small in-line filter and into the optical cell. 0.22 μm, 0.45 μm or 1.2 μm syringe filters were used.

The weight-average molecular weight, M_w, and the second virial coefficient, A_2, were computed from scattered light intensities using a plot of the excess Rayleigh factor as a function of polymer concentration. The Rayleigh factor, R_θ, contains the ratio of the scattered light intensity to the incident intensity. The excess Rayleigh factor, $\bar{R}_\theta$, is the difference in R_θ for the polymer solution and the solvent. At very low scattering angles, this quantity is related to the polymer concentration, C, by[8]

$$KC/\bar{R}_\theta = 1/M_w + 2A_2C$$

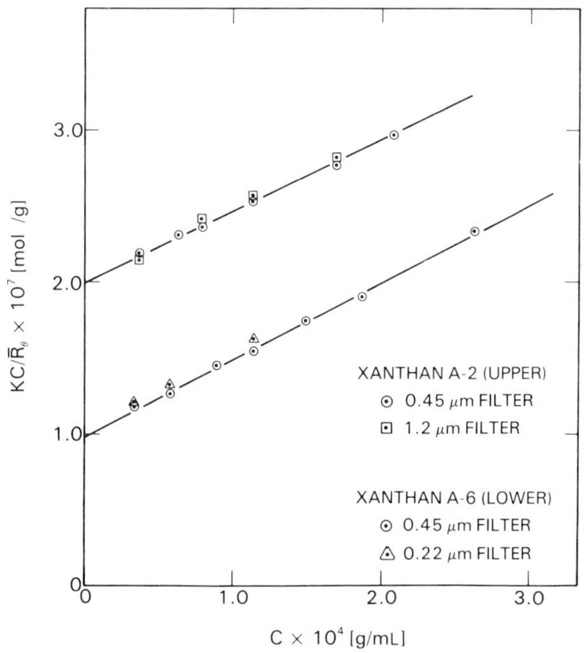

Figure 1. Low angle light scattering data for two xanthans in Brine 1. T = 25°C.

where the parameter K contains the refractive index increment, the solvent refractive index and several geometrical factors. The quantity $KC/\bar{R}_\theta$ should vary linearly with concentration, with a slope of $2A_2$ and an intercept of $1/M_w$. As an example, the data from two xanthan analyses are plotted in this manner in Figure 1. In each case, the data can be fit by a linear relationship. The similarity in the slope of the two lines reflects similarity in the second virial coefficients, but the large difference in the intercepts indicates a significant difference in M_w between the samples.

No effect of the in-line filter type was observed. The results of analyses with different in-line filter types are also presented in Figure 1. Filtration steps are a concern since this process can bias the average molecular weight, either by removing a portion of the distribution or by allowing strongly scattering debris into the optical cell. Good agreement was observed between analyses of Xanthan A-2 with 0.45 μm and 1.2 μm in-line filters. Similarly, good agreement was observed between analyses of Xanthan A-6 using a 0.45 μm and a 0.22 μm filter.

The effects of a different solution preparation scheme were tested in another experiment. After clarification, a xanthan solution was prepared "microgel-free" by the slow, serial filtration method of Chauveteau and Kohler.[9] This solution and the corresponding clarified, but otherwise untreated, xanthan solution were analyzed. Both solutions were coarsely filtered (1.2 μm) in the light scattering experiment to avoid biasing M_w. As shown in Figure 2, the two data sets form parallel lines with similar intercepts, indicating similar values of M_w. The small difference in M_w between the samples is within the reproducibility of the measurement. These observations again indicate that the measured M_w is representative of the dissolved polymer components.

The refractive index increment (dn/dc) for xanthan was measured using two differential refractometers. First, a manually-operated refractometer was used with xanthan solutions in the concentration range from 1.0-3.0 g/L. A halogen lamp coupled with a narrow band-pass filter simulated the HeNe laser light for the measurement. Second, the refractive index increment at

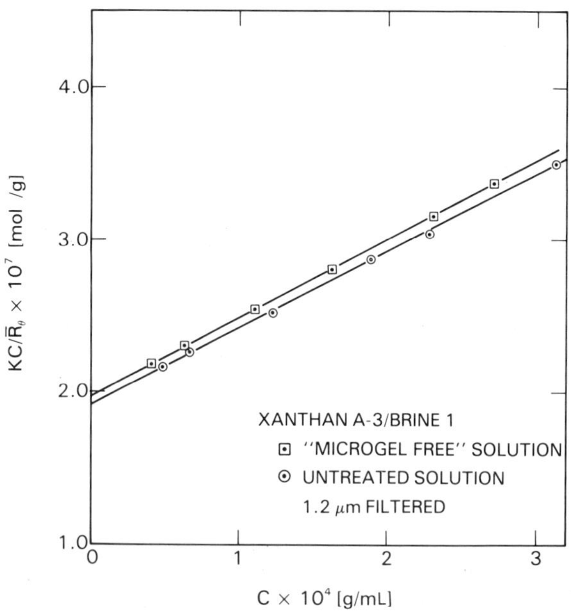

Figure 2. Comparison of a "microgel-free" solution and a clarified, but otherwise untreated xanthan solution in Brine 1. T = 25°C.

lower xanthan concentrations (0.2-1.0 g/L) was measured accurately using an HPLC refractive index detector. This instrument contained a white light source, so a small correction to dn/dc for wavelength was applied. This correction was based on measurements made with the first instrument using a white light source. Dialyzed xanthan solutions were used, and the reference solution was the final, matching dialyzate. The average refractive index increment for five Xanthan A samples in Brine 1 was 0.129 (±0.006) mL/g. The dn/dc values for these xanthans in 0.1 M NaCl fell within the same range, and one dn/dc value was used for all broth xanthans in all solvents. A lower value of 0.118 (±0.006) mL/g was measured for the Xanthan D sample in Brine 1 and 0.1 M NaCl.

The dynamic light scattering apparatus consisted of an argon ion laser (λ=488 nm), a photomultiplier mounted on a turntable, and an autocorrelator. The dilute polymer solution was filtered directly into a cylindrical glass cell, and the cell was suspended in a thermostatted bath during the analysis (25°C). An apparent average translational diffusion coefficient, $\bar{D}$, was computed from the autocorrelation function using the method of cumulants.[10,11] Polystyrene latex standards (58 nm diameter) were examined periodically to check the instrument operation. The instrument alignment was tested by measuring the diffusion coefficient of a small protein, (BSA), in a buffered saline solution at multiple angles. The scattering angle, θ, was varied between 30° and 90° in the xanthan studies.

RESULTS

Low Angle Light Scattering

A three-fold variation in xanthan molecular weight was observed. The results of the low angle light scattering experiments in Brine 1 are presented in Table 2. For the Xanthan A samples, the weight-average molecular weight varied between 4.2 x 10^6 g/mol and 10.1 x 10^6 g/mol. A variation in M_w was also observed among Xanthans B, C, and D, and the molecular weight range is similar to that observed for the Xanthan A samples.

The second virial coefficient for polyelectrolytes is influenced by the charge density on the molecule and by the solution ionic strength.[8] The charge density is the ratio of the total number of charges on the polymer to the molecular weight. The ionizable groups on xanthan are the glucuronic acid and pyruvic acid moieties on the side chain.[1] The narrow range in A_2 observed among the Xanthan A samples may indicate a similar chemical composition in these polymers. By the same reasoning, the greater variation in A_2 among the Xanthans A, B, C, and D may reflect significant variations in chemical composition among the samples, perhaps due to variations in the pyruvate content.

A pronounced effect of solvent on M_w was not observed. Xanthan D and three samples of Xanthan A were studied in 0.1 M NaCl. The results of these experiments are presented in the lower portion of Table 2 for comparison. The M_w of all xanthans in the dilute saline solution were nearly equal to or greater than the values measured in brine. This trend is opposite of the effect that would be expected if the calcium ions of the brine promote aggregation of the macromolecules. The change in the second virial coefficients reflects the change in solvent ionic strength. Xanthan A-3 was also studied in the high-salinity, high-hardness Brine 2. The weight-average molecular weight varies somewhat among the three solvents for this sample, but the highest M_w was measured in the dilute saline solution. These results also indicate that the divalent cations of the brines have not induced a significant aggregation of the macromolecules.

Table 2. Low Angle Light Scattering Results

Sample	$M_w[10^6$ g/mol]	$A_2[10^{-4}$(mL·mol)/g^2]
Solvent: Brine 1		
Xanthan A-1	4.2	2.0
A-2	5.0, 5.0*	2.3, 2.5*
A-3	5.4, 5.3*	2.4, 2.5*
A-3 ("microgel-free")	5.1	2.6
A-4	8.1	2.7
A-5	8.2	2.5
A-6	10.1	2.5
Xanthan B	4.1	0.95
Xanthan C	12.2	1.5
Xanthan D+	5.2	2.1
Solvent: 0.1 M NaCl		
Xanthan A-2	5.1	3.6
A-3	6.3	3.0
A-6	12.5	3.8
Xanthan D+	5.8	3.5
Solvent: Brine 2		
Xanthan A-3	4.7	2.4

\+ 0.22 μm filtered before analysis to remove particulates.
* 1.2 μm in-line filter used. Unless noted, a 0.45 μm in-line filter was used in the remaining analyses.
pH = 6-7

In general, the xanthan molecular weight estimates obtained in this study agree well with previous reports. The following three reports are representative of many recent determinations. Sato and coworkers[12] have studied a powdered xanthan dispersed in 0.1 M NaCl by light scattering. The weight-average molecular weight measured by Sato et al. was 7.4×10^6 g/mol, and the second virial coefficient of the sample was 3.2×10^{-4} (mL·mol)/g^2. Both values fall within the range observed in the present study. Müller and coworkers[13] have determined a lower M_w value of 1.8×10^6 g/mol for a broth xanthan in 5 g/L NaCl. This value is outside the range observed herein, but the sample was later described as a dissociated or single-stranded form.[14] Finally, Holzwarth[15] also found a range of xanthan molecular weight, from 4×10^6 to 12×10^6 g/mol, using a band sedimentation-light scattering technique.

The xanthan samples with higher solution viscosities were characterized by higher M_w. The viscosity measurements for the six Xanthan A samples in Brine 1 are presented in Table 3. The xanthan concentration for these measurements was 1.0 g/L. This concentration is commonly used for polymer evaluations in EOR applications. The shear rate was 0.03 s^{-1} in the first Newtonian region of the rheogram. As a group, the higher molecular weight xanthans displayed significantly higher solution viscosities.

The cause of the variations in M_w is not clear. Since M_w is an average quantity, the difference in molecular weight among the xanthans may reflect differences either in the polymer molecular weight or in the degrees of aggregation between the macromolecules. One possibility is that the higher molecular weight xanthans are simply dimers of the lower molecular weight

Table 3. Comparison of Xanthan Viscosities

Sample	Viscosity [cP]	M_w [10^6 g/mol]
A-1	53	4.2
A-2	27	5.0
A-3	44	5.4
A-4	274	8.1
A-5	310	8.2
A-6	164	10.1

T = 25°C; Xanthan concentration = 1.0 g/L in Brine 1; Shear rate = 0.03 s^{-1}

form. A "dissociation" experiment with two Xanthan A samples was conducted to test the latter idea. This experiment is outlined in Figure 3. First, each xanthan was dispersed into 0.1 M NaCl, as done for the M_w analyses. The xanthan concentration was 1.0 g/L. The solution was then dialyzed against distilled water to remove as much salt as possible. NaCl was added directly to a portion of this dialyzed solution (Treatment #1). Milas and Rinaudo[3] have suggested that this treatment sequence should induce an order-disorder conformation change in the polymer. The remainder of the polymer solution in distilled water was heated briefly to 96°C under anaerobic conditions. The solution was cooled, and NaCl was reintroduced (Treatment #2). Holzwarth and Prestridge[2] have suggested that the heat treatment in distilled water can denature the xanthan, or separate the strands of the native double helix. The double-stranded form should be regained upon salt addition. Similarly, Sato and coworkers[16] have interpreted this treatment sequence as inducing a dissociation and reaggregation of the xanthan strands.

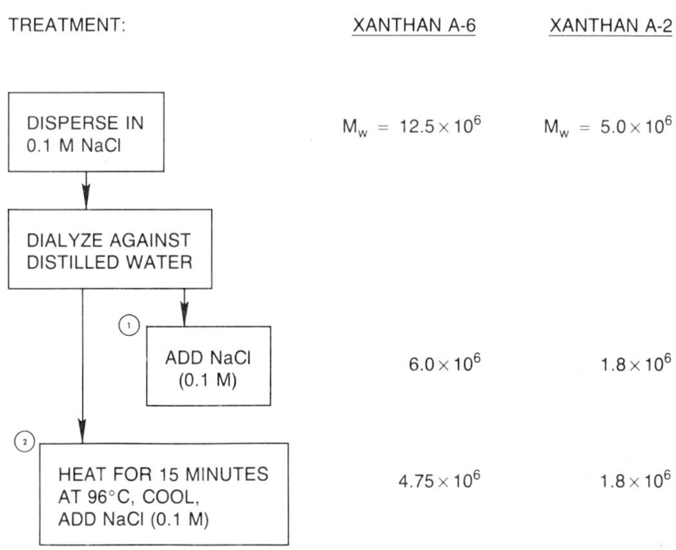

Figure 3. Outline of "dissociation" treatments and results for two Xanthan A samples.

The two Xanthan A samples behaved similarly in this experiment. The results of the light scattering analysis after each step are presented on the right side of Figure 3. The M_w of each sample decreased to less than half of the original value as a result of Treatment #1 or #2. These results suggest that the higher M_w xanthans are not simply dimers of the lower molecular weight form. Rather, the higher M_w xanthans must be composed of higher molecular weight subunits or be more resistant to the "dissociation" treatment. Interestingly, the average molecular weight of the A-2 sample after the treatments is comparable to the lowest M_w reported for xanthan in the literature. Finally, the observed decrease in molecular weight is contrary to some previous interpretations on the effects of these treatments. Differences in xanthan concentration, chemical composition or post-fermentation broth processing could be responsible for the differing outcomes reported with these treatments. In a recent study, however, Lecourtier and coworkers[17] have also observed a large, irreversible decrease in M_w upon exposing xanthan to a low ionic strength environment.

Dynamic Light Scattering

The aim of the dynamic light scattering experiments was to determine a hydrodynamic size for each xanthan by measuring the translational diffusion coefficient. Ideally, these experiments would be conducted at low polymer concentrations with a low scattering angle, such that only translational diffusion would be detected. In practice, true translational diffusion coefficients could not be measured at the available scattering angles ($\theta \geq 30°$) for the xanthans used in this study. The following results and discussion are presented to illustrate the difficulty.

The apparent diffusion coefficients measured for four xanthans in Brine 1 are presented in Table 4. A 40° scattering angle was used in these experiments to match the conditions of an earlier study.[5] $\bar{D}$ was a strong function of scattering angle over the range from 30° to 90° and a very weak function of xanthan concentration in Brine 1 (see Appendix A). The xanthan concentration used for most experiments was 0.4 g/L. The higher average molecular weight samples displayed lower $\bar{D}$ values in Brine 1. Similar values of the apparent diffusion coefficients were measured in 0.1 M NaCl.

These apparent diffusion coefficients are significantly higher than the values obtained by Southwick and coworkers[5] at comparable conditions of polymer concentration, wavelength and scattering angle (40°). The sample used by Southwick et al. had an average molecular weight of 2.2×10^6, or a fraction of the molecular weight of the xanthans used in this study. The discrepancy in the $\bar{D}$ values may have been caused by the difference in xanthan molecular weights. The higher molecular weight of the xanthans used in the present study could have two consequences. First, the overlap concentration, C*, may be below 0.4 g/L for many samples, so that dilute solution conditions were not attained. Second, intramolecular modes of motion may have been detected in the present study. The use of dynamic light scattering to detect these modes has been described by Pecora.[18] A translational diffusion coefficient is obtained directly from the autocorrelation function when the molecule acts as a point scatterer. A flexible macromolecule will appear as a point scatterer when the product of the scattering vector and the radius of gyration is much less than 1. When the product is greater than 1, the autocorrelation function can contain contributions from other faster modes of motion such as flexing of the chain. The magnitude of the scattering vector, k, is equal to $(4\pi/\lambda)(\sin(\theta/2))$. The radius of gyration, R_g, measured by Sato and coworkers[12] for a xanthan with M_w of 7.4×10^6 was 378 nm. At the 40° scattering angle with an argon laser, the product $k{*}R_g$ is greater than 3. Although xanthan may not be a flexible coil, this simple calculation suggests that detection of intramolecular motion was possible. A lower scattering angle will be required to measure

Table 4. Dynamic Light Scattering Results

Sample	$\bar{D}[10^{-8}cm^2/s]$
Xanthan A-2	3.8 (±0.4)
A-6	3.1 (±0.1)
Xanthan C	3.3 (±0.1)
Xanthan D	4.1 (±0.2)

Xanthan concentration = 0.4 g/L in Brine 1;
Solutions were 0.45 μm filtered;
θ = 40°, T = 25°C.

the translational diffusion coefficient for these higher molecular weight xanthans.

SUMMARY AND CONCLUSIONS

The classical low angle light scattering technique was useful in characterizing xanthans. Results indicate that xanthan average molecular weight is insensitive to the Ca^{++} and Mg^{++} ion concentration in typical reservoir brines. Apparently, the calcium ions do not induce aggregation of the molecules in these solvent environments. This observation is consistent with the insensitivity of solution viscosity to brine composition.

The commercial xanthans were markedly diverse in molecular weight. M_w varied between 4 x 10^6 and 12 x 10^6 g/mol. The variations were observed among samples from one manufacturer and among samples from several manufacturers. The origin of these differences in molecular weight is not clear. Solution history can affect xanthan molecular weight, but these variations in the polymer must have arisen from the fermentation or post-fermentation processing. As one idea, the higher M_w samples may contain higher molecular weight strands or be aggregated forms. Clearly though, the large difference in molecular weight among some samples cannot be explained by single-stranded and double-stranded forms. Additional "dissociation" studies and careful chemical characterization of these samples may provide further insights.

APPENDIX A

The autocorrelation function, $G(\tau)$,[10] obtained in a typical xanthan analysis at θ = 40° is presented in Figure A-1. The time scale, Δt, for the analyses at θ=40° was typically 30-40 μs. Evidence of additional relaxation phenomena was not detected at longer timescales.

The apparent diffusion coefficients were a weak function of xanthan concentration. Results for one xanthan sample in Brine 1 are presented in Figure A-2. In these experiments, the polymer concentration was varied between 0.2 g/L and 0.65 g/L. Below 0.2 g/L, the scattered light intensity was low, and poor reproducibility was observed. The dynamic light scattering analyses were typically conducted using a concentration range from 0.3 g/L to 0.4 g/L.

The apparent diffusion coefficients varied with scattering angle. Results for one xanthan sample and the small protein, BSA, are presented in Table A-1. The diffusion coefficient for the protein decreases slightly with decreasing angle. This effect can be attributed to greater interference from dust and debris at the lower angles, or to minor misalignment of the instrument. The apparent diffusion coefficients for the xanthan sample decrease sharply and steadily with decreasing angle. A change in $\bar{D}$ with scattering angle is also expected if the molecules are large with respect

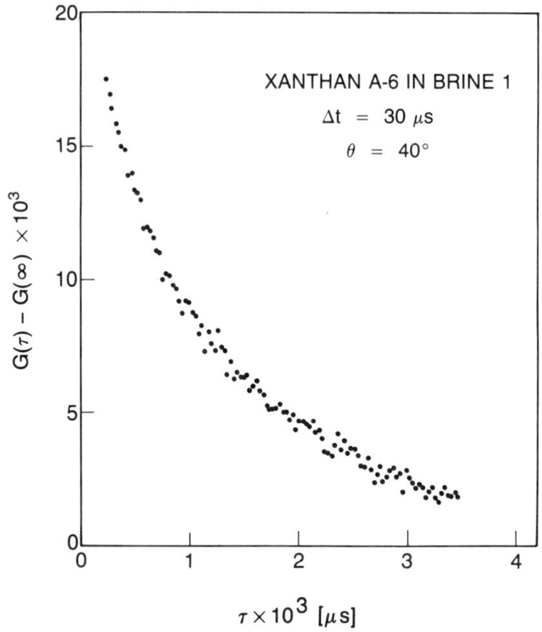

Figure A-1. Typical autocorrelation function. T = 25°C. Xanthan concentration = 0.27 g/L. $\bar{D} = 3.2 \times 10^{-8}$ cm^2/s and polydispersity index $\mu/\bar{\Gamma}^2 = 0.34$ from the method of cumulants analysis.

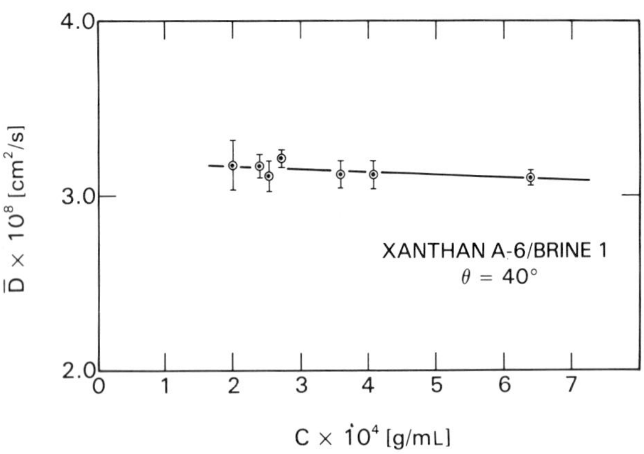

Figure A-2. Effect of xanthan concentration on $\bar{D}$. T = 25°C. Solutions were 0.45 μm filtered.

to the wavelength of light. The inconsistency of these values with previous reports[5,13] suggests that large polymer size is the main cause of the variation in $\bar{D}$ with θ.

TABLE A-1. Effect of Scattering Angle on $\bar{D}$

θ	$\bar{D}$ [10^{-8} cm^2/s]	
	BSA in a Buffered Saline	Xanthan A-6 in Brine 1
90°	60.5	5.6
60°	59.1	-
45°	52.6	3.7
40°	-	3.1
35°	-	2.8
30°	-	2.6

T = 25°C

ACKNOWLEDGMENT

The author wishes to thank B. E. Boeger for his assistance in the dynamic light scattering experiments, and Exxon Chemical Americas, Houston Plant, for use of the KMX-6 instrument. B. J. Henrici (EPRCo.) developed the clarification scheme.

REFERENCES

1. Jansson, P. E., Kenne, L., and Lindberg, B., Carbohydr. Res., 45, 275-282 (1975).
2. Holzwarth, G., and Prestridge, E. G., Science, 197, 757-759 (1977).
3. Milas, M. and Rinaudo, M., Polym. Bull., 12, 507-514 (1984).
4. Seeger, B., Die Nahrung, 25 (7), 655-666 (1981).
5. Southwick, J. G., Jamieson, A. M., and Blackwell, J., Carbohydr. Res., 99 117-127 (1982).
6. In retrospect, this procedure of dispersing the powdered Xanthan D in distilled water may have affected the M_w compared to direct dispersal into brine.
7. Dubois, M., Gilles, K. A., Hamilton, J. K., Rebers, P. A., and Smith, F., Anal. Chem., 28, 350-356 (1956).
8. Tanford, C., "Physical Chemistry of Macromolecules", Wiley, New York (1961).
9. Chauveteau, G, Kohler, N., Soc. Pet. Eng. J., 361-368, June 1984.
10. Ford Jr., N. C., in "Dynamic Light Scattering", Pecora, R., Ed., Plenum, New York (1985) pp. 7-57 and
    Weiner, B., in "Modern Methods of Particle Size Distribution Analysis", Barth, H., Ed., Wiley, New York (1984) pp. 93-116.
11. Koppel, D. E., J. Chem. Phys., 57, 4814-4820 (1972).
12. Sato, T., Norisuye, T., and Fujita, H., Polym. J., 16, 341-350 (1984).
13. Müller, G., Lecourtier, J., Chauveteau, G., and Allain, C., Macromol. Chem. Rapid Commun., 5, 203-208 (1984).
14. Müller, G., Anrhourrache, M., Lecourtier, J., and Chauveteau, G., Int. J. Biol. Macromol., 8, 167-172 (1986).
15. Holzwarth, G. Dev. Ind. Microbiol., 26, 271-280 (1985).
16. Sato, T., Kojima, S., Norisuye, T., and Fujita, H., Polym. J. 16, 423-429 (1984).
17. Lecourtier, J., Chauveteau, G., and Müller, G., Int. J. Biol. Macromol., 8, 306-310 (1986)
18. Pecora, R., in "Scattering Techniques Applied to Supramolecular and Nonequilibrium Systems", Chen, S. H., Chu, B., and Nossal, R., Ed., Plenum, New York (1980) pp. 161-172.

CONFORMATIONAL ANALYSIS OF XANTHAN AND WELAN USING ELECTRON MICROSCOPY

B.T.Stokke, O. Smidsrød, A.B.L. Marthinsen and A.Elgsaeter

Norwegian Biopolymer Center, Division of Biophysics and Biotechnology
University of Trondheim, N-7034 Trondheim - NTH, Norway

ABSTRACT

Electron microscopy of xanthan indicates that this biopolymer can exist as a single- or double-stranded or partly dissociated double-stranded structure depending on the ionic strength prior to preparation for electron microscopy. Welan appears as a uniform thick, convoluted structure with contour length varying from molecule to molecule. The two-dimensional spatial correlation of the tangent direction indicates that the persistence length of double-stranded xanthan in 100mM ammonium acetate is 150 nm; the persistence length of single-stranded xanthan in 2 mM ammonium acetate is observed to be 60 nm and the persistence length of welan in 100 mM ammonium acetate is observed to be 80 nm.

INTRODUCTION

The potential market for utilization of biopolymers in enhanced oil recovery as mobility control or permeability modifying agents has led to continued interest in exploring the molecular mechanisms making these technologically important applications possible. Xanthan and welan (or S-130) are both extracellular microbial polysaccharides that give rise to viscous and pseudoplastic aqueous solutions at relatively low polymer concentrations (1). This is a consequence of the polymer conformation. The primary structure of xanthan, the extracellular polysaccharide produced by the bacterium Xanthomonas campestris, consists of a pentasaccharide repeat unit made up of two (1→4) linked β-D-glucose residues in the main chain and substituted with a trisaccharide, (1→3) linked to every second glucose residue (2). The terminal side chain mannose is substituted with pyruvate, and the mannose residue adjacent to the main chain with acetate. The degree of substitution varies for xanthans of different origin. X-ray fiber diffraction data (3-4) indicate that the axial repeat distance in crystalline xanthan is 0.94 nm. This corresponds to a mass per unit length of about 1000 Dalton/nm for a single-stranded xanthan chain or 2000 Dalton/nm for a double-stranded xanthan chain. There is substantial evidence that xanthan at high ionic strength is dissolved as a double-stranded, semi-flexible structure with an estimated persistence length of 100-150 nm (5-8). The extracellular heteropolysaccharide welan produced by an Alcaligenes specie, ATCC 31555 (9), is reported to have a primary structure consisting of a pentasaccharide repeat unit, →3)-β-D-Glcp-(1→4)-β-D-GlcpA-(1→4) -β-D-Glcp -(1→4)-α-L-Rhap-(1→, in the main chain and substituted with an α-L-Rhap or α-L-Manp (1→3) linked to the glucose residue linked (1→4) to the rhamnopyranosyl residue in the main chain (10). X-ray fiber diffraction data indicate that the axial repeat distance in crystalline welan is 1.83 nm (11), which corresponds to a mass per unit length of about 500 Dalton/nm.

In this communication we report on electron microscopic visualization of xanthan and welan. Analysis of the electron micrographs provides, firstly, mass per unit contour length of polymers with known molecular weight as the ratio between the polymer molecular weight and the proper average of the contour length distribution (12), secondly, spatial correlation in the tangent direction of the polymer molecule, and thirdly, end-to-end distance relative to contour length. These quantities are related to the strandedness and stiffness of the polymer chain.

MATERIALS AND METHODS

Preparation of biopolymer solutions

Xanthan powder (Kelzan XCD, Kelco, a Division of Merck & Co., Inc.) was dissolved (0.3 mg/ml) in 20 mM NaCl, pH 7, and stirred overnight at room temperature. The solution was centriguged at 90 000 x g (Beckman 50 Ti rotor, 35 000 rpm.) for 90 min., to remove cell debris an unsolubilized polymer. Aliquots of the supernatant were dialyzed against 5 x 1000 ml aqueous solution containing from 2 to 100 mM NaCl for viscometry and 2 to 100 mM ammonium acetate for electron microscopy. Xanthan obtained as a preserved fermentation broth (Flocon 4800, Pfizer Inc.) was diluted directly with 100 mM ammonium acetate, and stirred overnight at room temperature. Welan obtained as a powder (Kelco, a Division of Merck & Co., Inc.) was dissolved (0.5 mg/ml) in distilled water, stirred overnight at room temperature, and centrifuged at 15 000 x g (Sorvall GSA rotor, 12 000) rpm) for 90 min. The pellet consisting of unsolubilized polymer and cell debris was discarded, and the polymer in the supernatant was precipitated using iso-propanol. The precipitated polymer was dissolved in distilled water, dialyzed against 5 x 1000 ml, 100 mM NaCl to convert it into the sodium form, and then extensively dialyzed against distilled water and freeze-dried. Freeze-dried welan was dissolved in 10mM NaCl (0.3 mg/ml) stirred overnight, centrifuged at 90 000 x g (Beckman 50 Ti rotor, 35 000 rpm.) and subsequently dialyzed as described for xanthan. The concentrations of xanthan and welan were determined using the colorimetric method of Dubois et al. (13).

Electron microscopy

Biopolymer solutions were mixed with glycerol, ammonium acetate, and distilled water to a final polymer concentration of 4 - 30 μg/ml, 50 % glycerol and the desired ionic strength. An aliquot of this mixture was sprayed onto freshly cleaved mica (14). The samples on the mica discs were vacuum-dried at 10 $^{-6}$ torr for 1 h or more, and rotary replicated with 0.7 nm Pt from an angle of 5 ^{o} and with 7 - 8 nm C from 90 ^{o} to get mechanical support. Replicas were floated off on distilled water, picked up on 400 mesh copper grids and inspected in an electron microscope. Electron micrographs were obtained using a Philips EM 400 electron microscope at an electron optic magnification between 17 000 x and 60 000 x. The total final magnification was calibrated using a line grid (S 102, Agar Aids). The contour lengths, end-to-end distances and spatial correlation of the tangent direction of the visualized xanthan and welan molecules were determined by digitizing and analyzing the electron micrographs as already described (15). The viscosities of dilute xanthan and welan solutions were determined for shear rates ranging from 0.8 s^{-1} to 20 s^{-1} using a Cartesian diver viscometer (16). Intrinsic viscosities were obtained as the intercept of η_{red} versus c in the Newtonian limit.

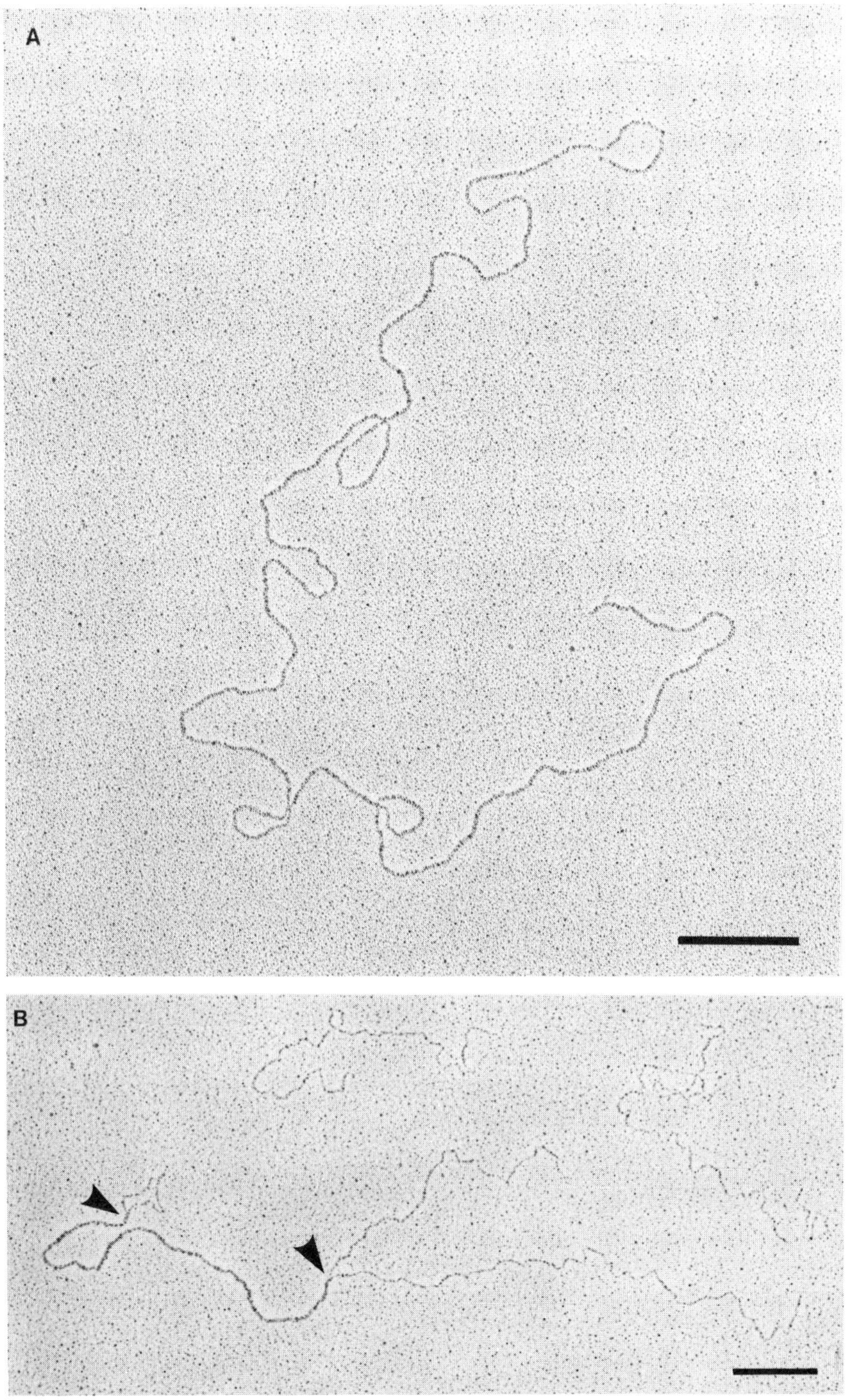

Figure 1. Electron micrograph of xanthan from a fermentation broth (Pfizer Flocon 4800) diluted in 100 mM ammonium acetate (A), and powdered xanthan (Kelco Kelzan XCD) in 2 mM ammonium acetate (B). Scale bar = 200 nm.

RESULTS AND DISCUSSION

Electron microscopy

Fig. 1. shows a representative electron micrograph of xanthan from the fermentation broth vacuum-dried from a glycerol solution containing 100 mM ammonium acetate, pH 7 (A) and powdered xanthan in 2 mM ammonium acetate (B). At high ionic strength, the xanthan obtained as a fermentation broth (Fig. 1A) appears as a uniformly thick, polydisperse and convoluted structure. This is similar to what we previously reported for powdered xanthan (15). At low ionic strength, the xanthan obtained as a powder is observed to branch from a uniformly thick structure into two thinner structures, also of uniform thickness (Fig. 1B, arrows). We find that xanthan obtained as a fermentation broth also displays the same branching phenomena at low ionic strength as the powdered xanthan. Powdered xanthan subjected two ultracentrifugation, sonication , and fractionation by gel-filtration and kept in 100 mM salt concentration during all preparation steps prior to vacuum drying shows similar electron optic contrast as the polymer chain in Fig. 1A, or the thick part of the molecule shown in Fig. 1B. Using the theory for the intrinsic viscosity of the helical wormlike coil (17) we found that the observed intrinsic viscosities and contour lengths of these xanthan fractions in high ionic strength solution were consistent with a double-stranded structure of xanthan (15). The branching observed at low ionic strength is therefore assumed to be the double-stranded xanthan branching into its two subchains. Although this double- to single-stranded conformational change may be related to the order- disorder conformational change of xanthan observed by measuring the optical activity (18,19), a quantitative correlation with data from electron microscopy is difficult because of the possible change in ionic strength during vacuum-drying and change in dimensionality when the three dimensional conformation in solution is changed to the essential two dimensional conformation on the mica disc. For a xanthan (from strain PX 061) with low degree of pyruvate substitution and doubly substituted with acetate on the mannose adjacent to the main chain, we observed previously (15) that the polymer appeared to be single- stranded when prepared from 2 mM ammonium acetate for electron microscopy. By adding ammonium acetate to an ionic strength of 100 mM, we observed single-stranded, perfectly matched double-stranded, and mismatched double-stranded species, along with different sizes and shapes of clusters containing several molecules in the electron micrographs (15). The contorted xanthan structures seen in the present electron micrographs contrast with both the stiff, rod-like appearance reported by Holzwarth and coworkers (20), and the collapsed globules observed by Wellington (21) using methanol in the preparation procedure for electron microscopy.

Welan appear as a convoluted, uniform thick, unbranched polymer with contour length varying from molecule to molecule (Fig. 2). We observed no chain splitting as for xanthan when the ionic strength in the solution prior to vacuum-drying ranged from 1 mM to 100 mM, however, chains of approximately twice the thickness were observed as a minor fraction (<2%). Contour lengths and molecular weight determinations of sonicated and fractionated welan to address the question of strandedness of welan were not obtained, and it is therefore not possible from the present data to determine whether welan is a single- or multi-stranded structure.

Spatial correaltion of the tangent direction

The spatial correlation of the tangent direction of the biopolymer contour was quantitated by evaluating the tangent vectors $\mathbf{v}_i$ and $\mathbf{v}_j$ at two points i and j on the polymer contour. The correlation between the two tangent vectors is expressed as the angle between them:

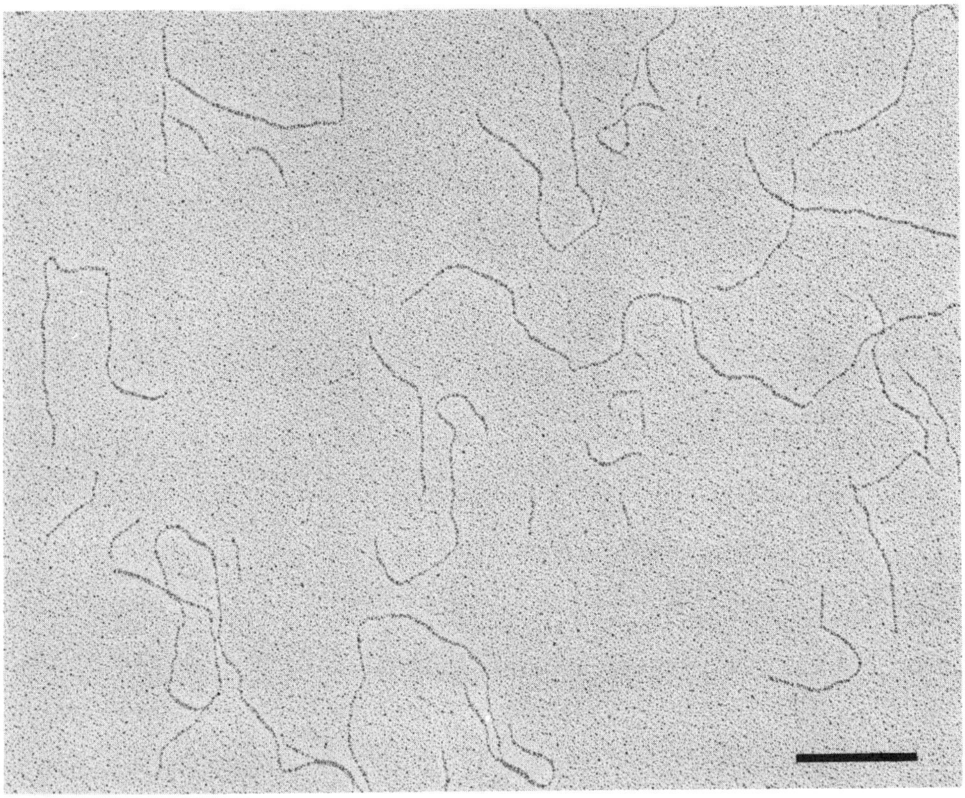

Figure 2. Electron micrograph of welan in 100 mM ammonium acetate. Scale bar = 200 nm.

$$\cos\theta = \mathbf{v}_i\,\mathbf{v}_j / |\mathbf{v}_i||\mathbf{v}_j| \tag{1}$$

Assuming that θ for hypothetical chain segments exhibits a Gaussian distribution around θ = 0, which corresponds to the rectilinear shape, Frontali and coworkers (22) showed that

$$\ln < \cos\theta > = - l / 2q \tag{2}$$

where the average of cos θ, designated by angle brackets, is taken for all combinations of i and j with constant contour distance l between them. In Eq. 2, q is a local flexibility parameter which is equal to the persistence length of the chain. The details of our implementation of this approach is given elsewhere (15). Except for the factor 2 due to the modification to two dimensions, Eq. 2 yields the correlation of the tangent direction versus the contour distance similar to that obtained from statistical mechanical calculations of polysaccharide chains (23). When $<\cos\theta>$ has been determined experimentally through an analysis of the electron micrographs, Eq. 2 yields an estimate of the persistence length. Determination of the persistence length using this method requires only that the total magnification is known. The assumption that θ for hypothetical segments has a Gaussian distribution around θ = 0 can be tested experimentally by evaluating $<\theta^4>/<\theta^2>^2$, taking into account that θ may be larger than π for large $|i-j|$ (18). Fig. 3 shows $\ln<\cos\theta>$ versus l for single- and double stranded

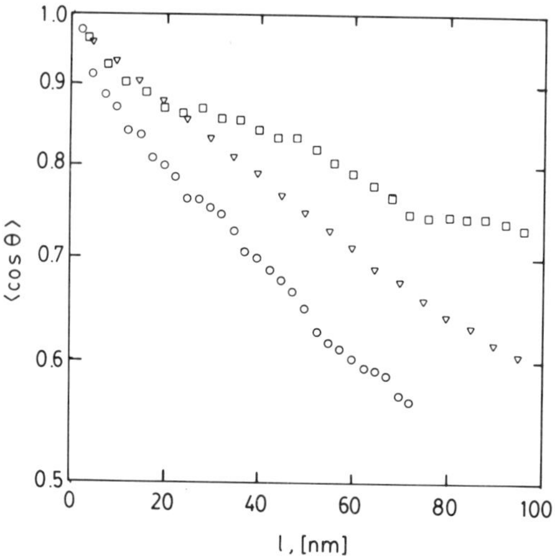

Figure 3. Ln <cos θ> versus l for double stranded xanthan in 100 mM ammonium acetate (□), single stranded xanthan in 2 mM ammonium acetate (O) and welan in 100 mM ammonium acetate (∇) .

xanthan as well as for welan. Digitized data from a total contour length of 142 μ m, 138 μm and 68 μm was included in the averaging procedure for double-stranded xanthan, single-stranded xanthan and welan, respectively. <cos θ> were averaged with an intervall of l = 4nm, 2.5nm and 5 nm as detailed elsewhere (15) for double-stranded xanthan, single-stranded xanthan, and welan, respectively. The contour distance included in the analysis was sufficient to validate the assumptions of the method proposed by Frontali and coworkers (22). Ln <cos θ> decreases approximately linearily with increasing l when l is less than q for the three polymers presented here. The slope of ln<cos θ> versus l yields q = 150 nm for double-stranded xanthan (Kelzan XCD, 100 mM ammonium acetate), q = 60 nm for single-stranded xanthan (strain PX061, 2 mM ammonium acetate), and q = 80 nm for welan (100 mM ammonium acetate). This value for double-stranded xanthan is in accord with a previously reported q = 120 nm based on solution measurements of xanthan modelled as wormlike, double-stranded xanthan chain (4,5). Solution measurements of a xanthan sample believed to be single-stranded yielded q = 50 nm (24).

Integrating the probability density function of θ along the polymer contour relates the end-to-end distance l_e and the contour length L_c with the persistence length q as the only other parameter (25):

$$l_e^2 = 4qL_c \{1 -(2q/L_c)(1-\exp(-L_c/2q))\} \quad (3)$$

Eq. 3 is the 2-dimensional analogue of the well known equation for the worm-like coil. Rewriting Eq. 3:

$$(l_e/L_c)^2 = 4q/L_c \{ 1 - (2q/L_c)(1-\exp(-L_c/2q))\} \quad (4)$$

By plotting the observed $(l_e/L_c)^2$ versus q/L_c for appropriate q, polymers having different stiffness should in principle follow the same general behavior for contour

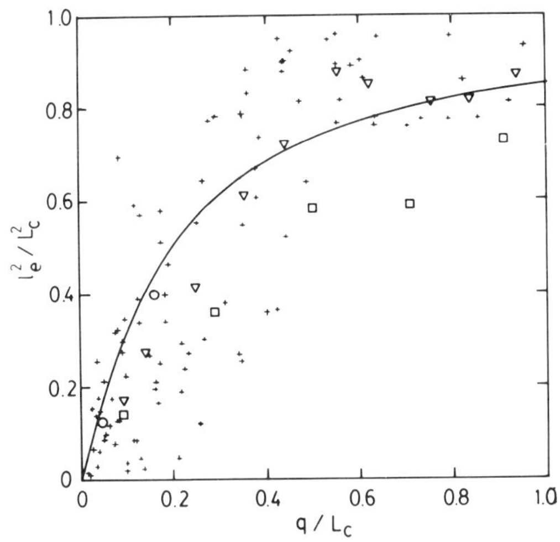

Figure 4. $(l_e/L_c)^2$ versus q/L_c for single welan molecules (+), and the average of $(l_e/L_c)^2$ in intervall of 0.1 of q/L_c for welan (∇) , for double stranded xanthan (□) and single-stranded xanthan (O). The solid curve is calculated from Eq. 4.

lengths shorter as well as longer than their respective persistence length. Fig. 4 shows $(l_e/L_c)^2$ versus q/L_c for individual welan molecules observed from electron micrographs using the the value q = 80 nm obtained for welan using Eq. 2 (+). Also plotted are $(l_e/L_c)^2$ averaged in intervals of $q/L_c = 0.1$ for welan as well as for single- and double-stranded xanthan using their respective q values determined from the electron micrographs with Eq. 2. Fig. 4 shows that there is adequate agreement between experimentally observed values and those predicted from Eq. 4 for a large range of L_c both for single- and double-stranded xanthan as well as for welan. This indicates that the method based on analysis of the spatial correlation of the tangent direction and the method based on the end-to-end distance relative to the contour length are consistent. However, using only the information based on the the end-to-end distance and contour length yields a more uncertain estimate of q than the spatial correlation method.

Intrinsic viscosities

To compare the persistence length for welan obtained from the electron micrographs with an estimate obtained from solution properties, we measured the intrinsic viscosity of welan at different ionic strengths (I) as a function of shear rate in solution (from 0.8 s^{-1} to 20 s^{-1}), and the results were then extrapolated to zero shear rate. Fig. 5 shows that the intrinsic viscosity of welan increases from 12 000 ml/g at I = 100 mM to 16 000 ml/g at I = 3 mM. When the intrinsic viscosity of welan was plotted against $1/\sqrt{I}$, a straight line of slope S = 2.20 was observed (Fig. 5). This slope divided by the intrinsic viscosity at I = 0.1M to the power of 1.3 yields the Smidsrød Haug B parameter (26) equal to B = 0.0043 for welan. The intrinsic viscosity of the powdered xanthan at different ionic strengths, all corresponding to the chiro-optically ordered state of xanthan (18,19) yields B = 0.0030 (Fig. 5). Using the empirical correlation (26) between B-parameters and Kuhn statistical segment lengths, A_m, these B parameter values

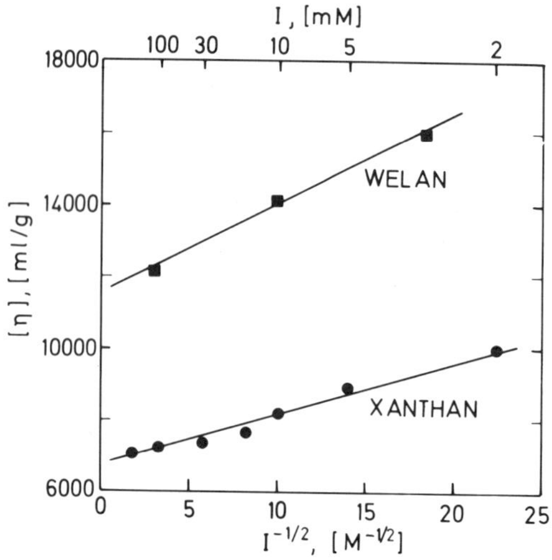

Figure 5. The zero shear intrinsic viscosity of xanthan and welan versus $1/\sqrt{I}$.

indicate that A_m = 130 nm for welan and A_m = 200 nm for double-stranded xanthan. The xanthan sample from strain PX 061, which displays aggregation behavior in the electron micrographs in preparations from 100 mM ammonium acetate (15) cannot be modelled as a linear polyelectrolytic chain at the salt concentrations used, and a B value for this xanthan was therefore not calculated. The values of A_m estimated from the empirical B - parameter are in good agreement with the persistence lengths obtained from the electron micrographs, bearing in mind that A_m = 2q for worm-like coils.

In conclusion, this study indicates that electron micrographs obtained from vacuum-drying of biopolymers in aqueous glycerol followed by Pt rotary shadowing, yields estimates of the persistence lengths for single- and double-stranded xanthan and for welan in adequate agreement with estimates obtained from solution properties. Xanthan and welan are well described in the terms of the worm-like model, and the persistence lengths obtained indicate that these polysaccharides are among the stiffest ones known. It is this structural chain stiffness that gives rise to the highly viscous and salt-tolerant solutions which are important in oil-field applications.

ACKNOWLEDGEMENTS

We gratefully acknowledge valuable comments from Dr. D.A. Brant on this manuscript. Welan was kindly provided by Kelco, a Division of Merck & Co., Inc., San Diego, California. This work was supported by grant T 6399 from the Exploration and Production Laboratories, The Norwegian State Oil Company, and fellowship 2823 from the Royal Norwegian Council for Scientific and Industrial Research to B.T.S.

REFERENCES

1. P.A. Sandford, I.W. Cottrell, and D.A. Pettitt, Microbial polysaccharides: New products and their commercial application, Pure and Appl. Chem., 56:879 (1984).

2. P.E. Jansson, L. Kenne, and B. Lindberg, Structure of the extracellular polysaccharide from Xanthomonas campestris, Carbohydr. Res. 45:275 (1975).

3. R. Moorhouse, M.D. Walkinshaw, and S. Arnott, Xanthan gum - molecular conformation and interactions, in: "Extracellular microbial polysaccharides", P.A. Sandford and A. Laskin, ed., ACS Washington, ACS symposium series 45:81 (1977).

4. K. Okuyama, S. Arnott, R. Moorhouse, M.D. Walkinshaw, E.D.T. Atkins and C. Wolf-Ullish, in: "Solution properties of polysaccharides", D.A. Brant, ed., ACS Washington, ACS symposium series 141: 411 (1980).

5. T. Sato, T. Norisyue, and H. Fujita, Double stranded helix of xanthan: Dissociation behavior in mixtures of water and cadoxen, Polymer J., 17: 729 (1985).

6. T. Sato, T. Norisyue, and H. Fujita, Double stranded helix of xanthan: Dimensional and hydrodynamic properties in 0.1M aqueous sodium chloride. Macromolecules, 17:2696 (1984).

7. G. Paradossi, and D.A. Brant, Light scattering study of a series of xanthan fractons in aqueous solution, Macromolecules, 15:874 (1982).

8. T. Sato, T. Norisyue, and H. Fujita, Double stranded helix of xanthan in dilute solution: Evidence from light scattering, Polymer J., 16: 341 (1984).

9. K.S. Kang, and G.T. Veeder, U.S. Pat. 4,342,866 (1982).

10. P.E. Jansson, B. Lindberg, G. Widmalm, and P.A. Sandford, Structural studies of an extracellular polysaccharide (S-130) elaborated by Alcaligenes ATCC 31555, Carbohydr. Res., 139: 217 (1985).

11. P.T. Attwool, E.T. Atkins, M.J. Miles and V.J. Morris, X-ray fibre diffraction results rom Alcaligenes (AATCC 31555) microbial polysaccharide S-130 and a comparison with gelan gum, Carbohydr. Res., 148:C1 (1986).

12. B.T. Stokke, A. Elgsaeter, G. Skjåk-Bræk, and O. Smidsrød, The molecular size and shape of xanthan, xylinan, bronchial mucin, alginate, and amylose as revealed by electron microscopy, Carbohydr. Res. 160:13 (1987).

13. M. Dubois, K.A. Gilles, J.K. Hamilton, P.A. Rebers, and F. Smith, Colorimetric method for determination of sugars and related substances, Anal. Chem., 28:350 (1956).

14. J.M. Tyler, and D. Branton, Rotary shadowing of extended molecules dried from glycerol, J. Ultrastruct. Res., 71: 95 (1980).

15. B.T. Stokke, A. Elgsaeter, and O. Smidsrød, Electron microscopic study of single- and double-stranded xanthan, Int. J. Biol. Macromol., 8: 217 (1986).

16. M. Troll, K.A. Dill, and B.H. Zimm, Dynamics of polymer solutions. 3. An instrument for stress relaxations on dilute solutions of large polymer molecules, Macromolecules. 13: 436 (1980).

17. H. Yamakawa, and T. Yoshizaki, Transport coeffecients of helical worm-like chains. 3. Intrinsic viscosity, Macromolecules, 13: 643 (1980).

18. I.T. Norton, D. M. Goodall, S. A. Frangou, E.R. Morris, and D.A. Rees, Mechanism and dynamic conformational ordering in xanthan polysaccharide, J. Mol. Biol., 175:371 (1984).

19. S. Paoletti, A. Cesaro, and F. Delben, Thermally induced conformational transition of xanthan polyelectrolyte, Carbohydr. Res., 123:173 (1983).

20. G. Holzwarth, and E. B. Prestridge, Multistranded helix in Xanthan polysaccharide, Science, 197:757 (1977).

21. S.L. Wellington, Xanthan gum molecular size distribution and configuration, Polym. Preprints, Prep. Div. Polym. Chem. Ass. Chem. Soc., 22:63 (1981).

22. C. Frontali, E. Dore, A. Ferrauto, E. Gratton, A Bettini, M.R. Pozzan, and E. Valdevit, An abolute method for the determination of the persistence length of native DNA from electron micrographs, Biopolymers, 18:1253 (1979).

23. B.A. Burton, and D.A. Brant, Comparative flexibility, extension and conformation of some simple polysaccharide chains, Biopolymers, 22:1769 (1983).

24. G. Muller, J. Lecourtier, G. Chauveteau, and C. Allain, Conformation of the xanthan molecule in an ordered structure, Makromol. Chem., Rap. Commun. 5: 203 (1984)

25. H. Hofman, T. Voss, K. Kuhn, and J. Engel, Localization of flexible sites in thread-like molecules from electron micrographs, J. Mol Biol., 172:325 (1984).

26. O. Smidsrød, and A. Haug, Estimation of the relative stiffness of the molecular chain in polyelectrolytes from measurements of viscosity at different ionic strengths, Biopolymers, 10:1231 (1971).

DYNAMIC UNIAXIAL EXTENSIONAL VISCOSITIES AND THEIR IMPORTANCE IN THE MECHANICAL STABILITY OF WATER-SOLUBLE CARBOHYDRATE POLYMER SOLUTIONS

David A. Soules and J. Edward Glass

Polymers and Coatings Department
North Dakota State University
Fargo, North Dakota 58105

ABSTRACT

The mechanical stabilities of a series of carbohydrate polymers are examined and correlated to aqueous solution dynamic uniaxial extensional viscosities. Both properties relate to the carbohydrate polymer's conformation in aqueous solution. As noted in the stability of guaran solutions this property is not reflected by Mark-Houwink coefficients determined under static conditions. Helical or rigid rod carbohydrate polymers exhibit the greatest mechanical stability in solution. Solution blends of the carbohydrate polymers with high molecular weight poly(oxyethylene) also were studied. Unique rheological responses were observed in Sclerotium glucanium polysaccharide/poly(oxyethylene) blends at equal weight percents (e.g., lower, low-shear-rate viscosities and the absence of a high extensional viscosity), suggesting complexation. Such complexation appears to effectively disrupt the three dimensional structure of Sclerotium glucanium polysaccharide in aqueous solution (storage modulus studies) and decrease the flexibility of poly(oxyethylene).

INTRODUCTION

The viscosity stability of water-soluble polymer solutions is critical to their performance in many application areas; this is particularly true in petroleum recovery processes. The importance of viscosity stability in mobility control buffer applications (i.e., in polymer augmented water flooding and as a low cost viscosifying medium behind more expensive chemical slugs in EOR) has been highlighted by Chang(1) and in drilling by Thomas(2). Maerker(3) has cited the mechanical instability of poly(acrylamide) aqueous solutions.

The mechanical degradation of synthetic polymers in organic solvents has generated a vigorous debate over the relationship between polymer conformation and mechanical stability. The most definitive study among these early investigations examined(4) poly(styrene) in solvents of good and poor solvating properties. The mechanical degradation of polystyrene in poor solvents, where the conformation is collapsed, occurs at a lower concentration than in a good solvent, where the conformation is that of a random coil. Above a critical shear stress at which the polymer degrades in both solvent types, degradation is faster in the poorer solvating media, in which the polymer is contracted and entangled.

Aqueous solution studies of sodium poly(styrene sulfonate), in which the electrolyte concentration was varied, again revealed(5-8) that the more extended the polymer conformation the greater its stability. The deformation was predominately extensional in these latter studies; gel permeation chromatography(5,6) and polarized light birefringent-pressure drop(7,8) techniques were used to follow degradation.

In petroleum applications the mechanical instability of synthetic polymers is well recognized. The current study has concentrated on water-soluble carbohydrate polymers in which the conformations of the polymers are different. With the exception of guaran, the carbohydrate polymers studied have a common glucopyranose repeating unit. Guaran differs only in the stereochemistry of the C-2 hydroxyl in the main chain and of the C-4 hydroxyl in the branched units. Since these polymers contain the same approximate repeating units, differences in mechanical stability in fresh water should be related to physical parameters such as the solution conformation of the polymer. The conformation is influenced by substituents attached to the backbone. The carbohydrate polymers prepared by fermentation processes, SGPS and XCPS, form helices in moderately concentrated solution(9-11). The former, SGPS, is also noted to form three dimensional networks in aqueous solution(12,13).

EXPERIMENTAL

The water-soluble polymers (W-SPs) were supplied by the manufacturers. The polymers were: Sclerotium glucanium polysaccharide, (SGPS, Actigum CS-11, Jetco Corp.); Xanthomonas campestris polysaccharide (XCPS, often referred to as Xanthan gum, Kelzan SS-4000, Kelco Corp.); guaran (Galactosol 211, a polygalactomannan from the Henkel Corp.); hydroxyethyl cellulose, (M.S. = 2.0, Cellosize QP100M, Union Carbide Corp. and M.S. = 2.5, Natrosol 250 HHR, Hercules Incorp.); Cellulose Sulfate Ester (M.S. = 0.7, an experimental polymer from Stauffer Chemical Co.); and poly-(oxyethylene), (POLYOX 301 Coag., Union Carbide Corp.).

All materials contained adsorbed water, inorganic salts and in some cases microorganism debris. The salts were extracted with an aqueous (25 volume %) t-butanol (75 volume %) mixture; however, little success was noted in lowering the salt content of the fermentation polymer XCPS. Adsorbed moisture was removed by vacuum drying. Microorganism debris when present were removed by filtration through Millipore polyacetate filters. Commercial guaran contained approximately 8 weight percent protein. Solvent extraction and precipitation techniques were only partially successful in removing the protein from the mannan polymer.

A. Preparation of Water-Soluble Polymer Solutions. Aqueous W-SP solutions were prepared by adding the required weight of polymer to 300cc of deionized water under vigorous agitation (Waring blendor) for approximately 20 seconds. The dispersed polysaccharide slurries were immediately added to 700cc of water and stirred by blade impellers until dissolution occurred.

B. Degradation Tests. Thickened solutions for mechanical degradation were prepared to 37.0-40.0 Pa·s viscosities ($\dot{\gamma} = 12s^{-1}$), and exposed to mechanical stresses in a Waring blendor (#4 speed). Water (10^oC) was passed through an exterior jacket to minimize heat buildup in the liquid under deformation.

C. Viscosity Measurements. Shear viscosities were measured with cone-and-plate viscometers(14). The dynamic uniaxial extensional viscosities were measured with a spinning fiber apparatus(15,16). Improved measurement sensitivity in this study was achieved by replacing the extension wheel with a suction device(17) and a rack and pinion stand for varying the length of the fiber. Additional improvements involved replacing the pressure transducer (recording the resistance to extensional deformation) with a Cahn Electrobalance (Model 7000), Hewlett-Packard XY recorder (Model 7047A) and the use of a pulse-free transfer pump.

RESULTS AND DISCUSSION

Chain Conformation and Solution Properties

To a first approximation the exponential constant, a, in the Mark-Houwink equation reflects the solution conformation of a macromolecule. The exponential constants for the carbohydrate polymers studied, excluding the cellulose sulfate ester, are listed in Table I. Generally a value of 0.7-0.8 reflects a random coil conformation; as the constant approaches or exceeds 1.0 the conformation approaches a rod-like shape.

Table 1. Solution Properties of Water-Soluble Carbohydrate Polymers[a]

Polymer	Mark-Houwink Exponent	Sol'n Viscosity (Pa·s) @ $100s^{-1}$ Deformation: Shear	Extension[f]
HEC	0.78[b]	0.71	1.22
GUARAN	0.80[c]	0.70	0.16
XCPS	0.96, 1.34[d]	0.56	0.24
SGPS	1.70[e]	0.55	0.04

(a) Concentration, Wt % = 1.0; Dynamic uniaxial extensional viscosity(DUEV) of POE = 3.6 Pa·s at $100s^{-1}$, Wt.% = 0.1.
(b) Brown,W., Henley,D., Ohman,J., Macomol. Chem.,(1963), 64,49; Ark.Kemi., (1962)19,227.
(c) Sharman,W.R., Richards,F.L., and Malcolm, G.N., Biopolymers,(1978) 17,2917.
(d) Holzwarth, G., Carbohydr. Res.,(1978) 66,173., higher value for lower molecular weight
(e) Norisuye, T., Yanake,T., and Fujita,H., J. Polym. Sci. Physics Ed. (1980),18,547.
(f) Fiber length were 2 cm for HEC, guaran, and XCPS; 1 cm for SGPS. The total extensional force in these experiments was 6 - 24 dynes.

In mechanical degradation, chain scission does not occur from individual attack on a specific atom of the macromolecular chain, but from the application of a critical stress. The extensional viscosities of high-molecular-weight, synthetic polymers have been studied for the past two decades, in attempts to relate the values with drag reduction behavior. To date the dynamic uniaxial extensional viscosities of aqueous carbohydrate solutions have not been reported.

Data comparing the relative relaxation times of naturally occurring carbohydrate polymers with the transitions in helical conformations observed in some fermentation carbohydrate polymers have not been reported under comparable conditions. Considering the difference in segmental flexibility between synthetic and carbohydrate polymers it might be expected that a multistranded helix would provide a lower dynamic uniaxial extensional viscosity than a conformationally flexible polymer (e.g., hydroxyethyl cellulose, HEC). This is observed; the aqueous solution data are reported in Table I. With one exception, guaran, the results are in accord with what might be expected based upon solution conformations reflected by the Mark-Houwink coefficient. The extensional viscosity of SGPS is particularly low. Extensional viscosities are proportional to the polymer's molecular weight. This fermentation polymer achieves high low-shear-rate viscosities via three dimensional networks (12,13) in aqueous solution. This polymer is relatively low in molecular weight ($5 \cdot 10^5$ g/mole) relative to the other polymers (10^6) used in this study.

Mechanical Stability

The viscosity stability under mechanical deformation as a function of time is illustrated in Figure 1 for solutions thickened with XCPS, SGPS and HEC. Under equivalent conditions high molecular weight synthetic polymers (e.g., poly(oxyethylene) or the 30% hydrolyzed acrylamide copolymer) lose 90 percent of their viscosity within 10 minutes. Carbohydrate polymers have greater mechanical stability. The greater stability of the XCPS solution is characterized by both a lower solution extensional viscosity at a moderate deformation rate ($\dot{\epsilon}$ = $100 s^{-1}$)and lower shear viscosity(18) at a high deformation rate ($\dot{\gamma}$ = $10^4 s^{-1}$).

A similar investigation of comparable viscosity HEC solutions is illustrated in Figure 2. The lower molecular weight HECs exhibit greater stability. This might be related to their greater rigidity(19), also reflected in lower characteristic ratios among cellulose acetates of variable molecular weights(20). The extensional viscosity of the lower molecular weight HECs could not be measured but the higher viscosities at high shear rate of the lower molecular weight polymers(21) reflect the lack of importance of high shear rate viscosities in mechanical degradation.

Recent studies(22,23) have indicated several methods for disrupting the hydrogen bonding that promotes helix formation in fermentation carbohydrate polymers (e.g., alkalinity in SGPS, urea in XCPS solutions). The influence of urea is less marked in SGPS solutions, probably due to the absence of an acid-base reaction. The addition of

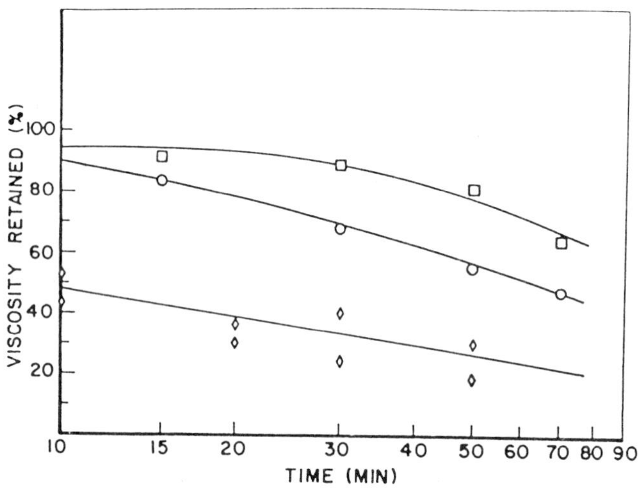

Figure 1. Time dependence of percent retained viscosity under mechanical deformation (Waring blendor, #4 speed);□, 0.28 wt.% XCPS;○, 0.28 wt.% SGPS;◇, 0.50 wt.% HEC. Initial viscosity 375 mPa·s at 37.8 s^{-1}

sodium hydroxide to SGPS solutions results in a rapid viscosity drop at ambient temperatures; a 0.04M NaOH level (a 0.3wt % SGPS solution with 0.2M caustic has a very low viscosity) produces a very dramatic viscosity drop with increasing solution temperature. The stabilities of the fermentation polymers in aqueous solutions alone and in aqueous solutions containing these conformational transforming additives are illustrated in Figure 3. Destruction of the helical conformations results in greater mechanical degradation.

The viscosity loss with time of solutions thickened with derivatized natural carbohydrate polymers is illustrated in Figure 4. Cellulose sulfate ester, whose anionic units are the salts of a strong acid, would be in an extended conformation due to electrostatic repulsions. In fresh water solutions this polymer would, therefore, be expected to have greater stability than HEC, which is observed. One polymer, guaran, does not appear to fit the above generalization. Guaran exhibits greater mechanical stability (Figure 4) than would be expected from the exponential constants in Table 1.

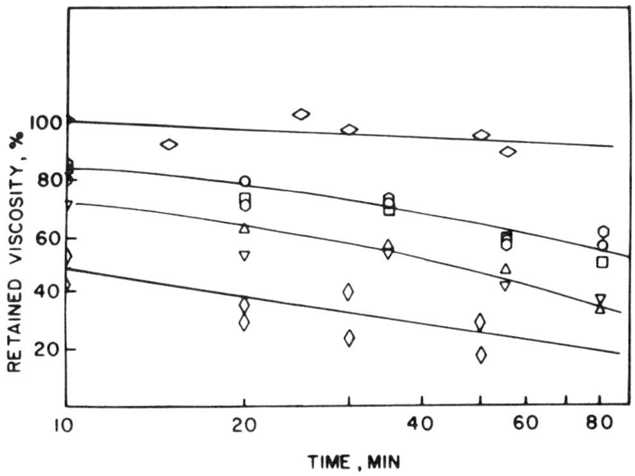

Figure 2. Time dependence of percent retained viscosity under mechanical deformation (Waring blendor, #4 speed) HEC molecular weights (10^{-5}): ◇,3.0; ○,4.0; ⬡,4.9; □,5.1; △,6.4; ▽,8.2; ◊,12.0. Initial viscosity 375 mPa·s at 37.8 s^{-1}

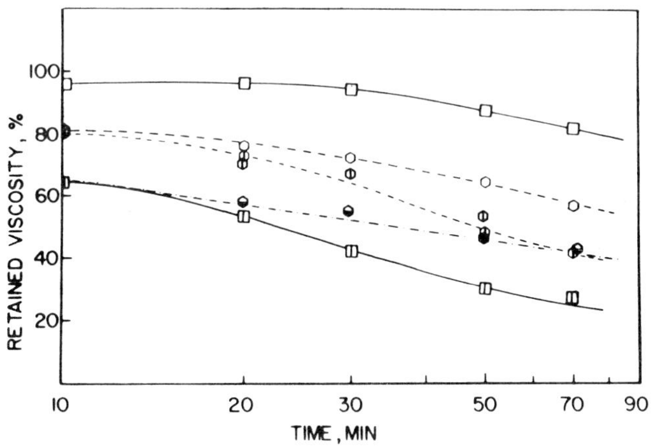

Figure 3. Time dependence of percent retained viscosity under mechanical deformation (Waring blendor, #4 speed) □, XCPS, 0.15 wt.%; ◫, XCPS, 0.15 wt.% in 4M urea; ⬡, SGPS, 0.3 wt.%; ⦶, SGPS, 0.3 wt.% in 4M urea; ⬢, SGPS, 0.3 wt. % in 0.04M sodium hydroxide.

Structurally guaran resembles a fermentation polysaccharide more than a modified cellulose ether (e.g., guaran contains pyranosyl branches from the main chain, in a 1:2 ratio, but the branching is not uniformly distributed as is the case in XCPS and SGPS). The extensional viscosity of the guaran solution is lower (Table 1) and greater stability is observed (Figure 4) in the mechanical degradation studies than is recorded in the 10^6 molecular weight HEC solution (Figure 1). Hyper-entanglement whereby distinct associations in guaran solutions have been proposed(24) is based on both specific viscosity and storage modulus (Figure 5) dependence on concentration measurements. Mechanistically, hyperentanglements with distinct associations lead to a slightly "stiffened conformation"(24) which would correlate with the greater stability of guaran and the lower dynamic uniaxial extensional viscosity observed in this study. The hyperentanglements are probably similar to the super molecular order observed in poly(vinyl alcohol) solutions(25).

Carbohydrate/Poly(oxyethylene) Aqueous Blends

The fermentation polymers, XCPS and SGPS, have found use in various petroleum applications, in part because of their greater mechanical stability. The major disadvantage with both polymers has been their injectivity due to microorganism debris and due to intermolecular associations. An observation made in carbohydrate polymer blends with 10^6 M_w poly(oxyethylene) could provide improved SGPS injectivity. At low shear rates synergistic increases in viscosities were observed in XCPS/POE and HEC/POE blends and increased extensional viscosities above that of POE were observed. The blend containing the most flexible carbohydrate polymer(HEC) gave the highest dynamic uniaxial extensional viscosity readings, consistent with the single component data in Table 1. The rheological response was dramatically different in the SGPS/POE blends. At low to moderate uniaxial extension rates, the high extensional viscosity of POE solutions was not observed in equivalent weight blends with SGPS.

The lower, low-shear-rate viscosities and the absence of a high extensional viscosity in POE/SGPS aqueous solution blends suggest complexation. Such complexation appears to effectively disrupt the three

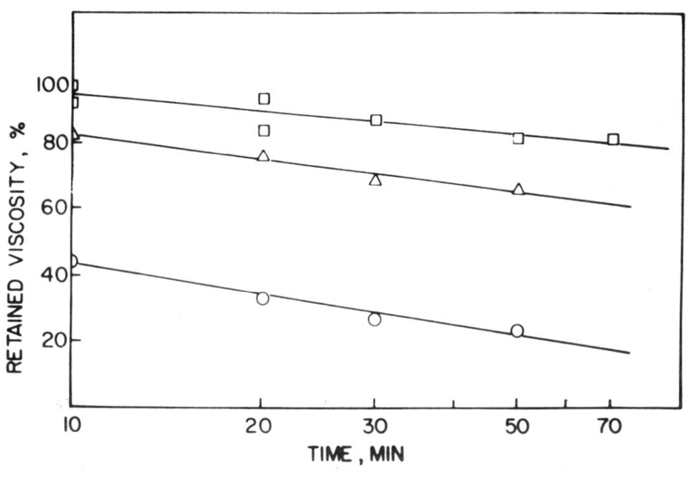

Figure 4. Time dependence of percent retained viscosity under mechanical deformation (Waring blendor, #4 speed) □, Guaran; △, CSE; ○, HEC. Initial solution viscosities approximately 400 mPa·s; $\dot{\gamma}$ = 12s^{-1}

dimensional structure of SGPS in aqueous solution (storage modulus studies, Figure 6, support this postulate) and decrease the flexibility of POE. The complex should allow greater injectivity into tight subterranean formations; however, utilization of these blends may be affected by other factors. For example, competitive adsorption(26) within the wellbore would probably lead to disruption of the complex and possibly reformation of the SGPS associations, with improved profile modification. Many aspects of this investigation require additional study.

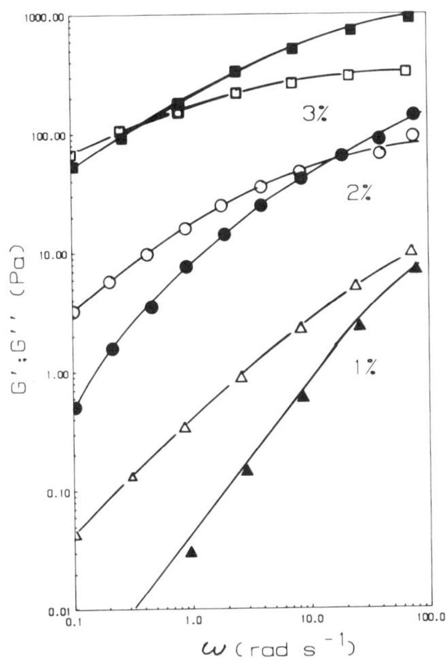

Figure 5. Moduli dependence on angular frequency of variable concentration guaran solutions. Open symbols: loss modulus; closed symbols: storage modulus. Reproduced with permission from reference 24. Copyright 1982 Elsevier.

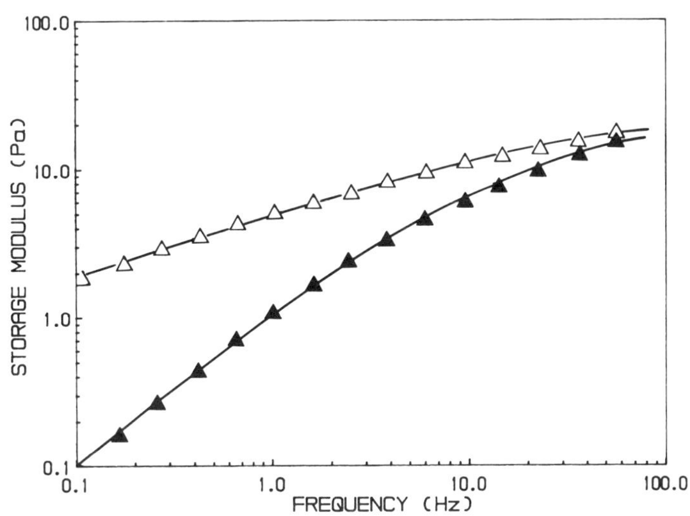

Figure 6. Angular frequency dependence of storage modulus. Open symbol: SGPS at 0.50 wt.%; closed symbol: SGPS/POE (0.50/0.90 wt.%).

REFERENCES

1. Chang, H.L. J. Pet. Tech. (1978), 113-1128.
2. Thomas, D. Soc. Pet. Eng. J. (1982), 171-180.
3. Maerker, J.M. Soc.Pet. Eng. J. (1975), 311-322.
4. Zakin, J.L.; Hunston, D.L. J. Appl. Polym. Sci. (1978), 22,1763.
5. Merril, E.W.; Leopairat, P. Polym. Eng. Sci. (1980), 20(7),505.
6. Leopairat, P., Ph.D. Thesis, Massachusetts Institute of Technology (1982).
7. Ferrel, C.J.; Keller, A.; Miles, M.J.; Pope, D.P. Polymer (1980), 21,1292.
8. Miles, M.J.; Tanaka, K.; Keller, A. Polymer (1983), 24,1801.
9. Holzwarth, G. Biochemistry (1976), 15(19),4333.
10. Southwick, J.G.; Lee. H.; Jamieson, A.M. Carbohydr. Res. (1980), 84,287.
11. Sato, T.; Norisuye, T.; Fujita, H. Macromolecules (1984), 17,2696.
12. Norisuye, T.; Yanake, T.; Fujita, H. J. Polym. Sci. Physics Ed. (1980), 18,547.
13. Kashiwagi, Y.; Norisuye, T.; Fujita, H. Macromolecules (1981), 14,1220.
14. Van Wazer J.R.; Lyons, J.W.; Kim, K.Y.; Colwell, R.E. "Viscosity and Flow Measurement; a Laboratory Handbook of Rheology"; Interscience Pub., New York, N.Y., 1963.
15. Weinberger, C.B.; Goddard, J.D. Int. J. Multiphase Flow (1974), 1,465.
16. Hudson, N.E.; Ferguson, J.; Mackie, P. Trans.Soc.Rheol. (1974), 18,541.
17. Khagram, M.; Gupta, R.K.; Sridhar, T. J. Rheol. (1985), 29(2),191.
18. Fernando, R.H.; Soules, D.A.; Glass, J. E. J.Rheo. in press
19. Foweraker, A.R.; Jennings, B.R. Makromol. Chem. (1977), 178,505.
20. Flory,P.J., "Principles of Polymer Chemistry", Cornell University Press, Ithaca, New York, 1953, Chap.14.
21. Glass, J.E. J. Coatings Technol. (1978), 50(640),61.
22. Lambert, F.; Milas, M.; Rinaudo, M. Proceedings XIth International Carbohydrate Sym., Univ. British Columbia 1982 22-28, Vancouver, B.C., Canada.
23. Frangon, S.A.; Morris, E.R.; Rees, D.A.; Richardson, R.K.; Ross-Murphy, S.B. J. Polym. Sci.: Polym. Letters Ed. (1982), 20,531-538.
24. Robinson, G.; Ross-Murphy, S.B.; Morris, E.R. Carbohyd.Res. (1982), 107,17.
25. Gruber, E.; Soehendra, B.; Schurz, J. J. Polym.Sci. (1974), N-44,195.
26. Glass, J.E.; Ahmed, H.; Karunasena, A.; McCarthy, G.J. Colloids and Surfaces (1986),21,323.

THERMALLY PROMOTED HYDROLYSIS OF POLYACRYLAMIDE

J. J. Maurer, G. D. Harvey and L. P. Klemann

Corporate Research Laboratories
Exxon Research and Engineering Company
Route 22 East, Clinton Township
Annandale, New Jersey 08801

INTRODUCTION

Most low molecular weight (MW) amides will not hydrolyze in water alone. Hydrolysis can, however, be effected via acid or base catalysis, but elevated temperatures for an extended period of time may be required.(1) It has been shown that the nature of organic reactions involving macromolecules may differ significantly from those of their low MW analogs. This may be reflected in the kinetics and mechanism of the reaction as well as in the products obtained.(2,3) Such effects have been ascribed to steric or electrostatic interactions between different structural units of the polymer or between functional groups of the polymer and a given reagent.

The acid or base catalyzed hydrolysis of polyacrylamide (PAM) or its partially hydrolyzed counterpart (HPAM) in aqueous solutions has been the subject of numerous studies, in part because this system provides an opportunity for evaluating the influence of polymer composition, and steric and electrostatic effects on the course of a simple organic reaction in a convenient solvent medium.

Analysis of the acid catalyzed hydrolysis of PAM is complicated due to the possibility of intra- and inter-molecular imide formation.(4) Perhaps for this reason, studies of the base catalyzed hydrolysis of PAM are more numerous. These investigations indicate that the base catalyzed hydrolysis of PAM may be influenced by a number of factors including electrostatic interactions (2,5), nearest neighbor effects (6), and imide formation associated with specific polymer chain microstructure(s).(7)

Aging of aqueous or saline solutions of PAM at elevated temperature in the absence of catalyst can also lead to polymer hydrolysis (8-10), herein termed thermally promoted (TP) hydrolysis. Many studies have been concerned either directly or indirectly with the influence of TP hydrolysis on the phase behavior of hydrolyzed PAM in brine solutions which contain sufficient divalent ions (i.e. Ca^{2+} and/or Mg^{2+}) to lead to phase changes in the system at higher degrees of hydrolysis. The TP hydrolysis of PAM has recently been treated by a general kinetics approach with some simplifying assumptions regarding the mechanism of hydrolysis(11). In general, studies of the TP hydrolysis of PAM have been of relatively limited scope.

This contribution will consider two approaches for examining the extent and nature of the TP hydrolysis of PAM. One involved conversion of carboxylate groups on the hydrolyzed polymer to carboxyl groups and subsequent titration of the acid. The other approach consisted of analyzing the ammonium ion formed during TP hydrolysis. One of our objectives will be to consider the relative merits of these characterization procedures for elucidating the hydrolysis of PAM. Hydrolysis data obtained by these techniques will then be used to explore several aspects of the TP hydrolysis of PAM including the influence of salt and the possibility of imide formation.

EXPERIMENTAL

Polymers: PAM was synthesized by conventional free radical polymerization techniques.

Polymer Solutions: Solutions were made up to approximately 0.2 wt% with deionized, distilled water, or NaCl containing aqueous brine as indicated. Oxygen was not excluded since similar PAM hydrolysis characteristics were observed in previous aerobic vs. anaerobic aging studies.

Aging Conditions: Polymer solutions were aged in a 93°C air oven in sealed glass ampoules, or in volumetric flasks fitted with a special stopcock (Ace Glass #8648-20) and taped for safety considerations.

Hydrolysis Determination: Aged solutions were analyzed by one or more of the following methods:

(a) Ion Exchange/Titration: The degree of hydrolysis of the polymers was determined for aqueous solutions by (1) conversion of the ionic acrylate groups to carboxylic acid moieties via treatment with cation (Dowex 50W-X4)- and anion (Dowex 1-X8)-exchange resins, followed by (2) titration of the carboxylic acid functionalities. Saline solutions of polymer were dialyzed to remove most of the salt before they were contacted with the ion exchange resins. Dialysis of the polymer solutions was performed using Spectrapor® Membrane Tubing #2 which had been pre-conditioned by boiling in 2% $NaHCO_3$ for two hours, followed by washing extensively with distilled water. Polymer solution (0.1-0.2 wt%; 0.04 L) contained in this membrane was dialyzed in 4 L of distilled water for 48 hours. The solution outside the membrane was replaced with fresh distilled water after 8, 24 and 32 hours. The precision of this method is about ±10 % of the measured value.

(b) Ion Chromatography (IC): IC characterization of aged, aqueous PAM solutions was conducted using a Dionex 2020i ion chromatograph operating at ambient temperature. The ammonium ion analysis was performed in an isocratic mode with 0.005 N HCl eluent (2.0 mL/min) and a 200 mm x 4 mm ID Dionex HPIC-CSI cation separator column. A standard cation fiber supressor which was continuously regenerated with 0.04 N tetramethylammonium hydroxide solution (3.0 mL/min) was used. Ion detection was by conductivity (nominal 8 μS background). Aqueous samples were charged to the injection loop after passage through a 0.4 μ filter. The samples examined by IC were typically 100-fold dilutions made up in distilled, deionized water. Standard ammonium fluoride solutions were used for calibration, and detector response was shown to be linear over the range of sample concentrations of interest here. In all other respects, the operating conditions for ammonium ion analysis were similar to those described elsewhere for analysis of sodium ion in water soluble polymers.(12) The precision of this method is about ±10 % of the measured value.

(c) Ammonia Analysis: Ammonium ion was converted to ammonia which was analyzed by an Orion Model 95-12 Ammonia Electrode coupled with an Orion Model 901 Ionalyzer. The precision of this method is about ±5 % of the measured value.

UV spectra were recorded on an IBM 9430 UV-VIS Spectrometer and aqueous samples were contained in 0.5 cm path length Quartz Suprasil cells (Hellma Cells).

Proton decoupled ^{13}C NMR spectra were obtained on 5% solutions of PAM in D_2O sealed in 10 mm precision bore tubes, on a Varian XL-300 spectrometer using dioxane as an internal reference (δ = 67.0 ppm). The following spectrometer conditions were used: gated decoupling, 45° pulse width, 0.5 second acquisition time, and 2.5 second cycle time. A minimum of 20,000 transients were acquired. Integration of carbonyl carbon and aliphatic carbons indicated that these conditions were satisfactory for quantitation. The degree of hydrolysis of the aged PAM solutions was determined by a comparison of the integrals of the amide carbonyls (δ_C = 180 ppm) and the carboxylate carbonyls (δ_C = 182 ppm).

The acidity values, ammonium ion contents and NH_3 concentrations determined by the approaches described above were used to calculate the degree of hydrolysis of the aged polymer. This treatment assumes that all the ammonium ion or ammonia arises solely from the primary hydrolysis reaction (i.e. the conversion of amide to carboxylate and accompanying ammonium ion). If, for example, some of the NH_4^+ ion derives from the formation of imide linkages, then the apparent degree of hydrolysis based on an analysis of carboxyl group concentration might differ from that based on ammonium ion or ammonia, depending on aging conditions and time.

RESULTS AND DISCUSSION

The first phase of this study involved measurement of the degree of hydrolysis via conversion of ionic carboxylate groups on the polymer molecule to carboxylic acid groups followed by titration of the acid groups. TP hydrolysis of aqueous vs. saline solutions of PAM, as determined by this route, is shown in Figure 1. In agreement with other studies (8-10), it is seen that substantial hydrolysis occurs in the absence of base catalyst, but at a slower rate than when a catalyst is present (2,7).

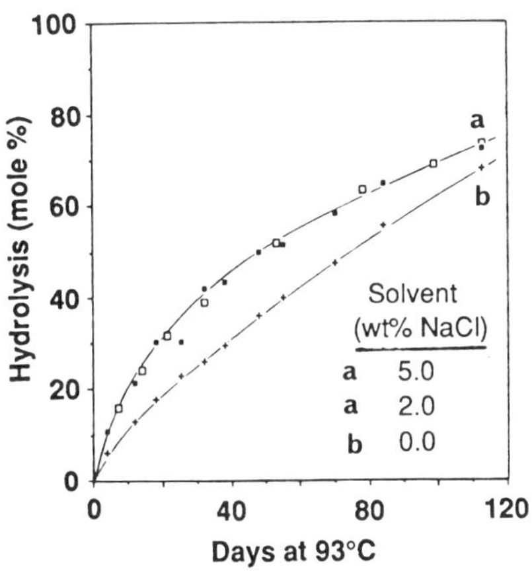

Fig. 1. Evaluation of PAM Hydrolysis (Titration Method).

The hydrolysis curve for the aqueous system indicates that the rate of hydrolysis is gradually decreasing as the degree of hydrolysis increases. This general trend was also noted by Moens and Smets for the base catalyzed hydrolysis of PAM and HPAM (2). The mechanism which they proposed was that the negative carboxylate ions partially shielded the remaining amide groups from the negatively charged hydroxyl ions (i.e. electrostatic repulsion of polymer bound carboxylate groups and OH^- ions). With regard to TP hydrolysis, the observed effect might also involve repulsive effects due to the carboxylate groups on the polymer chain. In this case, however, the mechanism may involve separation of the remaining amide groups from each other. Higuchi et al. (6) have concluded from a study of the base catalyzed hydrolysis of PAM and of low molecular weight analogs that an important consideration is how closely spaced the amide groups are, rather than the fact that they are specifically on a polymer chain. The latter mechanism was proposed by Smets and co-workers (2) based on an analysis of the base catalyzed hydrolysis of PAM and HPAM.

The rate of the TP hydrolysis of PAM in salt solutions is substantially faster than that in aqueous solutions. The reason for this effect has not yet been established; however, several possibilities can be suggested including changes in polymer-polymer vs. polymer-solvent interactions, and the effect of the salt on the polyelectrolyte (HPAM) formed during hydrolysis of PAM. As shown in Fig.1, virtually identical hydrolysis behavior was noted for 0.2 wt% PAM solutions in 2.0 vs. 5.0 wt% NaCl.

The second phase of this study involved determination of the degree of hydrolysis via analysis of the ammonium ion produced during TP hydrolysis. Two methods of ammonium ion analysis were investigated. The first of these involved Ion Chromatography (IC) of aged aqueous solutions of PAM. Saline solutions were not considered due to problems associated with the large amount of salt (3.5%) vs. polymer (0.2 %) in the solutions to be aged. The use of IC for analyses of the sodium counterion concentration in HPAM solutions has been described elsewhere(12). Similar procedures were used to analyze the ammonium ion in the HPAM solutions which result from the TP hydrolysis of PAM. The results of duplicate aging experiments, conducted at different times, are shown in Figure 2. This Figure compares the hydrolysis data obtained by IC vs that obtained by the titration method. Reproducibility of the data is considered good in view of the fact that the data were generated in separate aging studies.

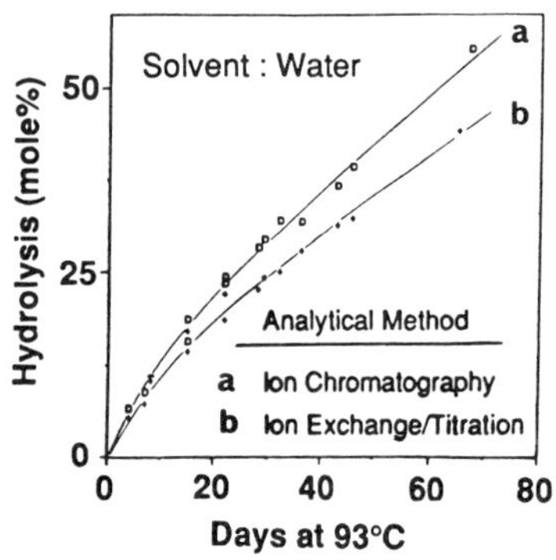

Fig. 2. Evaluation of PAM Hydrolysis (IC vs. Titration Methods).

An interesting aspect of Figure 2 is that the hydrolysis curves developed by the two methods diverge for samples aged longer than about 10 days. Two possible interpretations for this behavior have been considered. One is that there is a discrepancy in one of the analytical techniques. This appears unlikely based on IC calibration studies and data reported elsewhere for a series of HPAMs of different composition (12). In that study, good agreement was demonstrated betwen HPAM composition values determined by elemental analysis (for sodium), by the titration method, and by IC. Further, it is considered unlikely that resin treatment of aged PAM solutions (i.e. removal of ammonium ion and conversion of carboxylate groups to protonated carboxylic acid functions) would be any less effective than it is for HPAM solutions containing sodium counter ions.

An alternative possibility suggested by Figure 2 is that ammonium ions are being produced during aging of PAM solutions by a process other than conversion of amide units to ammonium carboxylate ion pairs. One such process would involve imide formation, occuring concurrently with TP hydrolysis. If imide formation does occur during TP hydrolysis of PAM, then one might observe complexity in the hydrolysis curve determined by titration, due to the possibility of two processes (amide vs. imide hydrolysis) occuring at different rates. Such complexity is not immediately evident in the titration derived hydrolysis curve. However, due to the precision of the method (ca ± 10% of the measured value) and the relatively small number of data points (due to the nature of the experiment which involves aging of solutions in sealed ampoules) it would be difficult to detect subtle features of the hydrolysis process by this method.

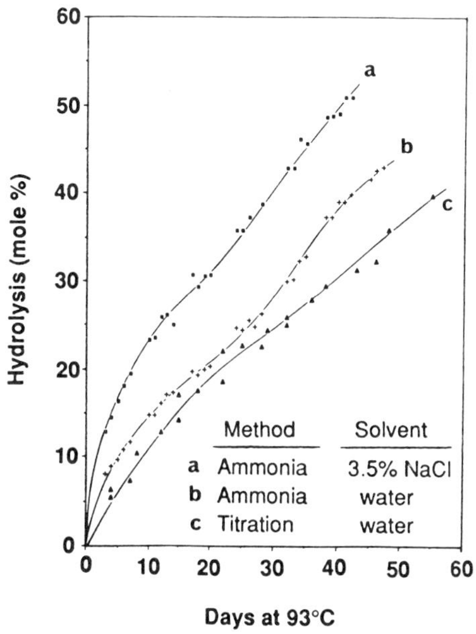

Fig. 3. Evaluation of PAM Hydrolysis (Ammonia vs. Titration Methods).

The second method employed to determine the ammonium ion content of aged PAM solutions involved treatment of the ammonium ions with a strong base and detection of the resultant ammonia. This method offers one or more of the following advantages vs. the titration and/or IC techniques: (a) small sample size, (b) rapid analysis, (c) measurements can be made in the salt solution, and (d) improved precision. Analysis of the TP hydrolysis of PAM by this method is shown in Figure 3. Note the expanded scale of Figure 3 vs. that of Figure 2.

The most interesting aspect of Figure 3 is the nature of the hydrolysis curve for the aqueous PAM solution. As indicated also by the other methods of analysis, the rate of hydrolysis was initially rapid but decreased significantly after about 20-25 days aging. Commencing at about 25 days, however, there is a significant inflection in the hydrolysis curve suggesting that (a) more than one process is leading to the formation of ammonium ions, (b) there may be a lengthy induction period associated with one of the processes, (c) different microstructures are hydrolyzing at unequal rates, and/or (d) some gradual change in the polymer-solvent and/or polymer-polymer interactions is influencing hydrolysis (e.g. electrostatic repulsion due to anionic carboxylate groups). The increased precision and other previously noted features of the ammonia method for determining hydrolysis appear to enable detection of subtle features of the aqueous TP hydrolysis of PAM. In saline solution, this feature is much less well defined, presumably due to the effects of the salt on polymer-polymer and/or polymer-solvent interactions.

Also shown in Figure 3 is TP hydrolysis of PAM in aqueous solution as measured by the titration method. The data shown were obtained via three separate aging experiments. For comparison with the aqueous hydrolysis curve determined by ammonia analysis, a similarly complex curve has been drawn through the composite data for the three titration experiments. Such a curve appears to give a reasonable fit to the data, but is admittedly arbitrary. Without benefit of the ammonia analyses (curve b), one would probably be inclined to represent the titration data as a smooth curve. A similar comment applies to curve a of Figure 3.

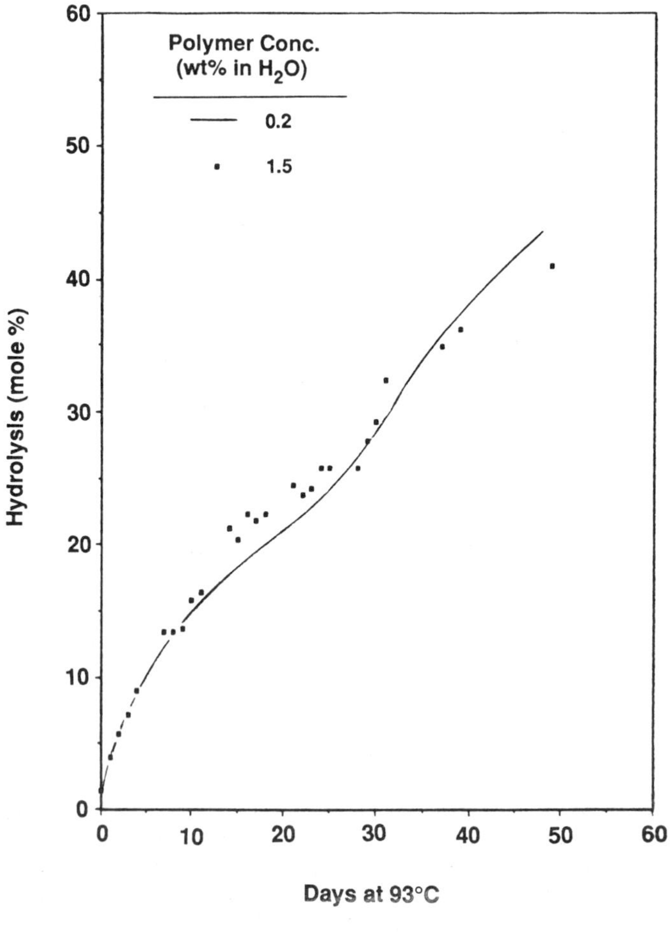

Fig. 4. Effect of Polymer Concentration on PAM Hydrolysis (Ammonia Method).

The final point to be considered is a comparison of the aqueous hydrolysis of PAM as measured by titration vs ammonia. As shown in Figure 3, the curve based on ammonia analysis indicates a higher degree of hydrolysis, compared with that determined from the titration data over the the entire aging period covered. It will be recalled however that comparison of the IC vs titration data (Fig. 2), based on duplicate experiments, indicates that the two hydrolysis curves are essentially identical up to about 7-10 days and diverge beyond this point. Additional ammonia analysis experiments will be required to enable firm conclusions regarding imide formation based on the comparison of ammonia vs. titration analyses. With regard to reproducibility of the inflection region of curve b in Figure 3, the hydrolysis data of Figure 4 are of interest. An inflection region in about the 15 to 35 day aging period is suggested in both systems shown in Figure 4.

Direct Analysis for Imide Formation during TP Hydrolysis of PAM

In order to obtain a reference polymer which contains imide functionality, acid catalyzed hydrolysis of PAM was conducted according to the general method of Minsk and coworkers. (13) A one weight percent PAM solution (400 mL) was contacted with acid (6.6 mL of 10 N HCl) and then aged at 93°C. Samples were withdrawn after 3 and 6 hours aging. A large quantity of precipitated polymer was noted after 24 hours aging. The six hour sample was precipitated in methanol and then extracted with methanol in a Soxhlet apparatus for 24 hours. The imide content of this sample (PAM-6) was then assessed by UV (7) and NMR (14) analysis.

Examination of a one weight percent solution of PAM-6 by UV revealed a subtle shoulder at about 235 nm, the region in which imide functionality has been reported in samples of hydrolyzed PAM (7). However at a concentration of 0.2 weight percent, there was no discernable peak in this region. Similarly, ^{13}C NMR analysis of the more concentrated sample indicated a peak at 178.49 ppm, which has been attributed to the imide carbonyl functionality (14), but there was no detectable peak in a 0.2 weight percent solution of this polymer. Based on the intensity of the 178.49 ppm peak in the 1% solution of PAM-6, it is estimated that this polymer sample contains a low level of imide content (≤ ca 10 mole %).

The foregoing analysis indicates that low levels of imide functionality would be difficult to detect in the dilute (0.2 wt %) polymer solutions used in this TP hydrolysis study of PAM. For this reason, a 5 wt % solution of PAM in D_2O was sealed in an NMR tube and was aged at 93°C. The imide content of this sample was assessed periodically by NMR. There was no discernable peak at 178.49 ppm in the ^{13}C NMR spectrum of this sample after 62 days aging. It appears that if there is any imide functionality in this TP hydrolyzed PAM, it is less than that of PAM-6 (i.e. ≤ ca. 10 mole %).

The general form of the hydrolysis curve derived from this NMR study (Fig. 5) is in agreement with that obtained via the other methods described above (cf. Fig 1) and also reported by Moens and Smets (2). The actual hydrolysis values shown in Fig. 5 however, are subject to considerable uncertainty due to the nature of the NMR experiment. Therefore, absolute differences between the NMR and the other methods are not considered significant.

UV analysis of various aged PAM solutions (0.2 wt % in water or NaCl solution) has not indicated any well defined absorption near ≈ 235 nm which may be associated with imide. Hence, it is assumed that if imide formation occurs during TP hydrolysis of PAM under these conditions, it must produce imide at a concentration less than about 10 mole %. Interestingly, this is about the level which would account for the difference in the hydrolysis curves for the titration vs. the ammonia or IC methods.

The UV spectra recorded for a PAM solution which had been permitted to age for varying periods at 93°C are shown in Figure 6. Clearly evident in these spectra is an absorbance, centered around 265 nm, whose relative intensity increases linearly with aging time (correlation coefficient = 0.99). The chemical functionalitiy responsible for this absorbance feature is not well defined at present, and one can only speculate on its nature. It seems quite unlikely that this absorbance is the result of imide groups within the polymer structure, since the observed absorbance is significantly red-shifted from that proposed for the imide moiety.

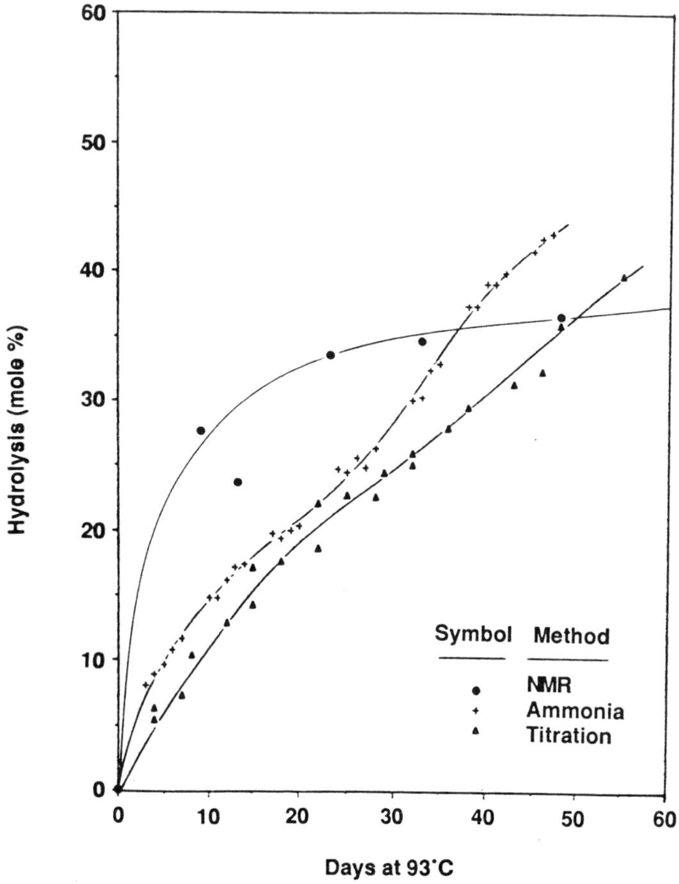

Fig. 5. Evaluation of PAM Hydrolysis
(0.2 wt% polymer in water)

Possibly the absorbance observed near 265 nm can be accounted for by the introduction of unsaturated centers, i.e. vinylic groups, either along the polymer backbone or in lower molecular weight, soluble decomposition fragments. The introduction of vinylic groups might be the reasoned end product of an initial oxidative attack on the polymer occuring in parallel with TP hydrolysis. If this were the case then samples aged under anerobic conditions might be found to be essentially devoid of the absorbance around 265 nm. In part to test this hypothesis, an investigation of TP hydrolysis under anerobic conditions is planned. It is interesting to note that PAM samples aged at elevated temperature in the presence of aqueous brine, appear to develop a significantly less intense absorbance near 265 nm. This suggests that the reaction process(es) which lead to this UV feature may be strongly influenced by the tertiary structure of the polymer in solution.

SUMMARY AND CONCLUSIONS

This study has examined three methods for analyzing the TP hydrolysis of PAM. The ammonia analysis method is the most interesting since it appears to enable resolution of subtle details of the hydrolysis process, which are not readily detectable by the IC or the titration methods. Comparison of hydrolysis data from the ammonia or IC methods, vs. the titration method, suggests that a low level of imide formation may occur during the TP hydrolysis of PAM. Direct analysis by UV and NMR does not indicate any evidence of imide functionality in 0.2 % solutions of TP hydrolyzed PAM. However, analysis of a known imide-containing PAM sample indicates that low imide contents (≤ ca. 10 mole %) would not be detectable in the dilute PAM solutions used in this study. The overall conclusion regarding imide formation during TP hydrolysis is that if it occurs under these conditions it must be at a level less than about 10 mole percent. Finally, data obtained by the ammonia analysis technique suggests that more than one process is operative during the TP hydrolysis of PAM. Additional studies are needed to determine whether this is due to a low level of imide formation or to some other reaction pathway.

ACKNOWLEDGEMENTS

The authors wish to express their appreciation to Drs. George E. Milliman and James Brown of the Exxon Corporate Research Analytical Science Laboratory for assistance with the acquisition and interpretation of portions of the spectroscopic data.

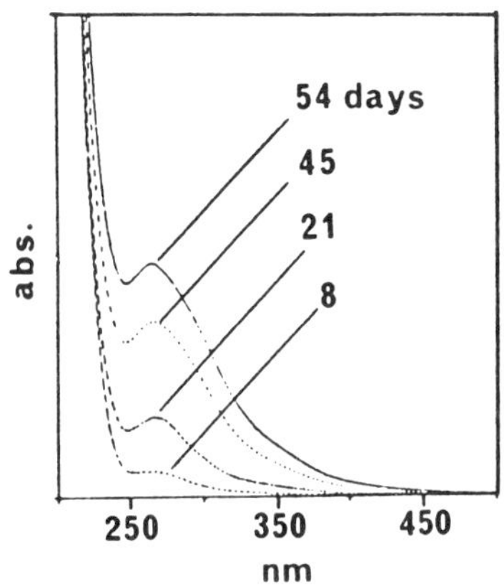

Fig. 6. Effect of polymer solution aging time at 93°C on UV absorbance.

REFERENCES

(1) J. March, "Advanced Organic Chemistry", 2nd. Ed., McGraw - Hill (1977) pp. 353-356.

(2) J. Moens and G. Smets, J. Polym. Sci., 23 (1957) 931-948.

(3) R. W. Lenz, "Organic Chemistry of Synthetic High Polymers", Interscience, NY (1967) pp. 694-698.

(4) H. C. Haas and R. L. MacDonald, J. Polym. Sci., A-1 (1971) 3583-3593.

(5) K. Nagase and K. Sakaguchi, J. Polym. Sci. Part A, 3 (1965) 2475-2482.

(6) M. Higuchi and R. Senju, Polymer Journal, 3 (1972) 370-377.

(7) S. Sawant and H. Morawetz, J. Polym. Sci., Polymer Letters Ed., 20 (1982) 385-388.

(8) P. Davison and E. Mentzer, Soc. Pet. Eng. J., (June 1982) 353-362.

(9) G. Muller, Polymer Bulletin, 5, (1981) 31-37.

(10) A. Zaitoun and B. Potie, International Symposium on Oil Field Geothermal Chemistry, Denver, CO, June 1-3, 1983, SPE paper number 11785.

(11) A. Moradhi-Araghi and P. H. Doe, 59th Ann. Tech. Conf. Soc. Pet. Eng., Houston, TX, September 16-19, 1984, SPE paper number 13033.

(12) J. J. Maurer and L. P. Klemann, J. Liq. Chrom., 10 , 83-94 (1987).

(13) L. M. Minsk, C. Kotlarchik, G. N. Meyer and W. O. Kenyon, J. Polym. Sci.: Polym Chem. Ed., 12 (1974) 133-140.

(14) F. Halverson, J. E. Lancaster and M. N. O'Connor, Macromolecules, 18 (1985) 1139-44.

ROLE OF IMIDIZATION IN THERMAL

HYDROLYSIS OF POLYACRYLAMIDES

A. Moradi-Araghi, E.T. Hsieh and I.J. Westerman

Phillips Petroleum Company
Bartlesville, OK

INTRODUCTION

Polyacrylamides (PAm), including copolymers having acrylamide as the major component, have been a popular model system in many recent studies of water-soluble polymers[1-13]. In the literature there is general agreement that longer heating of dilute aqueous PAm solutions at higher temperatures favors the formation of carboxylates. Intermolecular imidization resulting from heating solid PAm also is widely reported[1,3,6,7,13]. However evidence varies as to whether intramolecular imidization occurs to any detectable extent in aqueous PAm solutions[1,3,4,13,14]. Only Sawant and Morawetz report imidization as a reaction pathway in alkaline hydrolysis of PAm[4]. Guerrero, et al.[10,13], on the other hand, report seeing no intramolecular imide structure using a variety of characterization techniques. Generally speaking, the opinions have been widely divided in the literature on this particular issue.

Our interest in intramolecular imidization of PAm is for the many potential benefits of having the non-ionic, rigid closed ring structure in the molecular backbone. For example, imidization would lessen chain coiling. This report summarizes our recent study of molecular changes of PAm resulting from thermally promoted hydrolysis reactions. In particular, we will focus our discussion on reaction conditions involving intramolecular imide formation.

EXPERIMENTAL

Two commercially available PAm's with reported carboxylate contents of 0 and 30 mole percent were used in this study. To reduce the difficulties of removing bubbles from the NMR sample solutions, polymers were presheared to lower molecular weights. This was accomplished by forcing 5,000 ppm aqueous PAm solutions through stainless steel tubing (0.07 cm I.D. and 8.9 cm in length) at a shear rate of 75,000 per second and room temperature. A 75% reduction of solution viscosity was achieved with no noticeable molecular structural changes (as seen by C-13 NMR analyses). Solid polymers were recovered from the sheared solutions by freeze-drying.

NMR sample solutions were prepared in 10 mm tubes at concentrations of about 5 to 7 weight % in distilled water at 40°C (104°F). D_2O and p-dioxane were used as lock and reference solvents respectively, and were separated from the sample solution by using 4 mm diameter inserts.

All NMR experiments were performed on a Varian XL-200 spectrometer at 50.31 MHZ. Relevant instrument settings include 90 degree pulse angle, 1.0 second acquisition time, 0.5 second pulse delay, 238.5 ppm spectral width, and broad band proton decoupling. About 40,000 transients were collected for each spectrum. Temperature was maintained at 40°C. Spin-lattice relaxation time (T1) and Nuclear Overhauser Enhancement (NOE) values for all C-13 NMR resonances were carefully measured to determine the optimum NMR experimental conditions. The spectral intensity data thus obtained were assured of having quantitative validity.

Two C-13 NMR experimental parameters were found to be particularly important. One is the temperature, for which the effects have been well illustrated in this report. We have found that 40°C is low enough for neutral solutions of PAm and yet still different enough from room temperature to allow close regulation. The other is the use of lock solvent. If the lock solvent is directly mixed into the solution, proton-deuterium exchange reactions are known to take place that could cause NOE differences[2]. A better way is to use lock solvent in an insert.

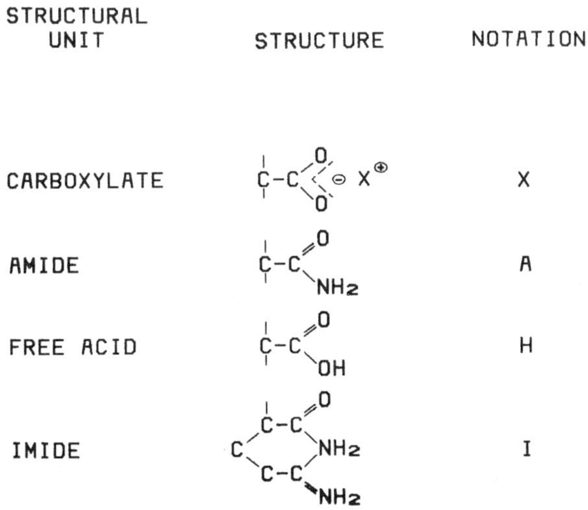

Figure 1. Identification of various structural units.

RESULTS AND DISCUSSIONS

Peak Assignments

Figure 1 shows the basic structural units and the corresponding notation for each unit. The assignments for the three types of peaks, namely, the carbonyl carbon, methine carbon and methylene carbons are listed in Table I. The chemical shift values are based on p-dioxane at 67.4 ppm. The peak assignments are in good agreement with the literature[2,5,12].

TABLE I

C-13 NMR Peak Assignments for Polyacrylamide and its Hydrolyzed Structures

Unit Sequence	Range (ppm)	Maximum (ppm)
Carbonyl Carbon Peaks		
C=O (XXX)	186.2 - 184.9	185.1
C=O (XXA)	184.9 - 184.0	184.3
C=O (AXA)	184.0 - 182.9	183.5
C=O (XAX)	182.5 - 181.5	181.6
C=O (AAX)	181.5 - 180.8	181.1
C=O (AAA)	180.8 - 179.9	180.4
C=O (XHX)	182.0 - 180.0	180.8
C=O (HHH)	180.0 - 179.0	179.5
C=O (IX)	179.5 - 178.7	178.9
C=O (IA)	178.7 - 178.0	178.5-178.2
Methine Carbon Peaks		
CH (X)	48.6 - 44.1	Varies
CH (A; rr)*	44.1 - 43.1	43.2
CH (A; mr)*	43.1 - 42.6	42.8, 42.7
CH (A; mm)*	42.6 - 41.6	42.4
CH (XIX)	41.6 - 41.0	41.2
CH (XIA)	41.0 - 40.6	40.8
CH (AIA)	40.6 - 39.6	40.3, 39.8
CH (H)	Overlap with CH(A)	---
Methylene Carbon Peaks Unit Sequence not resolvable	39.0 - 34.0	(rrr - mmm)*

* r = racemic and m = meso.

The Effect of pH on Thermal Hydrolysis of Polyacrylamides

Carbon-13 NMR spectra for samples (a) through (h) for the carbonyl carbon region are presented in Figure 2 with the sample solution environment and key structural features labelled. The spectrum for sample (b) shows that starting PAm was about 4% hydrolyzed. The spectrum of a similar sample after 10 days at 40°C (104°F) showed no detectable additional structural change, indicating that an overnight NMR run at 40°C is mild enough to have not altered the original polymer structure. Under the same conditions, the acidic solution resulted in about 10% imidization as shown in spectrum (a). The next four spectra, i.e., spectra (c) through (f), show clearly that the heat treatment of the neutral solution has led to formation of both carboxylate and imide with carboxylate favored, while heating the acidic solution leads predominantly to imidization. A look at spectra (f) and (h) indicates that the carboxylates can be converted to imide by treatment with acid. The reverse also can be accomplished by treating the imide with a base as shown in spectra (e) and (g).

Thermal hydrolysis of PAm under alkaline conditions is very rapid and carboxylate is the dominant product. As shown in Table II, after 3 days

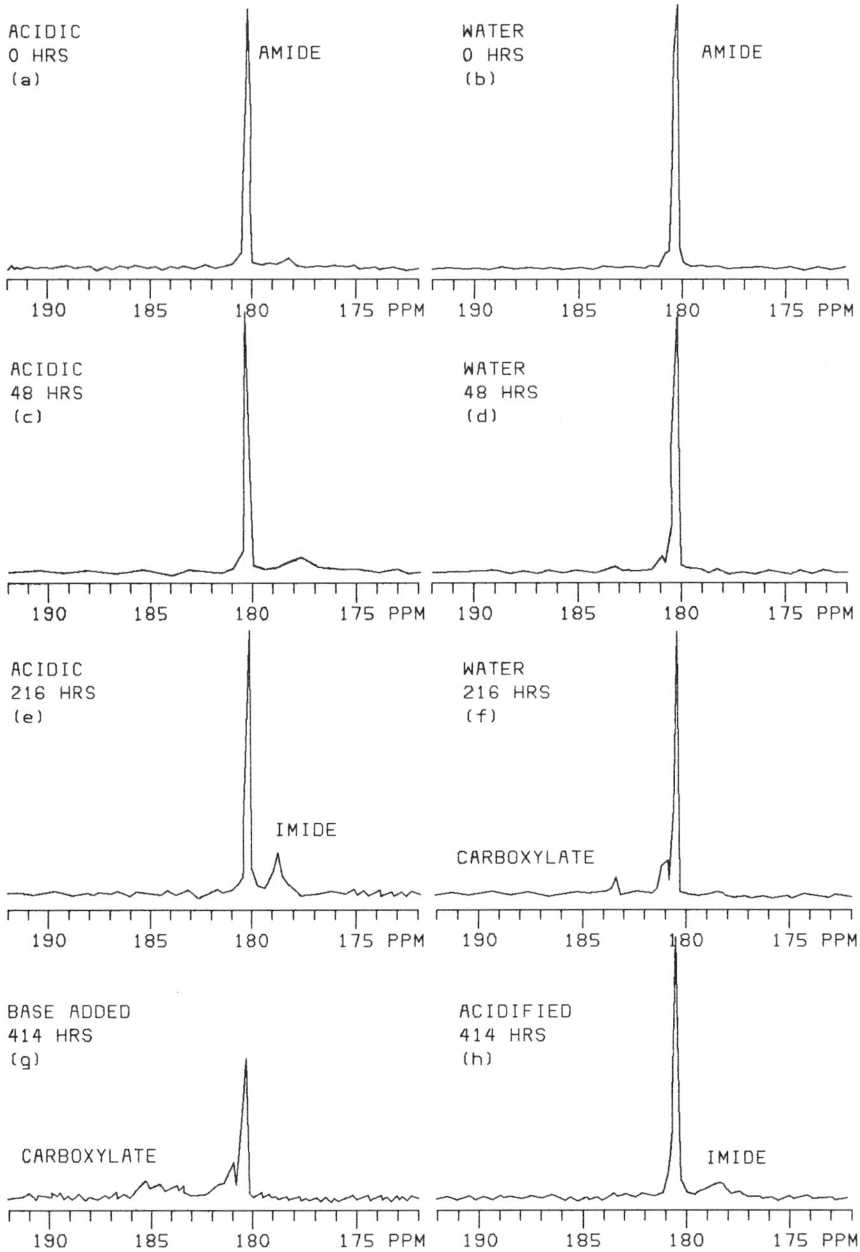

Figure 2. C-13 NMR spectra (carbonyl carbon region; 172 ppm to 192 ppm) of dilute aqueous PAm solutions heated at 90°C in acidic (spectra (a), (c) and (e)) and neutral (spectra (b), (d) and (f)) environments. Spectra (g) and (h) show the structural changes when the solution pH is changed.

aging at 90°C (194°F), a solution of originally unhydrolyzed PAm contained 25.3% carboxylate and 1.6% imide. Under neutral conditions, thermal hydrolysis of PAm is slower and the dominant product is the carboxylate. After 9 days of aging at 90°C, the solution contained 7.6% carboxylate and 2.9% imide. Extended aging at 110°C (230°F) resulted in the formation of

TABLE II

The Effect of Thermal Hydrolysis and pH on Relative Population of Various Structures in Originally Unhydrolyzed Polyacrylamide

pH Condition	Temperature (^oC)	Aging Time (Days)	Percent Population: Amide	Carboxylate	Imide
Dilute Base	90	0	84.5	14.2	1.3
		3	73.1	25.3	1.6
Neutral	90	0	96.0	1.9	2.1
		2	92.4	5.0	2.6
		9	89.5	7.6	2.9
	110	Short	86.8	7.4	5.8
		Long	17.4	70.5	12.1
Dilute Acid	90	0	95.2	0.6	4.2
		3*	89.0	3.0	8.1
		0*	91.9	5.0	3.1
Concentrated Acid	90	0	94.4	0.8	4.7
		2	79.0	2.4	18.6
		9	71.7	2.5	25.8

* Dilute base added.

70.5% carboxylate and 12.1% imide. Under mild acidic conditions, imide formation is the preferred route. After 3 days of aging at 90^oC, the solution contained 3.0% carboxylate and 8.1% imide. Addition of a small amount of base to this solution destroyed 5% of the imide structures and produced about 2% each of carboxylate and amide. In concentrated acid imide is once again the dominant product. After 9 days of aging, the solution contained 2.5% carboxylate and 25.8% imide.

As discussed above, low levels of carboxylates can be converted into imides by acid treatment (Figure 2f and 2h). To determine whether this treatment can be extended to polymers with high levels of hydrolysis, we

TABLE III

The Effects of Thermal Hydrolysis and pH on Relative Population of Various Structures for a Commercial Polyacrylamide Originally at 30% Hydrolysis Level

Thermal Aging	pH	Percent Population: Amide	Carboxylate	Imide
None	10	55.5	43.4	1.1
	7	60.6	38.1	1.3
	4	(73.1)*	24.6	2.3
3 Days @ 90 C	7	54.0	43.5	2.5
	4	(61.2)*	35.8	3.0

* includes acid form produced from protonation of carboxlates.

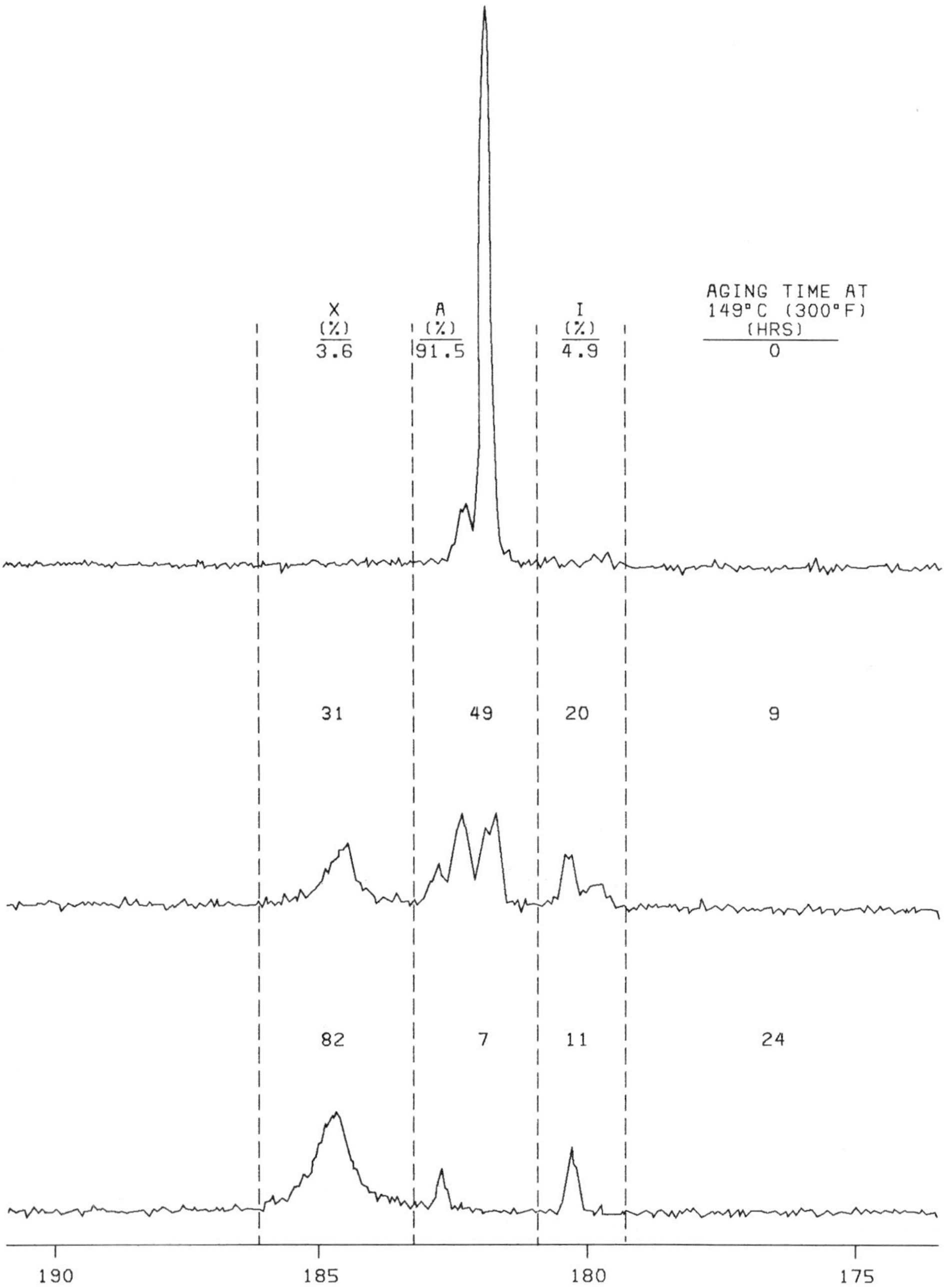

Figure 3. The effect of aging on relative population of imide, I, amide, A, and the carboxylate, X, in thermal hydrolysis of PAm.

chose a PAm that according to the manufacturer was 30% hydrolyzed. Table III shows the effect of pH on relative population of amides, carboxylates and imides for this polymer. In alkaline solutions, as expected, the polymer will further hydrolyze. Under acidic conditions, the carboxylates are protonated back to the acid form, which is indistinguishable from the amide form. No appreciable amounts of imides were produced as a result of this acid treatment.

Thermal Stability of the Imide

Thermal stability of the produced imides is illustrated in Figure 3. A solution of PAm with a nominal extent of hydrolysis, whence heated for 9 hours at 149°C (300°F), resulted in a 20% imide content. Additional 15 hours of aging still reduced the amide level by 42% while 11% of the imide survived this severe thermal aging. These results indicate that the imide structures are moderately stable at elevated temperatures.

CONCLUSIONS

1. C-13 NMR was successfully applied to determine the relative amounts of each carboxylate and imide in thermally promoted hydrolysis of polyacrylamides. While minute amounts of imide have been observed under alkaline conditions, 2% to 26% of this material are formed under neutral and acidic conditions, respectively.

2. Depending on the pH level, imide or carboxylate can be the dominant hydrolysis product. In alkaline solutions carboxylate is preferred. Under neutral conditions, while the carboxylate is still the preferred product, imide formation becomes significant. In acidic solution imide is the dominant hydrolysis product of polyacrylamides.

3. The produced imide can survive weak alkaline solutions. Under strong base or prolonged heating it predominantly produces carboxylate. In acidic environments the imide is stable, although its long term stability has not been assessed. Under neutral conditions imide is relatively stable and outlasts the amide during extended thermal aging.

ACKNOWLEDGMENTS

The authors wish to thank Mrs. K. H. Carney and Mr. F. W. Blanton for their laboratory assistance, Mr. R. M. Infield for freeze-drying of the samples and Mr. B. E. Jones for obtaining the NMR data. We would also like to thank the management of Phillips Petroleum Company for permission to publish this information.

REFERENCES

1. Minsk, L. M., Kotlarchick, C., Meyer, G. N. and Kenyon, W. O., Imidization During Polymerization of Acrylamide, J. Polym. Chem. Ed., 12:133 (1974).

2. Cais, R. E. and Stuk, G. J., Copolymerization of Acrylamide with Sulfur Dioxide. Determination of the Effect of Copolymerization Temperature on the Monomer Sequence Distribution by 13C N.M.R., Polym., 19:179 (1978).

3. Shepitka, J. S., Case, C. E., Donaruma, L. G., Hatch, M. J., Kilmer, N. H., Khune, G. D., Martin, F. D., Ward, J. S. and Wilson, K. V., Partially Imidized, Water-Soluble Polymeric Amides. I. Partially Imidized Polyacrylamide and Polymethacrylamide, J. Appl. Polym. Sci., 28:3611 (1983).

4. Sawant, S. and Morawetz, H., Reaction of Polyacrylamide Provides Evidence for Some Head-to-Head Addition of Monomer Residues, J. Polym. Sci., Polym. Lett. Ed., 20:385 (1982).

5. Lancaster, J. E. and O'Connor, M. N., High Field 13C-NMR Study of the Stereoregularity of Polyacrylamide, J. Polym. Sci., Polym. Lett. Ed., 20:547 (1982).

6. Kishore, K. and Santhanalakshmi, K. N., Studies on Thermal Polymerization of Acrylamide, J. Macromol. Sci., Chem. Ed., A16(5):941 (1981).

7. McCormick, C. L., Chen, G. S. and Hutchinson, B. H., Water-Soluble Copolymers. V. Compositional Determination of Random Copolymers of Acrylamide with Sulfonated Comonomers by Infrared Spectroscopy and C13 Nuclear Magnetic Resonance, J. Appl. Polym. Sci., 27:3103 (1982).

8. Inoue, Y. Fukutomi, T. and Chujo, R., Carbon-13 NMR Analysis of the Tacticity of Polyacrylamide, Polym. J., 15:103 (1983).

9. Panzer, H. P., Halverson, F. and Lancaster, J. E., Carboxyl Sequence Distribution in Hydrolyzed Polyacrylamide, Polym. Mater. Sci. Eng., 51:268 (1984).

10. Zurimendi, J. A., Guerrero, S. J. and Leon, V., The Determination of the Degree of Hydrolysis in Poly(acrylamides): Simple Methods using C13 N.M.R. and Elementary Analysis, Polym., 25:1314 (1984).

11. Leung, W. M., Axelson, D. E. and Syme, D., Determination of Charge Density of Anionic Polyacrylamide Flocculants by NMR and DSC, Coll. Polym. Sci., 263:812 (1985).

12. Halverson, F., Lancaster, J. E. and O'Connor, M. N., Sequence Distribution of Carboxyl Groups in Hydrolyzed Polyacrylamide, Macromolecules, 18:1139 (1985).

13. Guerrero, S. J., Boldarino, P. and Zurimendi, J. A., Characterization of Polyacrylamides Used in Enhanced Oil Recovery, J. Appl. Polym. Sci., 30: 955 (1985).

14. Maurer, J. J., Harvey, G. D. and Klemann, L. P., Thermally Promoted Hydrolysis of Polyacrylamide, Polym. Mater. Sci. Eng., 55:705 (1986).

EVALUATION OF POLYMERS FOR OILFIELD USE: VISCOSITY DEVELOPMENT, FILTERABILITY AND DEGRADATION

Chris Franklin Parks*, Bonnie L. Gall**, and Peter E. Clark+

*Kelco Oil Field Group Inc. Tulsa, Oklahoma
**National Institute for Petroleum and Energy Research/IIT Research Institute Bartlesville, Oklahoma
+Department of Mineral Engineering, The University of Alabama, Tuscaloosa Alabama

BACKGROUND

Water soluble polymers are used extensively throughout the oil producing industry for a variety of applications. Such applications begin with the drilling of a well, progress through chemical and hydraulic treatments designed to stimulate production, include workover operations which maintain production, and are significant in secondary and tertiary recovery operations.

Polymer requirements are dictated by the specific application. In drilling, the polymers must be shear stable and provide a means of suspending and removing formation cuttings. For use in hydraulic fracturing, the polymeric fluid must also suspend solid proppants, help control fluid loss, reduce friction and provide viscosity for the creation of fracture width. However, after the treatment the polymer containing fluid must be degradable, so that it can be readily returned from the formation and not interfere with oil and gas production.

In those applications where it is desirable for the polymer solution to penetrate the rock matrix, it is of uptmost importance that the polymer not be filtered out on the formation face, thereby causing impairment of formation permeability. These applications include secondary and tertiary recovery as well as other operations such as gravel packing for sand control purposes.

As a result of these various requirements there are four specific concerns dealing with polymer selection and usage. These are:

> <u>solvation</u> which is concerned with the complete solubilization of the polymer so that the polymer not be physically separated during its intended application,

hydration when used in this context refers to the solubilization of the polymer (hydration is used because some of the polymers used in the oil field are not truly water soluble but form hydrocolloids),

functionality which, in this discussion, relates to the specific purposes for which the polymer was selected, and not to the classical chemical term, and

degradation which assures that the fluid has lost those physical properties initially contributed by the polymeric material and which are no longer desireable.

Historically, apparent viscosity* has been the single most important physical measurement used to characterize fluids for use in petroleum related applications. Viscosity was an important factor in hydraulic fracturing, and this is where most of the water soluble polymers were first used in the oil field. The advent of an easy to use rotational viscometer helped to further entrench the use of viscosity as the primary metric for fluid performance. Viscosities could be easily measured both in the lab and in the field, while other tests were more difficult to carry out. Thus, viscosity measurements emerged over a period of time to be the accepted test for characterizing a fluid.

The purpose of this discussion is to draw attention to some of the shortcomings of relying on viscosity measurements, and to suggest additional tests that can be used to more completely define polymer solution properties. Specifically, this text will focus on polymer solvation (hydration), functionality, and degradation. The discussion will also relate to the shortcomings of relying solely on viscosity measurements, and how additional tests can lead to the selection of the optimum polymer for a specific application.

POLYMER SOLVATION

Introduction

For many workers in the industry, a polymer is fully dissolved when the fluid is visually uniform and free of undissolved solids. In the absence of more specific procedures, the procedure outlined in API Specicfication 13A "Oil Well Drilling Fluid Materials", which was designed primarily to describe a technique to totally disperse silicate minerals, is often used. This procedure suggests mixing for 20 minutes in a Hamilton Beach mixer at 11,000 rpm. If additional mixing of polymer solutions showed that viscosity measurements had reached a plateau after the initial 20 minutes of mixing, then the procedure suggested that there was little

* There is a great deal of confusion about the meaning of the term apparent viscosity. The apparent viscosity is really the apparent Newtonian viscosity which is determined by finding the slope of a line drawn between the origin and the point on the data curve at the shear rate of interest. The apparent viscosity should be reported along with the shear rate at which it was determined.

need to mix for a longer period of time. In fact, there is some justification to limit mixing based on a fear of shear degradation of the polymer.

Laboratory testing shows that visual examination and viscosity measurements are not sufficient to fully define polymer solvation. In this work, the solvation of hydroxyethyl cellulose (HEC) and xanthan has been studied. These polymers are both widely used in various petroleum applications. HEC is used in many workover and completion applications, while xanthan has its most wide spread uses in drilling and enhanced oil recovery (EOR) applications. Solublization of both polymers results in fluids with pseudoplastic (or shear thinning properties). Even though the polymers both exhibit pseudoplastic behivior, the polymers vary considerably as to their molecular size and physical properties.

It was found that complete solvation of both polymers was not obtained when the apparent viscosity plateau was used as the sole criterion for solvation. A more definitive standard for determining polymer solvation should also include tests that would assure that partially dissolved polymer agglomerations could not be removed from the fluid during its application. In the work described here, viscosity measurements were supplemented by filtration testing and particle settling experiments.

Filtration Testing

Data comparing viscosity to API fluid loss test results are presented in Tables 1 and 2. Under both low and high shear mixing conditions the HEC fluids developed similar apparent viscosities when measured at six customary shear rates. However, there was a large difference in the filterability of the fluid as determined by using a filter press procedure described in API Recommended Procedure 13B. Even though the apparent viscosities of the fluids were similar, the fluid prepared using high shear mixing had a fluid loss rate significantly greater than fluids prepared using low shear mixing. When using low shear mixing conditions the amount of HEC fluid leaking through the filter paper was 75 mls in 30 minutes. When using high shear mixing conditions the amount of HEC fluid was 300 mls (the capacity of the fluid loss cell) in 8 minutes. Qualitatively, there was also a considerably thicker layer of partially dissolved polymer present on the paper through which the fluid prepared under low shear conditions was passed. Less polymer was visible on the filter paper through which the fluid prepared under high shear conditions passed.

Another set of tests was made using xanthan polymer (shown in Table 2) and produced a similar set of results. In these tests, the xanthan containing fluid was circulated through a centrifugal pump for different periods of time. After 10 minutes of circulation, the fluid appeared to be free of visible solids. Samples were taken and viscosity measurements made at 10 minute intervals for a total of 40 minutes. It was found that the fluid apparently reached maximum viscosity at each of the six shear rates after only 10 minutes of circulation.

Table 1. Hydration of Hydroxyethyl Cellulose* (80 lbs/1000 gallon, in 2% KCl Brine)

Low Speed Mixing - 30 Minutes on a Magnetic Stirrer

Viscosity at 80 °F Shear Rate (s^{-1})	Centipoise	API Fluid** Loss (80 °F)
1022	95	75 mls/30 min
511	154	
340	202	
170	324	
10.22	1800	
5.11	2600	

High Speed Mixing - 30 Minutes on a Waring Blendor (30% Low Speed)

Viscosity at 80 °F Shear Rate (s^{-1})	Centipoise	API Fluid Loss (80 °F)
1022	88	300 mls/8 min, 15 s
511	148	
340	198	
170	321	
10.22	1800	
5.11	2600	

* HEC-10 produced by Union Carbide Company.

** Pressure differential was 100 psi; No. 50 Whatman filter paper was used.

Table 2. Hydration of Xanthan Polymer* (48 lbs/1000 gal, 2% KCl Brine)

	Viscosity (cp) (Circulation Time Through a Centrifugal Pump)			
Shear Rate (s^{-1})	10 min	20 min	30 min	40 min
1022	17.2	17.4	17.3	17.1
511	27.4	27.6	27.8	27.6
340	36.3	36.9	36.9	36.9
170	60.0	61.2	61.2	61.2
10.22	550.0	540.0	540.0	540.0
5.11	960.0	940.0	940.0	940.0

Circulation Time (Centrifugal Pump)	API Fluid Loss** (80 °F - 100 psi)
10 min	148 mls/30 min
20 min	300 mls/4 min, 30 s
30 min	300 mls/2 min, 30 s
40 min	300 mls/1 min, 47 s

* KELZAN XCD Polymer produced by Kelco Division of Merck & Co. Inc.

** Pressure differential was 100 psi, No. 50 Whatman filter paper was used.

Additional viscosity measurements made after a total of 20, 30, and 40 minutes of circulation showed almost no viscosity variations. Again, the filterability of the fluids varied significantly. The fluid removed after 10 minutes of circulation had a filtrate volume of 148 milliliters in 30 minutes. The filtration rates increased as the fluid circulation time increased. Ultimately, the filtration volume reached 300 mls (the fluid cell volume) in under two minutes. The filter papers removed from these tests also showed a wide variation in the amount of polymer that was removed. The fluid circulated for only 10 minutes had between 1/16 and 1/8 inch accumulation on the paper. While fluid circulated for 40 minutes showed no measurable amount of polymer present on the paper.

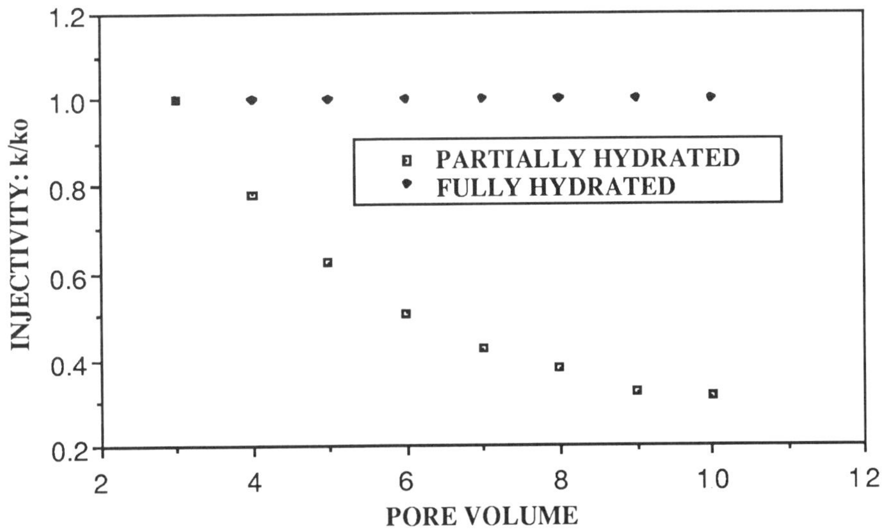

Figure 1. Injectivity of hydroxyethyl cellulose.

Additional tests were made to determine the effect of full polymer hydration on the injectivity of both HEC and xanthan solutions through formation core plugs. These results generally followed the filtration tests. Where fully hydrated fluids were used, there was little change in fluid injectivity, at constant injection pressure (Figures 1 and 2). In contrast, when the partially hydrated fluids were injected, injectivity decreased greatly at constant injection pressures. It is quite apparent that polymer solids were filtering out on the face of the core plugs, and causing the decrease in injectivity. The message is clear, relying on viscosity measurements alone may lead to choosing a fluid which causes formation damage and high injection pressures as well as poor polymer utilization and waste.

To assure full polymer solvation longer mixing times and/or higher shear mixing conditions may be required. The specific set of mixing conditions required depends upon the polymer, and is best determined through laboratory testing. The first step in determining polymer solvation is to mix the polymer containing fluid until viscosity plateaus are reached. The second step is to apply a filtration criteria. Additional mixing is required if the polymer containing fluid fails to filter readily.

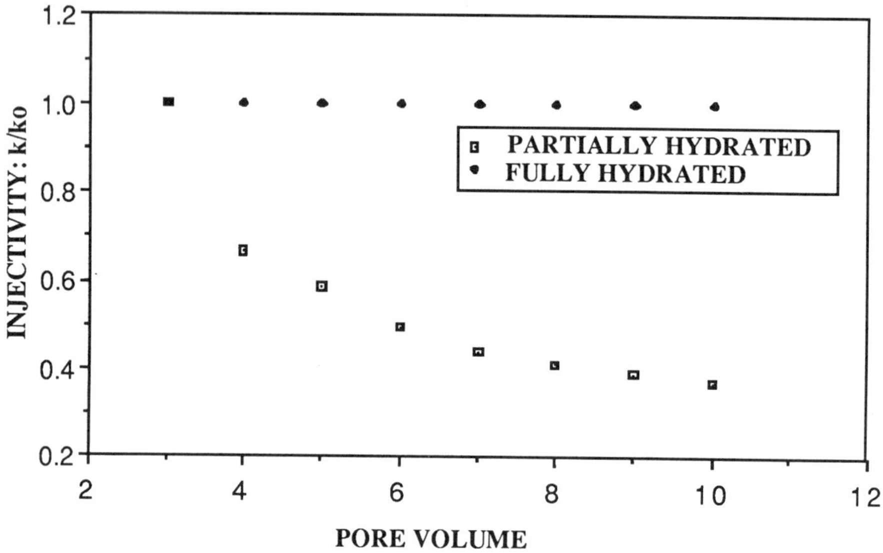

Figure 2. Injectivity of xanthan polymer solutions.

POLYMER FUNCTIONALITY

Introduction

The functional behavior of a polymer is defined here as the ability of a polymer solution to provide certain desirable properties under the wide range of conditions found in its oil field application. As mentioned earlier, there has been such a strong reliance on viscosity measurements, that the measure of the usefulness of a fluid is often completely defined by the apparent viscosity measured at two or three specific shear rates*. This total dependence on

* The industry has established three standard test shear rates based on the shear rates generated from the 100, 300, and 600 rpm speed settings of a Fann Viscometer equiped with a R1-B1 rotor/bob combination. Unfortunately, the resulting lowest shear rate is considerably above 100 s^{-1}. A better test protocol would involve taking several data points from each decade of shear rate for a wider range of shear rates (0.01 to 1000 s^{-1}).

viscosity measurement can be misleading. For example, polymer containing fluids are used in many petroleum producing applications to suspend and transport solids. These include drilling, to remove cuttings, hydraulic fracturing, to transport proppants, and gravel packing, to place gravel for the control of formation sand production. Despite widespread use of apparent viscosity as a metric, it is known[1,2] that there is not necessarily a good correlation between apparent viscosity and particle suspension.

Particle Suspension vs. Viscosity

To provide greater insight to the applicability of viscosity data for the characterization of polymer solutions, a series of tests were conducted to determine if viscosity measurements could be used to accurately define particle suspension and at what shear rate or shear rate range was correlation closest. This was accomplished by comparing the apparent viscosity measured at specific shear rates and particle settling velocities. The apparent viscosities of HEC and xanthan polymer solutions were determined at shear rates ranging from 0.01 to 1000 sec^{-1}. A Rheometrics Pressure Rheometer was used because of its ability to conduct long term tests without evaporative loss of fluid. Settling rates were determined by measuring the time required for suspended particles to settle a specified distance in the different polymer containing fluids. Two particle diameter ranges, (16-18 and 18-20 mesh) were used

These data are presented in Table 3. What is immediately evident from these data is that the closest correlation occurred when particle settling rates were compared to the apparent viscosity measured at shear rates of 1 sec^{-1}. The xanthan polymer solution showed a greater ability to suspend the solids and had a higher apparent viscosity in the low shear rate range. In contrast the HEC solution had higher apparent viscosities at shear rates above 10 sec^{-1} but exhibited lower particle suspension properties. The dependence on low shear rate apparent viscosity is not entirely unexpected. Roodhart[2] has shown emperically that the zero shear viscosity must be factored into a Stokes law type calculation before settling velocities can be calculated for HPG solutions. Thus, reliance on viscosity measurements at the customary shear rates would not have selected the more efficient fluid for particle suspension.

Likewise the use of the Marsh funnel viscosity to predict the best suspending fluid would not have led to selecting the best fluid. The Marsh funnel viscosity of the HEC fluid was 84 seconds as compared to 53 seconds for the xanthan polymer solution.

A fluid was also prepared using 48 lbs of xanthan per 100 gallons (0.58 percent by weight). Lithium hypochlorite was added to chemically degrade the dissolved polymer. At ambient conditions degradation occurs slowly and takes many hours. It was thus possible to monitor the changes occurring in the apparent viscosity and to compare these changes to changes in the particle suspension properties of the fluid. Viscosity changes occurring at shear rates between 0.01 and 1000 sec^{-1} were compared to the changes in the particle settling rates.

Figure 3 compares changes in settling rate to changes in viscosity measured at shear rates of 1, 10, 100, and 1000 sec^{-1}. As the polymer chemically degraded, the particle settling velocity increased from less than 0.01 to 0.7 centimeters/second. Likewise, there were corresponding decreases in viscosity. However the rates at which the apparent viscosity changed differed with shear rate at which the measurement was made. The most direct correlation with settling rates were obtained with viscosities measured at the shear rates about 1 sec^{-1}. Currently, viscosity measurements made to predict particle transport properties of fluids stresses those viscosities measured at shear rates of 170, 511, and 1022 sec^{-1}. It is assumed that viscosities obtained under these conditions do in fact control particle suspension properties. However, the data in Figure 3 (after 12 hours clearly shows that significant viscosities can exist at these shear rates without desired particle suspension.

Table 3. Comparison of Apparent Viscosity and Particle Settling.

	HEC*	Xanthan Polymer*
Shear Rate (s^{-1})	Apparent Viscosity (cp)	
0.1	**	23740.0
1.0	540.0	3712.0
10.0	311.0	569.0
100.0	114.0	84.0
1000.0	26.0	12.0

Particle Used	Particle Settling Velocity (cm/s)	
	HEC*	Xanthan Polymer*
16-18 Mesh Interprop®	0.33	0.0037
18-20 Mesh Interprop®	0.27	0.0029
16-18 Mesh UCAR®	0.2	0.0011
18-20 Mesh UCAR®	0.17	0.0009

* 0.48% by weight, room temperature, 2% KCl solution.

** Torque was too low for a reliable reading.

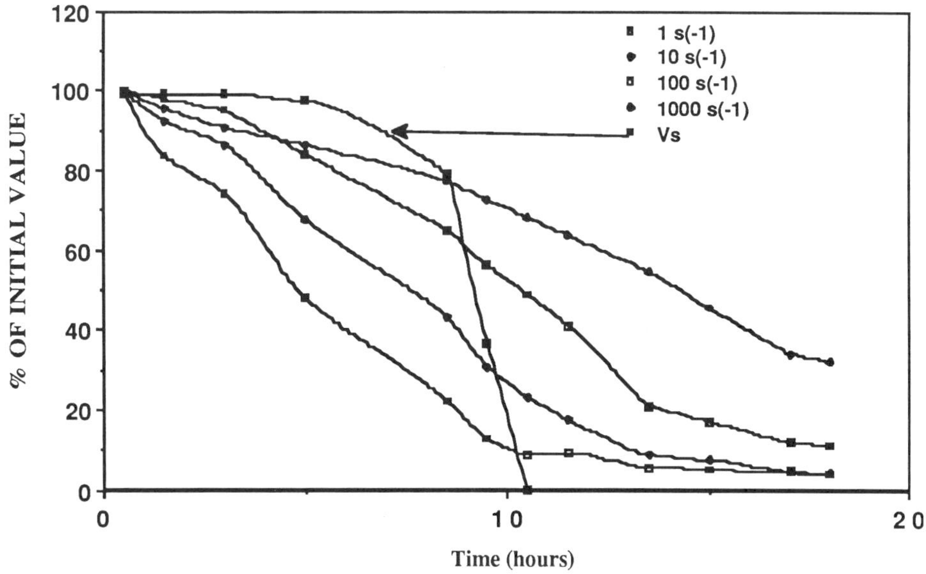

Figure 3. Correlation of Sand Falling Rate to Viscosity at Various Shear Rates.

Relying solely on the viscosity measurements currently used by the industry to predict particle transport can lead to either the use of a less efficient polymer system and/or to possibly the use of a polymer system before it is fully hydrated. A particle suspension test would help assure that the best polymer fluid is used.

POLYMER DEGRADATION

Introduction

Water-soluble polymers are commonly used to increase fluid viscosity of hydraulic fracturing fluids for improved proppant placement and fluid loss control. After completion of the fracturing treatment fluid viscosity should decrease to allow proppant placement and rapid fluid return through the fracture. It is important to control the time at which the "viscosity break" occurs. In addition the degraded polymer should produce little residue to restrict the flow of fluids through the fracture. The polymer may cause significant flow restriction depending on the degree of degradation (molecular size reduction) and insoluble residues produced during degradation.[3,4]

Molecular Weight vs. Viscosity

Loss of solution viscosity has often been used as a measure of polymer degradation. Solution viscosity however is a complex physical parameter which can depend on polymer interactions in solution as well as polymer size (molecular weight) and concentration. Loss of viscosity may be caused by a decrease in polymer interparticle attractions as well as a decrease in molecular size of the polymer. Low viscosity fluids that are considered "broken (less than 10 cp) may contain molecules with relative molecular weights greater than 5 million. Even solutions with viscosities less than 3 cp can have as much as 2 percent of the polymeric material with relative molecular weights greater than 3 million.[5]

Breaker schedules that are designed on the basis of viscosity measurements may leave many large polymer molecules in the proppant pack to retard fluid return. However, using sufficient breaker concentration to efficiently degrade polymer molecular size also reduces solution viscosity too rapidly to maintain a useful fracturing fluid. For example, Table 4 compares short term viscosity with long term molecular weight reduction for the most commonly used fracturing fluid, hydroxypropyl guar (HPG), degraded with ammonium persulfate oxidant breaker. Solution viscosity therefore should not be used as a measure of fluid flow damage.

TABLE 4. HPG Broken at 158 °F Using Ammonium Persulfate Oxidant Breaker

Breaker concentration (lb/1000 gal)	Viscosity 1 hour (cP) at 29 sec^{-1}	Molecular Weight at 72 hours
2	6.5	$3x10^6$
4	5.8	$2x10^6$
8	1.7-2.5	$2\text{-}4x10^5$

Designed degradation depends on a number of variables including polymer/breaker type and concentration, reaction time, temperature, and the presence of fluid constituents (salts, crosslinkers, acids, etc.). Size exclusion chromatography (SEC) has become a useful tool to examine the progress of polymer breakdown for different reaction conditions. Progress of polymer degradation is reported as changes in polymer average molecular weight ($\overline{M}w$). These molecular weights are not absolute values but are relative to those of high molecular weight dextran standards. Details of the chromatographic system used in this study have been described previously.[3]

The degree of molecular weight degradation was then correlated with fluid-loss characteristics of the polymer solutions. Increased fluid-loss characteristics would imply greater ease of fluid cleanup from a fracture. The kinetics of polymer-breaker reactions was also monitored using changes in polymer molecular weight.

Polymer Molecular Weight vs. Fluid-Loss characteristics

Designed polymer degradation can produce two types of polymer residue visible precipitates (insoluble reaction products and impurities such as plant parts in the original gel and invisible residues (partially degraded polymer molecules). Fluid loss characteristics of three water-soluble polymers used by the oil industry have been investigated. The three polymers were chosen because they produce different amounts of insoluble residues as they degrade. The three polymers were hydroxyethyl cellulose (HEC), HPG, and guar gum. HEC is considered residue free while guar gum may produce 8 to 14 percent weight percent insoluble residues.[6,7]

Fluid-loss studies were conducted using a gas pressured filter apparatus. The partially degraded polymer solutions were filtered through specific pore size filters (Millipore MF 0.05 to 5.0 μm using a gas pressure of 100 psi. In addition to insoluble residues, the 0.05 μm filter retained polymer degradation products with molecular weights greater than 3.0 x 10^5 d. Larger filter sizes were used to retain degraded polymer with higher molecular weights and to examine the effect of the insoluble residues on filtration characteristics.

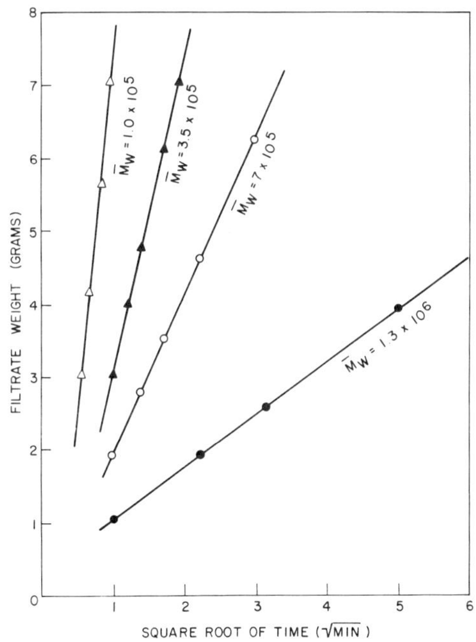

Figure 4. Static fluid loss curves: filtrate passing through a 0.05μm filter as a function of the square root of time for different HEC molecular weights.

These polymer solutions showed wall building characteristics so that fluid flow through the filter was a linear function of the square root of time.[8] Fluid loss coefficients were determined from the slopes of the square root of time plots. Figure 4 shows the change in fluid loss rate for degraded HEC solutions as molecular weight decreased. In oil field units the fluid loss coefficients (C_{III}) are calculated using the following equation:

$$C_{III} = 0.0164 \frac{M}{A} \left(\frac{ft}{\sqrt{min}}\right) \qquad \text{..............................(1)}$$

where: M = slope
A = filter area.

Figure 5 shows the relationship between polymer $\overline{Mw}$ and C_{III} for the three polymer types. Two phenomena are affecting the $\overline{Mw}/C_{III}$ relationship. If $\overline{Mw}$ exceeds that which corresponds to the pore size of the filter, C_{III} remains low and relatively constant. Larger polymer molecules build up a filtercake which then controls fluid-loss characteristics. As $\overline{Mw}$

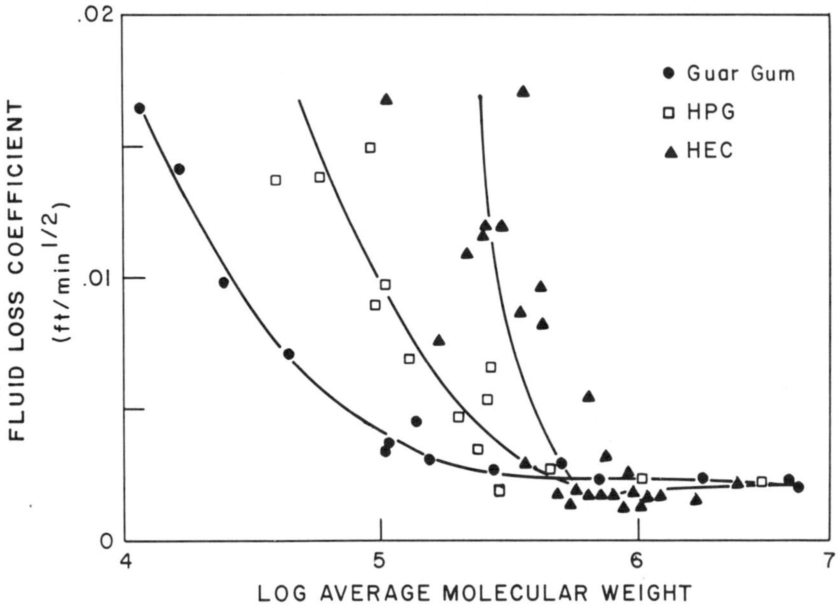

Figure 5. Fluid-loss coefficients as a function of degraded polymer molecular weight.

decreases, C_{III} eventually increases at the point when most of the degraded polymer molecules can pass through the filter. The increase in C_{III} is not the same for the three polymer types, however. For the polymers which produce insoluble residues, polymer molecular weight reduction must proceed to a lower value. The effect is greatest for guar gum which produces the greatest amount of insoluble residue. For example, to obtain a five-fold increase in C_{III} compared to the value for a non-degraded polymer solution, HEC $\overline{Mw}$ must be reduced to 2.5 x 10^5 d.; HPG $\overline{Mw}$ must be reduced to 1.0 x 10^5 d.; and guar gum $\overline{Mw}$ must be reduced to 2.5 x 10^4 d. when using the 0.05 µm filters. HPG and guar gum residues changed the pore size characteristics of the filter and retained degraded polymer of lower molecular weight than occurred with HEC degradation.

Using larger pore size filters, however, has a greater effect on HEC fluid-loss characteristics than on those of HPG. HEC passed without restriction through a 0.22 µm filter after only mild degradation. This is similar to the results of improved filterability after improved mixing and hydration experiments described earlier in this chapter. On the other hand HPG fluid-loss characteristics did not show marked improvement until significant $\overline{Mw}$ reduction had occurred. Results are shown in Table 5.

TABLE 5. Fluid-Loss coefficients for 40 lb/1000 gal HPG Solutions of Different Average Molecular Weights.

$\overline{Mw}$	Reaction Time at 343 K (158 °F) (lb/1000 gal)	Breaker conc.	Fluid Loss coefficient (ft/min 1/2) for Filters 0.22 µm	0.05 m
10 x 10^6	0	-	1.4×10^{-3}	1.4×10^{-3}
4.4×10^6	2	0.8	1.8×10^{-3}	1.5×10^{-3}
6.6×10^5	2	1.6	2.0×10^{-3}	1.8×10^{-3}
1.9×10^6	16	0.8	1.8×10^{-3}	1.5×10^{-3}
4.7×10^5	16	1.6	6.2×10^{-3}	3.2×10^{-3}

Similar results were obtained using 0.8 and 5.0 µm filters. The residues change filter pore sizes resulting in a polymer filtercake buildup that controls fluid loss. Viscosity measurements would not have been able to separate these different effects. Design of efficient breaker systems require information of this type to maximize fluid return and minimize fracture and formation damage.

Kinetic Studies - HPG Degradation Using Ammonium Persulfate

Degradation experiments have been performed to determine the effects of breaker concentration and temperature on polymer molecular weight reduction. Figure 6 shows HPG

$\overline{Mw}$ reduction at 70 °C for different initial concentrations of ammonium persulfate breaker as a function of reaction time. Molecular weight decrease is most rapid within the first 10 to 20 hours. The rate of polymer $\overline{Mw}$ degradation then slows considerably. As breaker concentration increases, lower values of $\overline{Mw}$ are reached as degradation terminates. Results of degradation experiments at 80 and 90 °C are similar. However, as temperature increases a lower value of $\overline{Mw}$ is reached for the same initial breaker concentration as shown in Figure 7.

The $\overline{Mw}$ decrease appears to have a faster initial component (breaker-initiated) and a slower component (possibly caused by hydrolysis, impurities, etc.). If the slower reaction rate results are extrapolated to zero time, the molecular weight decrease caused only by the breaker can be estimated. These values depend on the square root of the breaker concentration as shown in Figure 7. This relationship is found at all three reaction temperatures.

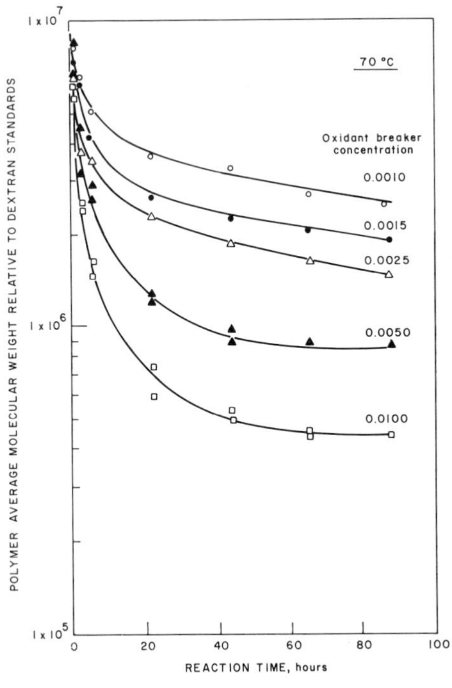

Figure 6. HPG molecular weight decrease a function of time and concentration.

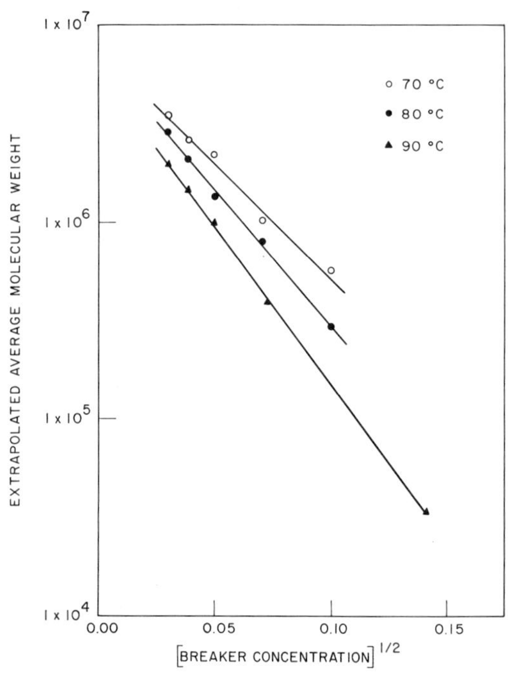

Figure 7. Temperature effect on maximum as decrease in HPG $\overline{Mw}$ caused only by oxidant breaker-initiated degradation reactions as a function of the square root of the breaker concentration.

If the molecular weight degradation rates are examined to determine possible reaction kinetics, the analysis shows that the data best fit a composite of two first-order rate curves. These two first-order rates correspond to the faster breaker initiated degradation and the slower secondary degradation reactions. To separate the breaker-initiated degradation molecular-weight values are determined by subtracting the extrapolated slow degradation portion of the curve ($\overline{M}w^*$) from the actual measured molecular-weight values, ($\overline{M}w$). If the reactions are first-order, a plot of log ($\overline{M}w$-$\overline{M}w^*$) versus time should be a straight line.

Figure 8 shows this relationship for the first 7 hours of reaction time at 70 °C. These data are a composite of at least three experiments using the same breaker concentration (0.005 percent). The straight line fit has a slope of -0.11 hour^{-1}with a correlation coefficient of 0.98. The slope corresponds to the negative value of an observed rate constant for the first order reactions.

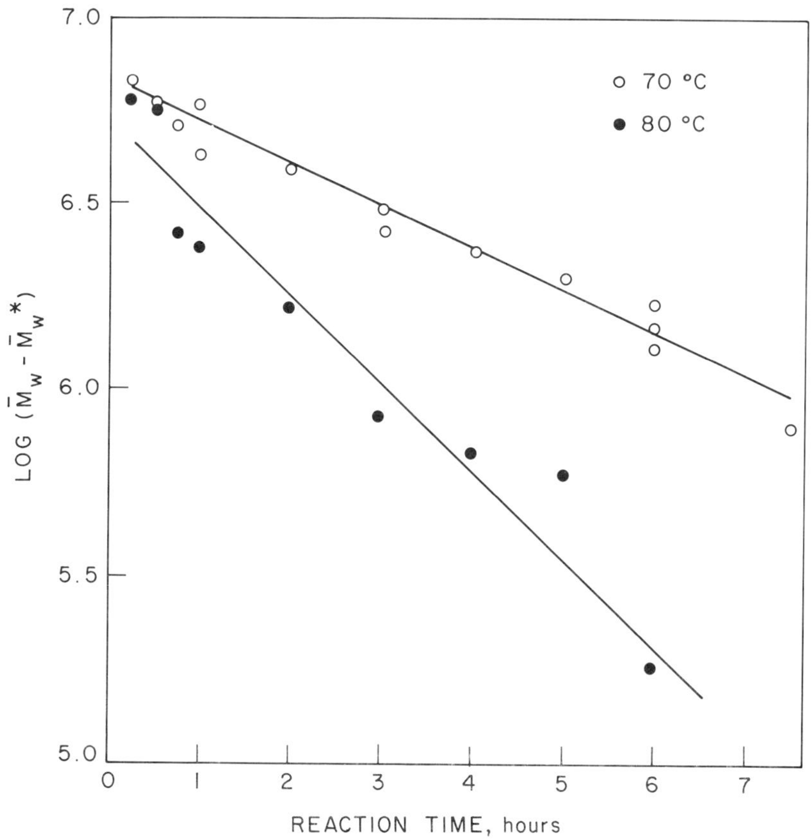

Figure 8 Kinetics analysis of HPG $\overline{M}w$ decrease caused by oxidant breaker-initiated reactions. Initial breaker concentration was 0.005 weight percent.

A similar analysis has been made for HPG degradation at 80 °C using 0.005 percent breaker to initiate degradation. The plot of log ($\overline{Mw}$-$\overline{Mw}$*) as a function of time is also shown in Figure 8. The slope of this curve is -0.23 hour^{-1} with a correlation coefficient of 0.94. The reaction rate has increased by a factor of 2 for the 10-degree temperature increase. This is often the case for reactions involving free radicals.[9]

Ammonium persulfate decomposes in solution to form free radicals $SO_4^{\cdot}$ and $OH^{\cdot}$ which then attack the polymer and start free radical chain reactions. The mechanisms of persulfate oxidations have been reviewed in detail.[10] Some of the possible chain reaction steps include:

Initiation

$$S_2O_8^{=} \rightarrow 2SO_4^{\cdot} \quad (2)$$

$$SO_4^{\cdot} + H_2O \rightarrow HO^{\cdot} + HSO_4 \quad (3)$$

Propagation

$$HO^{\cdot} + P \rightarrow P^{\cdot} + H_2O \quad (4)$$

$$SO_4^{\cdot} + P \rightarrow P^{\cdot} + HSO_4^{-} \quad (5)$$

Termination

$$R^{\cdot} + R^{\cdot} \rightarrow R\text{-}R \quad (6)$$

where $P^{\cdot}$ represents a free radical on the polymer and $R^{\cdot}$ represents any free radical in solution.

Kinetics analysis of reaction rates which involve free radical chains often result in a first order dependence on time. The observed rate constant is a combination of the rate constants of the various steps in the mechanism. The observed rate constant may also depend on concentrations of various components in solution such as the solvent or the polymer. Presently, studies to determine possible concentration dependence have been performed only for the initial breaker concentration. For reactions at 70 °C, the observed rate constant did not depend on the initial breaker concentration. For molecular weight decreases during the first six hours reaction time, the observed rate constants were 0.080 0.081, 0.11 and 0.089 hour^{-1} for initial breaker concentrations of 0.001, 0.0015, 0.0050 and 0.010 weight percent ammonium persulfate, respectively.

In summary, the rate of HPG $\overline{Mw}$ reduction cannot be adjusted by changing breaker concentration. However, the final of $\overline{Mw}$ reduction does depend on breaker concentration. In addition, the rate of $\overline{M}w$ reduction depends on temperature. The rate at which the fracturing fluid reaches formation temperature will also probably affect the initial rate of $\overline{M}w$ decline.

Design of better breaker systems requires a built-in delay mechanism which initially retard reactions which can reduce solution viscosity and polymer molecular weight but do not prevent polymer molecular weight reduction in the long run. SEC is a good analytical technique to monitor these changes with time. Figure 9 shows polymer molecular weight reduction for a commercial breaker/stabilizer system that shows some initial effect on molecular weight reduction. Ideally, however, the desirable shape of this curve would be a plateau at 100 percent maintenance of molecular weight for several hours followed by a rapid decline for ease in fluid removal after a hydraulic fracture treatment.

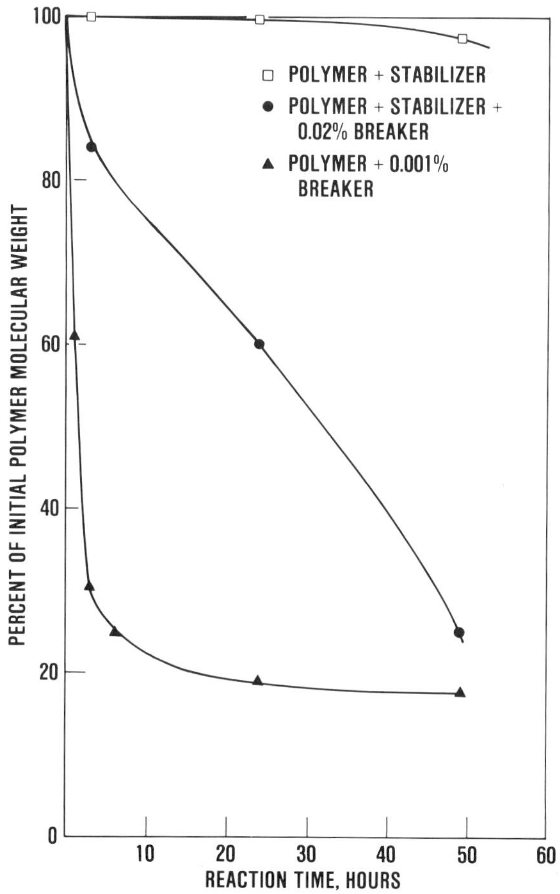

Figure 9. Comparison of HPG $\overline{M}w$ reduction using oxidant breaker and oxidant breaker with a gel stabilizer.

Conclusions

1. Size Exclusion chromatography can be used to study conditions which affect fracturing fluid polymer degradation. Comparison of the effects of different reaction parameters should aid the design and evaluation of different fracturing fluid breakers.

2. Retention of partially degraded polymer by a filtration medium can greatly reduce fluid flow through the medium. Polymers that produced insoluble residues (HPG, guar gum) require greater molecular weight reduction to minimize damage than those that were residue free (HEC CMHEC).

3. Low viscosity does not ensure rapid fluid return because low viscosity fluids can contain sufficient partially degraded polymer and residues to damage and restrict the fracture.

4. For oxidant breaker initiated degradation the initial rate of HPG molecular weight reduction is not dependent on the initial oxidant breaker (ammonium persulfate concentration. Adjusting breaker concentration is not an effective way to retard the initial degradation of the polymer.

5. Temperature increases the rate of molecular weight decrease. A 10-degree rise in reaction temperature increases the rate by a factor of 2.

6. Oxidant breaker initiated polymer degradation results are consistent with degradation via a free radical mechanism.

SUMMARY

It is believed that more beneficial as well as econcomical application of most polymers is entirely practical by supplementing viscosity measurements with additional testing. The tests suggested to assure full hydration and more complete functionality are available and can be made by persons with differing technical skills.

Specifically, it is suggested that a filterability evaluation be used as an added criteria to supplement viscosity to assure polymer hydration. Whereas the use of the API procedure is applicable in some oil field applications even more demanding filtration protocols may be appropriate. Similarly particle suspension considerations were used to define polymer solution functionality. It is of utmost importance to define the basic purpose of using water soluble polymers and to devise tests to assure that the polymer solutions selected be evaluated with those specific purposes in mind.

Finally polymer degradation standards should consider the use of size exclusion chromatography to supplement viscosity testing. Such testing will provide a more realistic standard for assuring proper polymer degradation.

With greater awareness benefits can be numerous. By stressing full hydration not only can polymer waste be controlled but concerns about polymer usage causing decreases in injectivity and possible formation damage can be minimized. By stressing functionality the proper polymer for specific application will more likely be selected. By recognizing that viscosity alone may not define polymer degradation more efficient breaker systems can be devised.

REFERENCES

1. Guler H. N. "Particle Motion in Non-Newtonian Fluids," M.S. Thesis The University of Tulsa 1982.
2. Roodhart L. P. "Proppant Settling in Non-Newtonian Fracturing Fluids," SPE/DOE 13905 presented at the SPE/DOE 1985 Low Permeability Gas Reservoirs Symposium in Denver May 19-22 1985.
3. Gall B. L. and Raible C. J. "Molecular Size Studies of Degraded Fracturing Fluid Polymers," SPE 13566, Phoenix Arizona April 1985.
4. Almond S. W. and W. E. Bland paper SPE 12485 presented at the Sixth SPE Symposium on Formation Damage Control, Bakersfield CA Feb. 13-14 1984.
5. Volk L. J. Gall B. L. Raible C. J. and Carroll H. B. "A Method for Evaluation of Formation Damage Due to Fracturing Fluids," paper SPE 11638 presented at the 1983 SPE/DOE Symposium on Low Permeability Denver CO March 14-16, 1983.
6. Githens C. J. and Burnham J. W. "Chemically Modified Natural Gum for Use in Well Stimulation, *Soc. Pet. Eng.J.*, 5-10 (Feb. 1977).
7. Chatterji J. and Borchardt J. K. "Applications of Water-Soluble Polymers in the Oil Field, *J. Pet. Tech.* 2042-2056 (Nov. 1981).
8. Howard G. C. and Fast C. R. "Hydraulic Fracturing," SPE of AIME Monograph, New York p.40 (1970).
9. Wellington L. *Soc. Pet. Eng. J.* **26**, 901-912 (1983).
10. House D. A. *Chemical Reviews* **62**, 185-203 (1962).

THE APPLICATION OF GELS IN ENHANCED OIL RECOVERY: THEORY, POLYMERS AND CROSSLINKER SYSTEMS

A. Moradi-Araghi, D.H. Beardmore and G.A. Stahl

Phillips Petroleum Company
Phillips Research Center
Bartlesville, Oklahoma

INTRODUCTION

A great deal has been written about the flooding of reservoirs with polymer solutions. The augmentation of a waterflood with water-soluble polymers permits a more controlled flow of fluids through the reservoir. The result is improved volumetric displacement efficiency, and more economical removal of the oil unproducible in the primary recovery step. The injection of a large volume aqueous polymer solution is capital intensive, and often requires a long time to produce the results. A commonly employed option is the placement of a gelable solution in the more permeable zones or fractures in the reservoir, away from the well bore, that is allowed to have sufficient time to gelation. This method, which is referred to as profile control, is relatively inexpensive and often is paid for by the first few months of incremental oil. Although in-situ gelation is often used for diverting the flow of driving fluids in injection wells, it is also used to stop or reduce the flow of water into producing wells to result in lower lifting, processing, and disposal costs. In this paper the application of gelled polymers in EOR will be discussed. Close attention will be given to polymers and crosslinking systems.

THEORY

The production of hydrocarbons from subterranean reservoirs results in a reduction of the driving force available to further produce the remaining oil. Residual oil constitutes about 60-90 percent of OOIP in most oil bearing formations. To sustain or increase this driving force, a fluid such as water, steam, or gas is injected into the reservoir. These fluids, which are in most cases less viscous than the oil, tend to follow the more permeable paths. In doing so, the fluids often by-pass oil bearing zones. Diversion of these fluids from the high permeability zones and fractures to the unswept oil-containing portions of the reservoir is desirable. The most commonly used chemical diversion is the treatment of the permeable zones with a gelled polymer. The following discussion covers a broad overview of related literature.

In gel treatment of a reservoir, a polymer solution is usually injected into the formation prior to crosslinking. The viscous, yet moveable, solution can be expected to flow into the more permeable streaks

- the same regions in which water was previously flowing. After injection, crosslinking is effected by either premixed crosslinkers or a subsequent crosslinker injection.

For a given reservoir the portion of the reservoir to be treated by a polymer gel will usually be less than that flooded by water or an uncrosslinked polymer solution. This means most gel treatments will be less, frequently considerably less, costly than polymer flooding. Since a smaller portion of the reservoir is treated, less incremental oil will be produced. The decision whether to use large or small treatment volumes will be determined by specific reservoir characteristics and economics will be considered in selecting a specific enhanced recovery technique.

A gel consists of a network of crosslinked polymer. The gels used in diverting the fluid path in EOR typically consist of about 0.5-3% crosslinked polymer forming a three dimensional network holding the 99.5-97% water together in an equilibrium state. Exposure of the gel to forces which might alter the nature or the extent of the crosslinking can disturb this equilibrium and result in shrinkage with expulsion of liquid from the gel[1]. This is a phenomenon called syneresis which is observed in some gels as a result of aging. For example, metallically crosslinked gels of polyacrylamide synerese in hard brines when exposed to elevated temperatures for extended times[2]. Syneresis can also occur when excessive amounts of crosslinker are used.

As stated earlier treatment usually includes injection of a low concentration of a polymer (ca 0.5-3%) and some desired level of crosslinker(s). In some treatments, the crosslinkers are injected sequentially. Other treatments involve in-situ polymerization by injecting the monomer(s) and crosslinker(s) into the formation. The crosslinker level is chosen so that the solution transforms into a gel once it reaches its desired destination. When the treatment volume for a given well is small and covers a radius of about 50-100 feet around the well-bore, it is referred to as a near-well treatment, profile modification, or profile control. In cases where the gelation time is slow enough and the gelling solution can be transported to a distance up to several hundred feet away from the wellbore prior to setting, it is called in-depth treatment. One of the true challenges facing the industry is to prepare novel crosslink systems which will delay the reaction for weeks. Such a long delay may make treatment of reservoirs with very large well to well distances practical.

Placement of a gel-forming composition into a high permeability streak of a reservoir is most successful in diverting the flow when a low level of crossflow exists between various layers of the reservoir[3]. At least one modeling study of a reservoir with relatively high vertical permeability has shown a significant incremental oil recovery[3]. The amount of incremental oil is sensitive to the level of permeability reduction. However, once the treated zone ends up with a permeability as low or lower than the adjacent reservoir layer, further reduction makes little difference.

TECHNIQUES OF LABORATORY EVALUATION

Determination of gelation rate or set-time for a given composition is a parameter of utmost importance to the success of a treatment. Premature setting of the gel will result in failure of the treatment and could be very expensive. Delay of gelation, on the other hand, could also be costly due to the need for additional shut-in time. A host of methods ranging from visual observation[4-6] to various physical measurements have been applied for laboratory evaluation of bulk gels[7-15]. These include time of flight measurement for a stainless steel ball to pass a preset distance through the solution[7-8], simple viscosity measurements[8-10], dynamic oscillatory measurements[10-13], and pulse shearometry[10]. While simple viscosity measurement as a function of time can estimate gelation time, it tends to shorten gel time, and it tends to break the gel once formed. Dynamic oscillatory measurement determines viscoelasticity and results in a parameter called the storage modulus, G', which is proportional to the cross-link density of the gel. This method disturbs the gel structure to a lesser degree. Pulse shearometry is based on measuring the elastic modulus of a fluid by determining the velocity of the shear waves through the material[10]. In pulse shearometry, since the collision rate is controlled only by diffusion rate, the gelation time determined is the least biased of the three methods. The variables which affect gelation time include temperature, concentration of polymer and crosslinker(s), pH, salinity and hardness of the brine, type and molecular weight of polymer, nature of crosslinker(s), chemical additives, dissolved oxygen level, and finally, the extent of shearing.

The final strength of a gel composition is another parameter of interest for determination of suitability of that gel for a given application. For example, a strong gel is a better candidate than a weak gel for plugging a large fracture. A storage modulus value of G'=10 dynes/cm^2, has been attributed to a weak gel, while a G' value of 100 dynes/cm^2 represents a strong gel[13]. Other methods proposed for evaluating gel strength include a bulk gel strength tester, which is based on determining the yield pressure which is required to extrude a gel through a 30 mesh stainless steel screen[14]. Measuring the yield pressure which is required to move a preset gel in a glass tube is another method proposed for gel strength evaluation[15]. This method, however, might be biased by the adhesion of the gel to the glass. It is important that the gels used for EOR applications exhibit long term stability under reservoir conditions, so that the need for retreating be minimized. The gel properties, therefore, should be checked in the laboratory prior to field application.

POLYMERS AND CROSSLINKERS

Twenty-two different metal ions have been shown to crosslink water-soluble polysaccharides[16]. Water-soluble synthetic polymers can also be crosslinked by several types of metal ions. This section will briefly summarize the state of knowledge regarding general aspects of metal ion crosslinked gels. Then, the most common EOR gel systems of this type, aluminum citrate, chromium redox, complexed chromium, and chromium/biopolymer, will be described in detail.

The process of gelling water soluble polymers with metal ions can be broken down into 3 basic steps: 1) production of active crosslinker, 2) attachment of active crosslinker to the polymer, and 3) crosslinking of two polymer segments[17].

The details of the first step, as they relate to each type of metal ion crosslinking system, will be discussed later in the appropriate section.

The second step of gelation was apparently thought to be a relatively rapid part of the overall gelation reaction in early studies[17,18]. However, other work indicated that the attachment of metal ions to the polymer and the subsequent crosslinking of polymer chains was not instantaneous, and was affected by temperature by an Arrhenius relationship[9,11,19-21].

Terry and coworkers[17] were able to form gels using a chromium redox system with hydrolyzed polyacrylamide, pure non-ionic polyacrylamide, acrylamide copolymerized with the anionic 2-acrylamido-2-methylpropane sulfonate, and polyacrylamide with unknown cationic groups copolymerized. Other workers present evidence for the formation of chromium complexes with either carboxyl or amide groups[11,22]. The hydroxyl groups on a variety of polysaccharides, such as xanthan gum, guar gum, hydroxypropyl guar, HEC, and CMHEC, can also react with metal ions to form gels[16,22,23].

The exact chemical mechanism of the third basic step in gelation, metal ion crosslinking of two water-soluble polymer molecules, is not well known. However, the finding that shearing a gelling solution increased the gelation rate implied that some step of the process was diffusion controlled[23-25]. It is sensible to assume that as shear increases the number of polymer/polymer collisions, the number of collisions resulting in crosslinks is also increased.

One detailed examination of the mechanism of gelation in a chromium redox/hydrolyzed polyacrylamide system determined that the rate of gelation depended on the crosslinker to the second power and on the polymer to the 2.7 power[21]. The authors suggest that in this particular system, the chromium crosslinker takes the form of a binuclear complex, hence the second power for the crosslinker in the rate equation. An exponent of 2 might be expected for the polymer term in the rate equation since crosslinking reasonably requires two polymer molecules to occur. The authors suggest that the 2.7 exponent on the polymer term is the result of some crosslinking reactions being ineffective in the formation of a gel structure, namely intramolecular crosslinks.

It has been found that a minimum polymer concentration, called the critical polymer concentration, is required for a gel to form[23]. This concentration represents the point where polymer/polymer entanglements become significant, in other words, where the hydrodynamic volumes of polymer molecules begin to overlap. Below this concentration, crosslinking would lead to polymer segregation.

A minimum number of effective (intermolecular) crosslinks per molecule is required to form a gel[11,16,23]. This number is variously reported in the literature as anywhere from 0.25 to 2, and probably depends on polymer concentration and other variables. Intuitively, 2 crosslinks per molecule would be needed to form an infinite polymer network if the polymer was at the critical polymer concentration.

The crosslinking bond may be more or less labile; in other words, the ligands associated with the metal ions may have a higher or lower exchange rate[16]. This concept applies to polysaccharides crosslinked by metal ions both in EOR gels and in fracturing fluids. For a weaker crosslinking bond, that is, a higher ligand exchange rate, the gel solution or gel will have a greater ability to "reheal" itself after it has been sheared. These weaker bonds are desirable in chromium/xanthan gels formed on the surface and in fracturing fluid gels for proppant transport into fractures. Higher temperatures increase the ligand exchange rate and weaken the crosslinking bonds, eventually to the point where no gel is detected[16]. Different metal ions exhibit different effective crosslinking temperatures and pH ranges.

Gelation rate, gel strength, and long term gel stability are the gel properties that describe gelation. As mentioned previously, there are many system variables that can affect gelation. Temperature, crosslinker properties, polymer properties, water properties, and shear all have major effects on gelation[1,2,11,14,16-19,21-28].

I. Aluminum Citrate Gels

The aluminum citrate process employs a fresh water solution of high molecular weight, hydrolyzed polyacrylamides and an aluminum citrate crosslinking solution injected in alternating slugs to achieve in-depth mixing and gelation, although near-well treatments have been performed with this system[9,29-31]. The citrate ion sequesters or protects the Al(III) so that the aluminum can penetrate deep into the formation and mix with previously (and subsequently) injected slugs containing polymer. Without the citrate complex, the aluminum would hydrolyze to an aluminum hydroxide precipitate, making the metal ion unavailable for crosslinking with polyacrylamide.

The form of the citrate complex is not well known. Models of complexes with a 1:1 Al:citrate ratio have been proposed, but it remains to be determined why solutions of aluminum citrate with greater ratios are stable[9].

The aluminum citrate process relies on alternating slugs for indepth permeability reduction. It was originally thought to be a layering process - a polymer slug adsorbs on the rock, aluminum from a citrate slug attaches to the adsorbed polymer, the next polymer slug attaches to the newly attached aluminum, and so on[29,32]. Evidence now points to a "bulk" gelation process; that is, pores are plugged by in situ gelation of a bulk mixture of the polymer and crosslinker[31]. If aluminum citrate is premixed with a polymer solution for pumping downhole its depth of penetration will be limited by the rapid rate of gelation.

Waters low in divalent cations are appropriate for the aluminum citrate process because their presence can cause an aluminum/polyacrylamide precipitate to form rather than a gel. This gel system is applicable in fresh water up to at least 85°C (185°F)[33].

II. Chromium Crosslinked Gels

Three main types of chromium crosslinked gel systems have been studied for EOR applications: dichromate redox/polyacrylamide, complexed chromium/polyacrylamide, and chromium/biopolymer.

The dichromate redox/polyacrylamide system consists of a mixture of three components: sodium dichromate, a reducing agent (such as sodium bisulfite), and polymer, in water. The chromium from the dichromate is in the +6 state, and as such cannot crosslink the polymer. The reducing agent, over time, reacts with the dichromate to form Cr(III), the active crosslinking species, and gelation commences through ionic bonds with anionic groups on the polymer. The chemistry of the redox reactions has been reviewed by Southard, et al.[19]. Controlling the rate of the redox reaction provides control over the gelation rate, which is desired for good EOR applicability. Much work has been published on this system[2,9,11,15,17-21,24-26]. Various reducing agents have been used, including bisulfite, thiourea, and thiosulfate. The ratio of reducing agent to dichromate has been found to affect the gelation rate[19]. The dichromate redox system can be used in field brines that are free of barium (barium dichromate is insoluble) and in brines that are free of ferrous iron (Fe^{++}) and H_2S (these species rapidly reduce dichromate and cause instantaneous gelation). This system can be used in brines containing significant hardness up to a temperature of about 66°C (150°F)[2].

Recently, complexed chromium/polyacrylamide systems have been developed as gels for EOR[34,35]. Chromium in the trivalent state can be complexed by anions, much as aluminum is by citrate, to provide delay to a gel system. Acetate, propionate, and citrate have been studied as chelators in this system[34]. A trimeric structure has been postulated for chromium acetate, which may also apply to the propionate structure[34,36]. An advantage that these systems have over the other metal ion/ polyacrylamide systems is their stability in the presence of the hardest of oilfield brines. Chromium propionate/polyacrylamide gels can be applied at temperatures up to about 71°C (160°F) in hard brines, above which the polyacrylamides are hydrolyzed and precipitated by reaction with the hardness ions[33]. Laboratory tests indicate that this system can be used premixed, as a near well gel treatment, or it may be applied sequentially, for in-depth treatments like aluminum citrate.

Like other polysaccharides, xanthan gums can be crosslinked with Cr(III) to produce a gel. It has been proposed that the crosslinking is accomplished by ionic bonding between the trivalent cations and the carboxylate groups present in the xanthan structure[37]. Another possible route could be complex formation between the hydroxyl groups of xanthan acting as ligands to the cation[16]. Based on crystal field energy calculations, these bonds are expected to be quite labile, and the resulting gels should be pseudo-plastic. The gelation network breaks down with shear and reconstructs once the shear is removed. This property allows mixing of the polymer and the crosslinker on the surface prior to injecting into the well. Based on this shear thinning property it has been proposed that xanthan-Cr(III) gelling system can be used for deep penetration into the formation. It is not clear, however, whether the lower shear experienced away from the well-bore is sufficient to make this gelling system flow.

The formation of a weak bond[37] between the metal and xanthan molecule is not expected to increase the thermal stability of the gel beyond that of xanthan. Since thermal stability of xanthan gum limits its application to temperatures below 80°C (176°F)[38,39], the effectiveness of this gelling system at temperatures above 80°C, is questionable[4].

III. Other Metal Ion Crosslinked Gel Systems

Two other metal ion crosslinking systems have been studied as profile control methods.

A dichromate gel system has been developed that uses lignosulfonate polymers[7]. Lignosulfonates are a by-product of the paper manufacturing process, and have an average molecular weight of 10,000. Dichromate crosslinker is reduced to the active crosslinking form, Cr(III), by reducing sugars that are present from the wood. The dichromate/lignosulfonate system is effective at temperatures of at least 107°C (225°F).

A modified polymer has been described that extends the effective temperature range of chromium ion crosslinked gel systems[12,40]. The NaA/NaAMPS^R copolymer, where NaA represents sodium acrylate and AMPS stands for 2-acrylamido-2-methylpropanesulfonic acid, produced stable dichromate/ redox gels at 121°C (250°F).

IV. Gel Formation Through Organic Crosslinking

Aldehydes can be condensed with materials such as phenols or alcohols to form gels. The history of condensation of formaldehyde with phenol dates back to 1872 by Adolf Baeyer. Later (1910), Leo H. Baekeland used this reaction to produce a product which he named Bakelite[41]. The condensation reactions of formaldehyde with phenolic compounds have recently been used for diversion of flow from high permeability regions in porous media[4-6,42-46]. A recent paper[6] describes formulation of a "phenoformaldehyde" gel which has been effective at concentrations of 2-10% in reducing the permeability of a carbonate core at 85°C (185°F). These gels have very low viscosities (1.5-2.5 cP) prior to setting, and can only be prepared in fresh water or low salinity brines. The mechanism by which these phenolic resins are formed include the following steps: (1) methylolation of the ortho and para positions on the benzene ring by formaldehyde, (2) condensation of the methylol phenol and a phenol to form a methylene bridge, and (3) condensation of two methylol phenols to form an ether bridge[42]. These reactions could produce a three-dimensional gel network.

The organically crosslinked gels have potentially an important advantage over metal ion systems. Since these systems gel through covalent and not ionic bonds, they can be designed to resist syneresis at high temperatures. A commercial gelling system, presumed to include resorcinol, formaldehyde, a pH buffer, and an accelerant[5], has been successful in diverting the flow of water in a 92°C (197°F) injection well. The rate of gelation for this system is very sensitive to salinity and especially the pH of the solution[5], and application requires careful monitoring of injection fluids to avoid premature setting or prolonging the shut-in time.

Naturally occurring polyphenolic compounds such as tannin extracts can be gelled with a source of formaldehyde such as paraformaldeyde and other additives[43-45]. At the levels of 5-15% concentration, these solutions showed viscosities in the range of 6-8 cP at 18000 sec-1 prior to in-situ setting. The gels have shown stability at 275°C (527°F) for at least several months[45].

Lower concentrations of phenolic compounds (ca. 1,000 ppm) and formaldehyde (ca. 2,000 ppm) can be used in conjunction with low levels of polyacrylamides (ca. 5,000 ppm) to form a gel which can be used for fluid diversion[46,47]. The gelation mechanism could include the three steps outlined above for phenoformaldehyde gels as well as condensation of the methylol groups with the amide groups present in acrylamide moieties[46]. The gelation of phenol, formaldehyde and polyacrylamide, however, requires temperatures above 70°C (154°F), a temperature beyond which the stability of polyacrylamides in most oilfield brines is questionable[48]. Substitution of resorcinol, which is more reactive than phenol, can produce gels at lower temperatures.

Aldehydes can also be used to gel solutions of polyvinyl alcohol[49]. A solution of 2.5% polyvinyl alcohol has been gelled with various mono- or dialdehydes. The aldehydes crosslink polyvinyl alcohol through acetal formation. These gels are insensitive to the hardness level of water, and are proposed for use under harsh conditions of high salinity, hardness, and elevated temperatures. To avoid syneresis, like other crosslinked gels, the level of crosslinker should be kept to a minimum. The preferred aldehydes include malonaldehyde and glutaraldehyde.

When an emulsion solution of an acrylic resin and an emulsified solution of an epoxy resin are added to a brine in the presence of a silane adhesion agent, a tough and rubbery gel, with sufficient strength for use as fluid diversion agent, is produced[50]. The silane facilitates adhesion of the gel to silica. Divalent cations commonly encountered in oilfield brine will precipitate the acrylic resin. About 200 ppm of hardness and 151°C (300°F) are the upper limits for this system. The initial viscosity of this system prior to gelation is low so it can be easily pumped down the well. This gel contains about 67-75% water in its formulation as compared to 90-95% water containing gels which are more common. Both of the ingredient resins are commercially available. Because of the high concentration requirement of the gel forming ingredients, the economic viability of this system is questionable.

V. Gel Formation by In-Situ Polymerization

An interesting variation of the polymer gel EOR technique involves in-situ polymerization[51-53]. In this method, a monomer (acrylamide), a crosslinking monomer (N,N' methylenebisacrylamide), an initiator (ammonium persulfate), and an inhibitor (potassium ferricyanide) are premixed on the surface and pumped downhole. The pumping is easier than with a viscous polymer solution because the viscosity of the monomer solution is 1.2 cP. The time delay before gelation is controlled by the catalyst concentration. The solution polymerizes into a gel downhole to provide profile modification. It is possible to place very hard gels with this technique at temperatures up to 200°F[52]. The concerns of employing in-situ polymerization include handling the often toxic and flammable monomer(s) and possible contamination of the ground with the same monomer(s). This bulk polymerization would not be expected to quantitatively convert monomers to polymer.

VI. Other Gel Systems

Silicate gels have been studied as a profile modification method[54-56]. A review of silicate gels for EOR applications notes that the silicate gels are composed of high ratio liquid sodium silicates, Na2O.nSiO2, where n>3[55]. Polymerization in silicate systems occurs in three ways: 1) condensation of monomer and dimer silicate species to form higher order oligimers, 2) intramolecular condensation of silanol groups within polymers leading to ring formation and eventual particle formation, and 3) aggregation of individual particles to form chains and microgels[55]. Gelation can occur under a wide range of pH and temperature [16-93°C (60-200°F)] conditions.

A recent patent describes the use of poly(diallylamine)-epichlorohydrin, which is a cationic resin, for crosslinking of nonionic or anionic polymers[13]. This cationic resin has been successfully used to crosslink anionic polymers of cellulose and guar, xanthan gum, and partially hydrolyzed polyacrylamides The nonionic polymers which can be crosslinked with this cationic resin include: polyacrylamide, polyacrylonitrile, and polyvinylpyrrolidone. The gels which are produced at elevated temperatures show a high level of strength as evidenced by a storage modulus of 300-500 dynes/cm2.

SELECTED FIELD APPLICATIONS

The application of gels in subterranean oil bearing formations as diverting agents for the flow of driving fluids is widely reported[29,37,53,57-69]. It is not within the scope of this paper to describe the details of these treatments. Following is a brief discussion of selective field treatments. The use of Cr(III) and Al(III) crosslinkers in gelation of polyacrylamides, polysaccharides, and lignosulfonates constitutes the majority of successful field treatments. The application of aluminum citrate for crosslinking polyacrylamides[29] was first reported in 1974, and since has been applied in a number of fields with impressive economic success[57-62]. The largest field treatment includes injection of four million pounds each of aluminum citrate and polyacrylamide in a 1,440-acre tract in the North Burbank Unit in Osage County, Oklahoma. This application was aimed at a more in-depth treatment of the field. Four years following this treatment, 854,000 bbls of incremental oil had been recovered with 2,700,000 bbls of total incremental oil projected within an

11 year period at a chemical cost of \$2.95/bbl of incremental oil[58]. Lower chemical costs of \$1.01-\$1.30/bbl of incremental oil have been reported for other fields[59-61].

The effectiveness of the xanthan-Cr(III) gelation system has been extensively tested by treating 205 wells in seven Oklahoma fields[69]. Six of the seven fields responded well to this treatment by producing a substantial amount of incremental oil. A recent paper reported on a successful treatment for diverting the flow for a CO_2 water-alternating-gas operation[53]. This field had a severe channeling problem prior to the injection of a low viscosity monomer solution containing an organic crosslinker. This treatment was an economic success with a payout time of 1.5 months. The effect of this treatment appeared to last about a year, producing a total of 25,000 bbls of incremental oil.

As a precautionary measure it is important to have a method at hand to destroy the gel and restore the permeability should the need arise. The most common method of destroying a gel is exposing it to an oxidant such as sodium hypochlorite (bleach) or hydrochloric acid.

SUMMARY

At the present time, a number of gelling systems exist which can be used to divert the flow of driving fluids to less permeable zones to recover more petroleum from subterranean formations. Several field treatments with polyacrylamide-aluminum citrate, polyacylamide-Cr(III), lignosulfonateCr(III), xanthan gum-Cr(III), and phenolic-formaldehyde gelling systems have been economically successful. An in-situ polymerization technique has also been successful in diversion of driving fluids in a CO_2 water alternating-gas operation.

The applicability of a majority of these gelling systems is limited to those reservoirs cooler than about 80°C (176°F). There is a pressing need for gelling compositions which can tolerate the higher temperatures commonly encountered in deeper oil bearing formations. High temperature stable Cr(III) systems have been reported.

One of the greatest challenges of the field is to prepare crosslinking systems which can form firm gels after several weeks delay. Such systems will allow less costly treatment of reservoirs such as those found in the North Sea where well to well spacing may be 1,000'-5,000'.

ACKNOWLEDGMENTS

The authors wish to thank Drs. P. H. Doe and N. A. Mumallah for reviewing this manuscript and the management of Phillips Petroleum Company for permission to publish this paper.

REFERENCES

1. T. Tanaka, Gels, <u>Scientific American</u>, 244:124 (1981).

2. P. M. DiGiacomo and C. M. Schramm, Mechanism of Polyacrylamide Gel Syneresis Determined by C-13 NMR, Paper SPE 11787, Presented at the International Symposium on Oilfield and Geothermal Chemistry, Denver, CO, June 1-3 (1983).

3. J. S. Tsau, A. D. Hill, and K. Sepehrnoori, Modeling of Permeability Reducing Vertical Conformance Treatments, Paper SPE 13569, Presented at the International Symposium on Oilfield and Geothermal Chemistry, Phoenix, Arizona, April 9-11 (1985).

4. P. W. Chang, I. M. Goldman, and K. J. Stingley, Laboratory and Field Evaluation of a New Gelant for High-Temperature Profile Modification, Paper SPE 14235, Presented at the 60th Annual Technical Conference and Exhibition of the SPE, Las Vegas, NV, September 22-25 (1985).

5. P. W. Chang, C. N. Meltz, G. D. Gruetzmacher, and R. A. Totino, Enhanced Hydrocarbon Recovery by Permeability Modification with Phenolic Gels, European Patent Application 0 177 324 A2, 09.04 (1986).

6. S. S. Nagra, J. P. Batycky, R. E. Nieman, and J. B. Bodeux, Stability of Waterflood Diverting Agents at Elevated Temperatures in Reservoir Brines, Paper SPE 15548, presented at the 61st Annual Technical Conference and Exhibition of the SPE, New Orleans, LA, October 5-8 (1986).

7. B. J. Felber and D. L. Dauben, Laboratory Development of Lignosulfonate Gels for Sweep Improvement, Soc. Pet. Eng. J., Dec.:391 (1977).

8. J. D. Allison, J. D. Purkaple, and L. E. Summers, Laboratory Evaluation of Crosslinked Polymer Gels for Water Diversion, Polym. Mater. Sci. Eng., 55:784 (1986).

9. G. P. Willhite, D. W. Green, T.-S. Young, J. L. Thiel, M. J. Michnick, S. Vossoughi, and R. E. Terry, Evaluation of Methods of Reducing Permeability in Porous Media by In Situ Polymer Treatment, DOE/BC/10354-16, Feb.:3-4 (1986).

10. S. Hubbard, L. J. Roberts, and K. S. Sorbie, Experimental and Theoretical Investigation of Time-Setting Polymer Gels in Porous Media, Paper SPE/DOE 14959, Presented at the SPE/DOE Fifth Symposium on Enhanced Oil Recovery, Tulsa, OK, April 20-23 (1986).

11. R. K. Prud'homme and J. T. Uhl, Kinetics of Polymer/Metal-Ion Gelation, Paper SPE/DOE 12640, Presented at the SPE/DOE Fourth Symposium on Enhanced Oil Recovery, Tulsa, OK, April 15-18 (1984).

12. R. G. Ryles and R. E. Neff, Thermally Stable Acrylic Polymer for Profile Modification Applications, in: "The Role of Polymers in Enhanced Oil Recovery," G. A. Stahl and D. N. Schulz, Eds., Plenum Press, New York, NY (1987).

13. C. Lukach and S.-T. Sun, Gelled Aqueous Composition, United States Patent 4,604,217, Aug. 5 (1986).

14. J. Meister, Gel Strength Tester, Paper SPE 13567, Presented at the International Symposium on Oilfield and Geothermal Chemistry, Phoenix, AZ, April 9-11 (1985).

15. J. P. Batycky, B. B. Maini, and G. A. Milosz, A study of the Application of Polymeric Gels in Porous Media, Paper SPE 10620, Presented at the SPE Sixth International Symposium on Oilfield and Geothermal Chemistry, Dallas, January 25-27 (1982).

16. M. W. Conway, S. W. Almond, J. E. Briscoe, and L. E. Harris, Chemical Model for the Rheological Behavior of Crosslinked Fluid Systems, J. Pet. Tech., Feb.:315 (1983).

17. R. E. Terry, C.-G. Huang, D. W. Green, M. J. Michnick, and G. P. Wilhite, Correlation of Gelation Times for Polymer Solutions as Sweep Improvement Agents, Soc. Pet. Eng. J., Apr.:229 (1981).

18. D. S. Jordan, D. W. Green, R. E. Terry, and G. P. Wilhite, The Effect of Temperature on Gelation Time for Polyacrylamide/Chromium(III) Systems, Soc. Pet. Eng. J., Aug.:463 (1982).

19. M. Z. Southard, D. W. Green, and G. P. Wilhite, Kinetics of the Chromium(VI)/Thiourea Reaction in the Presence of Polyacrylamide, Paper SPE/DOE 12715, Presented at the SPE/DOE Fourth Symposium on Enhanced Oil Recovery, Tulsa, OK, Apr. 15-18 (1984).

20. G. P. Wilhite, D. W. Green, and S. Vossoughi, Evaluation of Methods of Reducing Permeability in Porous Rocks by IN-Situ Polymer Treatment, in: "Contracts for Field Projects and Supporting Research on Enhanced Oil Recovery," B. Linville, ed., USDOE, Washington, D. C. (1985).

21. R. K. Prud'homme, J. T. Uhl, J. P. Poinsatte, and F. Halverson, Rheological Monitoring of the Formation of Polyacrylamide/Cr3+ Gels, Soc. Pet. Eng. J., Oct.:804 (1983).

22. G. B. Thurston, P. M. Ozon, and G. A. Pope, The Viscoelasticity and Gelation of Some Polycaylamide and Xanthan Gum Solutions, Paper #626, Presented at AIChE Spring National Mtg., Houston, TX, March 25-28 (1985).

23. J. A. Menjivar, On the Use of Gelation Theory to Characterize Metal Crosslinked Polymer Gels, Polym. Mater. Sci. Eng., 51:88 (1984).

24. C.-G. Huang, D. W. Green, and G. P. Wilhite, An Expermental Study of the In-Situ Gelation of Chromium(+3)-Polyacrylamide Polymer in Porous Media, Paper SPE/DOE 12638, Presented at the SPE/DOE Fourth Symposium on Enhanced Oil Recovery, Tulsa, OK, April 15-18 (1984).

25. S. Aslam, S. Vossoughi, and G. P. Wilhite, Viscometric Measurment of Chromium(III)-Polyacrylamide Gels by Weissenberg Rheogoniometer, Paper SPE/DOE 12639, Presented at the SPE/DOE Fourth Symposium on Enhanced Oil Recovery, Tulsa, OK, April 15-18 (1984).

26. T. S. Young, J. A. Hunt, D. W. Green, and G. P. Wilhite, Study of Equilibrium Properties of Cr(III)-Polyacrylamide Gels by Swelling Measurement and Equilibrium Dialysis, Paper SPE 14334, Presented at the 60th Annual Tech. Conference and Exhibition of the SPE, Las Vegas, NV, Sept. 22-25 (1985).

27. G. J. Rummo, Crosslinking Delay Achieved with Gel Additive, Oil & Gas J., Sept. 13:84 (1982).

28. M. W. Conway, S. W. Almond, and S. H. Shah, Polymer-Metal Interactions and Their Influence on Gel Performance in Fracturing Service, Polym. Mater. Sci. Eng., 51:7 (1984).

29. R. B. Needham, C. B. Threlkeld, and J. W. Gall, Control of Water Mobility Using Polymers and Multivalent Cations, Paper SPE 4747, Presented at the Improved Oil Recovery Symposium of the SPE of AIME, Tulsa, OK, April 22-24 (1974).

30. R. Parmeswar and G. P. Wilhite, A Study of the Reduction of Brine Permeability in Berea Sandstone Using the Aluminum Citrate Process, Paper SPE 13582, Presented at the International Symposium on Oilfield and Geothermal Chemistry, Phoenix, AZ, April 9-11 (1985).

31. H. A. Ghazali and G. P. Wilhite, Permeability Modification Using Aluminum Citrate/Polymer Treatments: Mechanisms of Permeability Reduction in Sandpacks, Paper SPE 13583, Presented at the International Symposium on Oilfield and Geothermal Chemistry, Phoenix, AZ, April 9-11 (1985).

32. C. P. Thomas, The Mechanism of Reduction of Water Mobility by Polymers in Glass Capillary Arrays, Soc. Pet. Eng. J., June:130 (1976).

33. N. A. Mumallah, Personal Communication (1987).

34. N. A. Mumallah, Chromium(III) Propionate, A Crosslinking Agent for Water Soluble Polymers in Real Oilfield Brines, Presented at the International Symposium on Oilfield Chemistry, San Antonio, TX, Feb. 4-6 (1987).

35. Y. Wu and N. A. Mummallah, Microbiocidal Anionic Sequesterants with Polyvalent Metal Cations for Permeability Correction Process, United States Patent 4,552,217, Nov. 12 (1985).

36. L. E. Orgel, Structure of Trinuclear Basic Acetates, Nature, 187:504 (1960).

37. M. R. Avery, L. A. Burkholder, and M. A. Gruenenfelder, Use of Crosslinked Xanthan Gels in Actual Profile Modification Field Projects, Paper SPE 14114, Presented at the SPE 1986 International Meeting on Petroleum Engineering, Beijing, China, March 17-20 (1986).

38. P. Davison and E. Mentzer, Polymer Flooding in North Sea Reservoirs, Soc. Pet. Eng. J., June:353 (1982).

39. R. S. Seright and B. J. Henrici, Xanthan Stability at Elevated Temperatures, Paper SPE/DOE 14946, Presented at the SPE/DOE Fifth Symposium on Enhanced Oil Recovery, Tulsa, OK, April 20-23 (1986).

40. R. G. Ryles and J. V. Cicchiello, New Polymers for EOR Applications, Paper SPE/DOE 14947, Presented at the SPE/DOE Fifth Symposium on Enhanced Oil Recovery, Tulsa, OK, April 20-23 (1986).

41. T. S. Carswell, History of the Development of Phenoplasts, in: "Phenoplasts - Their Structure, Properties, and Chemical Technology," Interscience Publishers, Inc., New York, NY (1947).

42. F. W. Bilmeyer, Jr., Thermosetting Resins , in: "Textbook of Polymer Science," 2nd Ed., Wiley-Interscience, New York, NY (1971).

43. A. J. Withworth, S. Y.-S. Tung, and E. A. Hajto, Soil Grouting Process and Composition, United States Patent 3,884,861, May 20 (1975).

44. M. Navratil, M. Sovak, and M. S. Mitchell, Diverting Agents for Sweep Improvements in Flooding Operations - Laboratory Studies, Paper SPE 10621, Presented at the SPE Sixth International Symposium on Oilfield and Geothermal Chemistry, Dallas, January 25-27 (1982).

45. M. Navratil, M. Sovak, and M. S. Mitchell, Formation Blocking Agents: Applicability in Water and Steamflooding, Paper SPE 12006, Presented at the 58th Annual Technical Conference and Exhibition of the SPE, San Francisco, CA, October 5-8 (1983).

46. D. O. Falk, Process for Selectively Plugging Permeable Zones in a Subterranean Formation, United States Patent 4,485,875, Dec. 4 (1984).

47. B. L. Swanson, Gelled Composition and Well Treating, United States Patent 4,440,228, Apr. 3 (1984).

48. A. Moradi-Araghi and P. H. Doe, Stability of Polyacrylamides in Hard Brines at Elevated Temperatures, Paper SPE 13033, Presented at the 59th Annual Technical Conference and Exhibition of the SPE, Houston, TX, September 16-19 (1984).

49. M. L. Marroco, Gel for Retarding Water Flow, United States Patent 4,498,540, Feb. 12 (1985).

50. R. H. Knapp and M. E. Welbourn, An Acrylic/Epoxy Emulsion Gel System for Formation Plugging: Laboratory Development and Field Testing for Steam Thief Zone Plugging, Paper SPE 7083, Presented at the Fifth Symposium on Improved Methods for Oil Recovery of SPE of AIME, Tulsa, OK, April 16-19 (1978).

51. H. C. McLaughlin, J. Diller, and H. J. Ayers, Treatment of Injection and Producing Wells With Monomer Solution, Paper SPE 5364, Presented at the Oklahoma City SPE Regional Meeting, Oklahoma City, OK, March 24-25 (1975).

52. W. D. Ford and W. F. U. Kelldorf, Field Results of a Short Setting Time Polymer Placement Technique, Paper SPE 5609, Presented at the 50th Annual Fall Meeting of SPE, Dallas, TX, Sept. 28 - Oct. 1 (1975).

53. P. Woods, K. Schramko, D. Turner, D. Dalrymple, and D. Vinson, In-Situ Polymerization Controls CO2/Water Channeling, Paper SPE/DOE 14958, Presented at the SPE/DOE Fifth Symposium on Enhanced Oil Recovery, Tulsa, OK, April 20-23 (1986).

54. J. O. Robertson, Jr., and F. H. Oefelein, Plugging Thief Zones in Water Injection Wells, J. Pet. Tech., Aug.:999 (1967).

55. P. H. Krumrine and S. D. Boyce, Profile Modification and Water Control with Silica Gel-Based Systems, Paper SPE 13578, Presented at the International Symposium on Oilfield and Geothermal Chemistry, Pheonix, AZ, April 9-11 (1985).

56. B. Vinot, R. S. Schecter, and L. W. Lake, Formations of Water-Soluble Silicate Gels by the Hydrolysis of a Diester of Dicarboxylic Acid Solubilized as Microemulsions, Paper SPE 14236, Presented at the 60th Annual Technical Conference and Exhibition of the SPE, Las Vegas, NV, Sept. 22-25 (1985).

57. D. D. Bruning, J. H. Hedges, and D. R. Zornes, Use of the Aluminum Citrate in the Commercial Scale North Burbank Unit Polymerflood, Presented at 5th Tertiary Oil Recovery Conference, Wichita, KS, March 9-10 (1983).

58. D. R. Zornes, H. Q. Long, and A. J. Cornelius, An Overview and Evaluation of the North Burbank Unit Block A Polymerflood Project - Osage County Oklahoma, Paper SPE 14113, Presented at the SPE 1986 International Meeting on Petroleum Engineering, Beijing, China, March 17-20 (1986).

59. T. E. Doll and M. T. Hanson, Performance and Operation of the Hamm Minnelusa Sand Unit, Campbell County, Wyoming, Paper SPE 15162, Presented at the Rocky Mountain Regional Meeting of the SPE, Billings, MT, May 19- 21 (1986).

60. H. Surkalo, M. J. Pitts, B. Sloat, and D. Larsen, Polyacrylamide Vertical Conformance Process Improved Sweep Efficiency and Oil Recovery in the OK Field, Paper SPE 14115, Presented at the SPE 1986 International Meeting on Petroleum Engineering, Beijing, China, March 17-20 (1986).

61. J. C. Mack and J. Warren, Performance and Operation of a Crosslinked Polymer Flood at Sage Spring Creek Unit A, Natrona County, Wyoming, J. Pet. Tech., July:1145 (1984).

62. J. E. Hessert and P. D. Fleming, Gelled Polymer Technology for Control of Water in Injection and Production Wells, Presented at 3rd Tertiary Oil Recovery Conference, Wichita, KS, April 25-26 (1979).

63. L. G. Schoeling, Implementation of Polymer Gel Treatments in Kansas, Presented at the Sixth Tertiary Oil Recovery Conference, Wichita, KS, March 6-7 (1985).

64. J. Frank and L. G. Schoeling, Injection Side Application of Polymer and Polymer Gels in Kansas, Presented at the Sixth Tertiary Oil Recovery Conference, Wichita, KS, March 6-7 (1985).

65. G. Fulco, Profile Modification in the Elrick Unit, Presented at the Sixth Tertiary Oil Recovery Conference, Wichita, KS, March 6-7 (1985).

66. O. R. Wagner, W. P. Weisrock, and C. Patel, Field Application of Lignosulfonate Gels to Reduce Channeling, South Swan Hills Miscible Unit, Alberta, Canada, Paper SPE 15547, Presented at the 61st Annual Technical Conference and Exhibition of the SPE, New Orleans, LA, October 5-8 (1986).

67. T. M. Garland, Selective Plugging of the Water Injection Wells, J. Pet. Tech., Dec.:1550 (1966).

68. W. F. Hower and J. Ramos, Selective Plugging of Injection Wells by In-Situ Reactions, J. Pet. Tech., Jan.:17 (1957).

69. M. K. Abdo, H. S. Chung, C. H. Phelps, and T. M. Klaric, Field Experience with Floodwater Diversion by Complexed Biopolymers, Paper SPE/DOE 12642, Presented at the SPE/DOE Fourth Symposium on Enhanced Oil Recovery, Tulsa, OK, April 15-18 (1984).

LABORATORY EVALUATION OF CROSSLINKED POLYMER GELS FOR WATER DIVERSION

L. E. Summers, J. D. Purkaple, and J. D. Allison

Conoco Inc., Petroleum Research and Development
P.O. Box 1267, Ponca City, OK, 74603, USA

INTRODUCTION

Many crosslinked gel treatments are currently available for injection profile modification in secondary oil recovery waterflood operations. The gels usually consist of a water-soluble polymer, such as polyacrylamide or polysaccharide, crosslinked with a metal ion (Al^{3+}, Cr^{3+}, Zr^{4+}) or organic compounds. Due to the large number of systems available and possible conditions which must be considered, a rapid testing procedure must be used to screen these systems on a timely basis.

Our approach is to evaluate gel systems with simple beaker tests and more complex flow tests in reservoir core samples. First, the beaker tests are conducted to determine gel time, gel strength, and gel stability. These parameters are related to the placement, the magnitude of permeability reduction and the longevity of a gel treatment. The beaker testing can be completed quickly. Thus, gel systems can be evaluated over a range of conditions, and the most promising systems selected for further testing. At this point, the gel evaluation is continued with flow tests in reservoir core. The core tests are the best indicators available of the potential reservoir performance with a given gel system and are therefore used to make the final selection of a polymer and a crosslinker for field testing.

OVERVIEW

The objective of this paper is to present Conoco's current methodology for the systematic evaluation of different gel systems for permeability modification. First, the beaker and core test techniques used in the evaluation of gel systems are reviewed. Then, the experimental approach followed for a complete gel evaluation is discussed. Laboratory results are included in this discussion to illustrate the importance of the various steps in the evaluation process. In addition to the general overview of our testing program, results are presented that demonstrate the usefulness of the beaker tests in identifying trends observed in core tests.

TESTING TECHNIQUES

Beaker Testing

Gelation is a complex process affected by many variables. There is a need for simple, yet quantitative screening tests to evaluate a large number of samples and to predict performance of gel systems prior to the more difficult and time consuming core tests. We have developed beaker tests for gel time, gel strength and gel stability.

Gel Time. A common means of determining gel time is by the measurement of solution viscosity [1,2]. The gel time is determined from the intercept of the extrapolations from the two straight line sections of a viscosity-versus-time curve (see Figure 1). We use two types of viscometers to monitor changes in viscosity.

For short gel times, a Brookfield viscometer is used. Solution viscosity is continuously logged using a laboratory computer and plotted to give a gel time curve as in Figure 1.

For long gel times, it is not efficient to continuously monitor solution viscosity with the Brookfield viscometer. A severe limitation is placed on the number of sample conditions that can be tested. These problems can be avoided with the use of an inexpensive ball viscometer [3]. The viscometer consists of a 5 mL Pyrex disposable pipet containing a 2.5 mm 440C stainless steel ball. This type of stainless steel has the advantage of being both highly corrosion resistant and magnetic. A schematic of the rolling ball viscometer is shown in Figure 2.

A viscosity measurement with the rolling ball method involves three steps. First, the pipet is placed at a fixed angle in a stand (see Figure 3). Then, the ball is picked up using a magnet and moved to the top of the gel solution where it is released. Finally, the time required for the ball to pass a preset distance through the solution is measured. A plot of the fall time versus the aging time of the solution determines the gel time of the solution.

With the ball viscometer, a large number of samples over a wide range of conditions can be studied simultaneously. The viscometers can be prepared anaerobically, flame sealed, stored up to 250°F safely, and discarded after use. In addition, the technique is ideal for monitoring gel times during a field treatment or on-site testing prior to a field test.

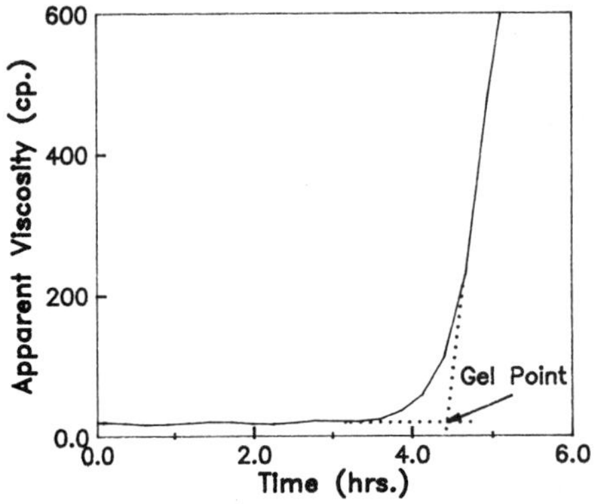

Figure 1 TYPICAL GEL TIME CURVE

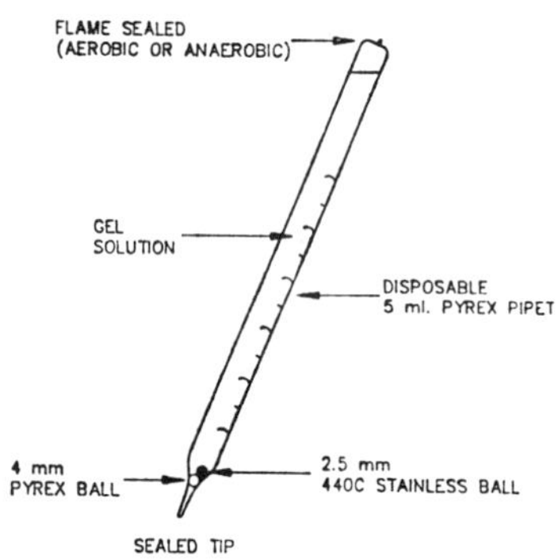

Figure 2 THE BALL VISCOMETER

Gel Strength. Gel strength is difficult to measure or even define in crosslinked polymer solutions. A transfer of the gel to any type of mechanical viscometer usually has a detrimental effect on the gel properties. Hence, unless a gel is allowed to set up and cure in the viscometer (a process that doesn't provide for quick measurement), the measured viscosity is usually not representative of the actual gel strength. An elaborate "Bulk Gel Strength Tester" has been devised by Meister [4] that eliminates the need for transferring the gel from containers to the apparatus.

We have devised a simple apparatus (see Figure 4) to determine the relative strengths of gels (relative to each other and the base polymer solution) that have been prepared and stored in glass containers. A glass capillary is attached to the apparatus by vacuum tubing and inserted into the sample. A constant vacuum is applied to the gel sample by adjusting house vacuum with a needle valve. The amount of time required to pull a specified volume of gel into the capillary is measured. The flowing time (FT) is an indirect measurement of the apparent gel viscosity. Since the apparatus makes use of a capillary to measure relative viscosities of solutions, it was termed the "capillary viscometer."

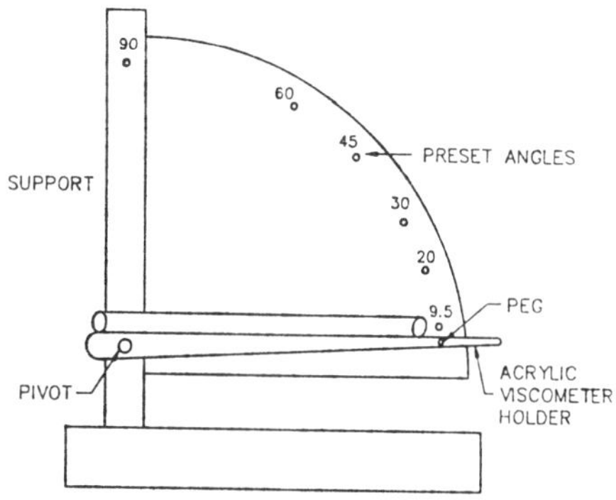

Figure 3 VISCOMETER STAND

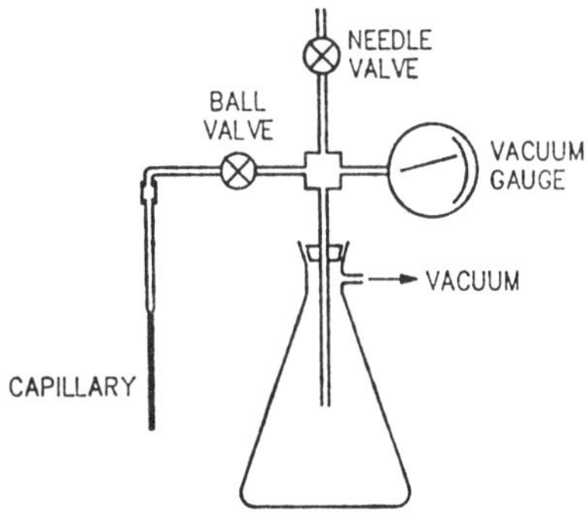

Figure 4 "CAPILLARY" VISCOMETER

Often gel strength measurements on different samples require variation in the vacuum level to make the readings in a timely fashion. Thus, a conversion of the flowing time values to a common vacuum is required to evaluate the gel strength results. As an approximation, the gel viscosity can be assumed to exhibit a power law behavior. Using this assumption, the conversion of the flowing time between different vacuum levels is given below:

$$FT_2 = FT_1 (V_1/V_2)^{1/n}$$

where

FT_1 = measured flowing time at V_1
FT_2 = estimated flowing time at V_2
V_1, V_2 = amount of applied vacuum
n = power law coefficient

The derivation of this equation and several examples of its application are found in Appendix A.

Gel Stability. A major cause of gel instability in crosslinked polymer systems is syneresis. Syneresis is defined as a shrinkage of the gel that forces out the liquid held in the lattice. The mechanism of syneresis is not well understood, but is thought to be due to an overcrosslinking of the polymer [5]. Two methods are used to monitor gel stability.

A simple measurement of gel syneresis is by visual observation. The volume of remaining gel is estimated and divided by the initial volume to give a percent syneresis. Plotting of the percent syneresis versus aging time gives a visual stability curve.

In some gels syneresis does not occur, but a steady decrease in gel strength is noted. In these cases it is necessary to measure the gel strength of the solution at periodic time intervals. For this, a series of ampules containing the gel solution is prepared. If required, the samples can be prepared anaerobically. The capillary viscometer is then used to measure the relative gel strength on the gel solutions at pre-established time intervals. Plotting these measurements against aging time gives a gel strength stability curve.

Core Testing

The second phase of the evaluation program tests the performance of gels under simulated flow conditions in reservoir core samples. The equipment used in core testing is discussed in this section.

A key element in the testing sequence is the core sample. If available, actual reservoir core samples are used for testing. However, frequently reservoir rock is not available and generic rock, like Berea sandstone, must be used.

Routine core testing of gel systems is conducted in two types of apparatus. A core screening panel is used to establish that gelation occurs in the core. The screening panel is designed for the rapid evaluation of gel systems in core. The apparatus is built to test four core samples simultaneously but independently at temperatures up to 250°F and confining pressures up to 2200 psig. The core samples used in this apparatus are typically 1.5 inches in diameter and 2 to 3 inches long. Rubber sleeves are used to confine the core samples during testing. Thus, only the total pressure drop across the core is measured, and the axial distribution of gel in the core can not be determined.

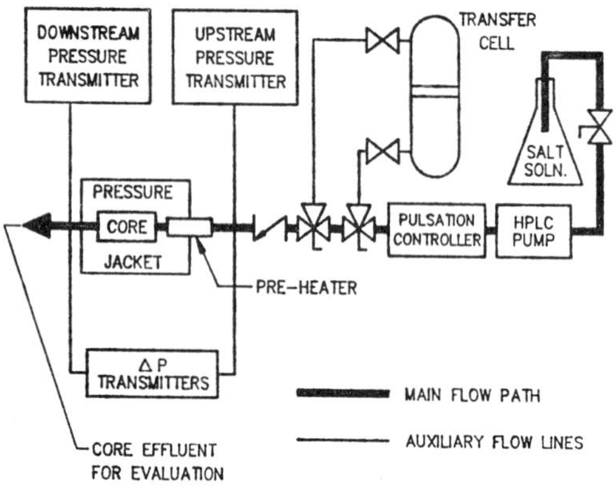

Figure 5 CORE TESTING CONFIGURATION

The second core test apparatus is designed for more detailed tests, such as injectivity/retention tests, gel placement evaluations, and gel stability tests under flowing conditions. Only one core sample at a time can be tested in this apparatus (see Figure 5). The temperature and confining pressure limitations are the same as those for the screening panel--250°F and 2200 psig. Core samples up to 2 feet long can be tested. The core is encased in epoxy so that multiple pressure taps along the length of the core sample can be used to differentiate between core entrance effects and in-depth gelation.

Regardless of which apparatus is used, the mechanics of the testing procedures are similar. First, headers are attached to the front and back face of the core to provide a flow path. The headers are designed to allow the removal of the gelled polymer solution with brine after treatment to ensure that gel does not plug the flow lines. Then, the core sample is placed in a rubber sleeve or coated with epoxy. Finally, the core assembly is placed in a vessel that provides temperature control and confining pressure.

The sequence in an injectivity/retention test typically involves four steps. First, the initial core permeability is measured with brine. Then multiple pore volumes of polymer are injected at a constant rate while the pressure drop is monitored and the effluent is fractionated for analysis. Next, brine is injected to flush unretained polymer from the core sample. Finally, the postinjectivity brine permeability is measured.

The sequence in the treatment of a core with a polymer gel system is similar to that just described and typically involves three steps. First, the initial permeability of the core sample is measured with brine. Then the core is treated with a polymer/crosslinker solution at a constant rate using a dual piston HPLC pump and a transfer cell (see Figure 5) while monitoring pressure with a differential pressure transducer. The core effluent is fractionated for analysis. This includes the determination of polymer and crosslinker concentration, the measurement of sample gelation tendencies, and the investigation of rock/polymer interactions. Finally, after a waiting period to allow gelation, the final permeability of the core is measured with brine to evaluate the effectiveness of the treatment.

The flow rates used in both injectivity and permeability testing with gelled polymer systems correspond to near wellbore shear rates. Typically, the rates have been between 0.5 and 2.0 feet/hour. Our testing has emphasized these conditions because of the economic and gel time delay limitations on the practical application of gelled polymer technology.

It is important to note that the proper evaluation of a polymer gel system must also include the evaluation of the polymer in addition to the gel system. Otherwise, it is not possible to differentiate whether the permeability reduction is due to the polymer alone or gel formation.

POLYMER/CROSSLINKER SYSTEM EVALUATION

Having described the mechanics of the beaker and core tests, the framework for a detailed evaluation of a gelled polymer system for a field application is discussed in this section. There are two categories of evaluation--beaker and core testing. An overview of the steps in a complete evaluation is shown in Figure 6.

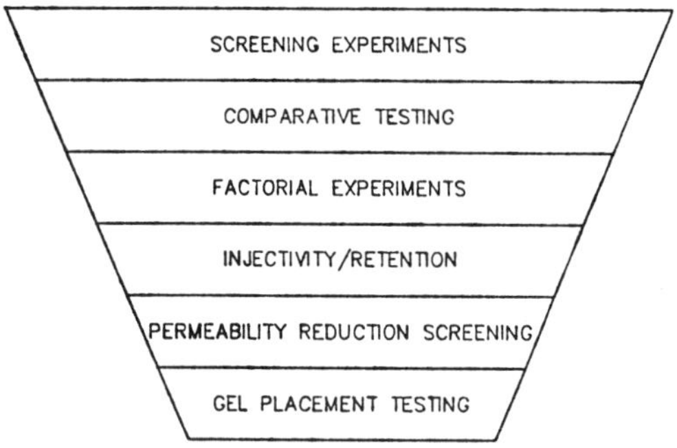

Figure 6 GEL EVALUATION APPROACH

The full methodology outlined in this section is required for new gel systems or for reservoir conditions not previously tested. However, once testing experience is developed with a gel system and in applications with similar reservoir conditions, an abbreviated evaluation is used. This type of evaluation would concentrate on factorial experiments in beaker tests and on placement testing in core.

Beaker Tests

The beaker testing phase of an evaluation involves three types of experiments that begin with broad variable screening and progress toward specific reservoir conditions. Initially, several polymer systems are screened to identify the major variables affecting the gel properties. Then, the trends these variables produce are defined in comparative testing. Finally, factorial experiments are used to investigate the variables that can be controlled in field application. It is important to note that although these tests, and subsequent statistical work up, give quantitative results, the point of these experiments is the comparative trends not the final regression equations.

Screening Experiments. The first step in the evaluation of a gel system is a screening experiment. As the name implies, the screening experiment is statistically designed to test a wide range of variables with a minimum number of samples. The purpose is to determine the major variables that influence the gelation process.

We use the Plackett-Burman screening approach [6] for initial gel system evaluations. The experiments determine the importance of six to eight variables using two levels in a 12-sample screening. The variables under consideration in a typical screening experiment are as follows: (1) reactant concentrations, (2) solution pH, (3) polymer related variables like molecular weight and/or percent hydrolysis, (4) salinity, and (5) temperature. A summary of a Plackett-Burman experiment is presented below.

Table 1 shows the variable levels that were used in the screening experiment. The experiment evaluated the gel reaction of two polyacrylamide polymers with identical levels of hydrolysis (20 percent). The reduction of potassium dichromate with thiourea was used to produce Cr(III) and crosslink the polymers.

The responses of gel time and gel strength were analyzed. The trends that the independent variables produce were identified by summing all of the responses associated with the high and low levels of a given

TABLE 1
VARIABLE CONDITIONS GENERAL SCREENING EXPERIMENT

Variable	High [+]	Low [-]
Polymer Conc. [ppm]	3500	2500
Dichromate Conc. [ppm]	1500	1000
Thiourea Mole Ratio*	12	6
pH (HCl Adjusted)	7.0	5.0
Polymer Molecular Weight in DI Water] [IV dL/g]	450	250
Salinity	Sea Water	DI Water
Temperature (deg F)	180	160

*Mole Ratio with respect to the dichromate concentration.

TABLE 2
RESULTS OF THE GENERAL SCREENING EXPERIMENT

Variable+	Gel Time	Gel Strength
Polymer Conc. [ppm]	Decreasing o	Increasing o
Dichromate Conc. [ppm]	Decreasing o	Decreasing o
Thiourea Mole Ratio	Decreasing o	No Trend++
pH (HCl Adjusted)	No Trend++	Increasing*
Polymer Molecular Weight	Decreasing**	Increasing*
Salinity	Decreasing**	Decreasing*
Temperature (deg F)	Decreasing*	Increasing o

+The trend in the response is based on increasing from the lower limit to the upper limit.

++No trend detected with the range of conditions tested.

The most significant variables in the statistical analysis are marked with the following:

o -- trend only
* -- to 90% confidence interval
** -- to 95% confidence interval

variable. The differences between high and low summations indicate the importance of the variable on the gel properties. The statistical analysis of the results is summarized in Table 2.

A review of actual experimental results shows the information that can be derived from a screening experiment. For the polymer/ crosslinker systems tested, the two most important variables affecting gel time and gel strength were polymer molecular weight and salinity.

Comparative Tests. The use of screening experiments to identify the major variables in a given gelation system is followed by detailed comparative testing. These tests often identify trends that single variables that can not be controlled in a field application, like reservoir temperature, produce on gel properties. With the change from broad variable screening to the testing of individual variable effects, the conditions investigated become very site specific. This includes testing the gelation reaction in the water that is to be used for mixing the chemicals in the field treatment.

Examples of comparative gel tests with a hydrolyzed polyacrylamide are presented below. The screening experiment discussed above indicated that polymer molecular weight and salinity are important in the gel reaction. The comparative tests are used to define the magnitude of these variable effects. In addition, the results demonstrate the usefulness of the simple beaker tests described earlier.

High molecular weight polymers have faster gel times and higher gel strengths than low molecular weight polymers. The increased rolling ball gel times associated with variations in polyacrylamide molecular weight are shown in Figure 7. Three conditions were tested with equivalent polymer and crosslinker concentration. Molecular weight was varied through shear degradation of the polymer solution, as demonstrated by the initial formulation viscosities. The effect of polymer molecular weight on gel times follows the trend observed in the screening experiment and predicted by theoretical arguments. In Flory's statistical crosslinking analysis [7], the number of crosslinks required for gelation in a linear polymer system is proportional to the inverse of the weight average molecular weight of the polymer. Given the equivalent availability of the crosslinking species, the higher molecular weight polymers gel faster.

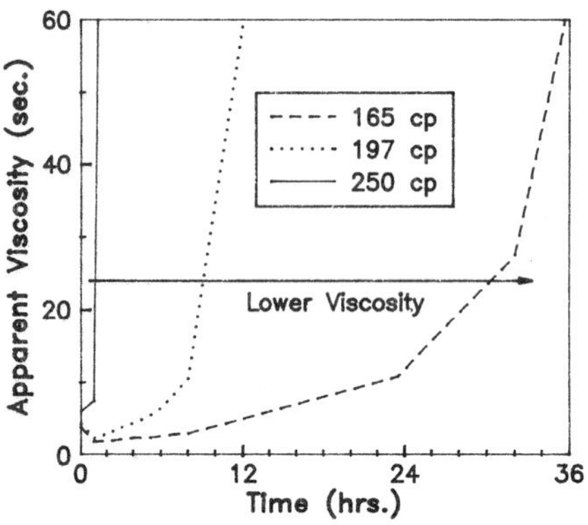

Figure 7 SHEARED POLYMER EFFECT
Gel Time Molecular Weight Dependence

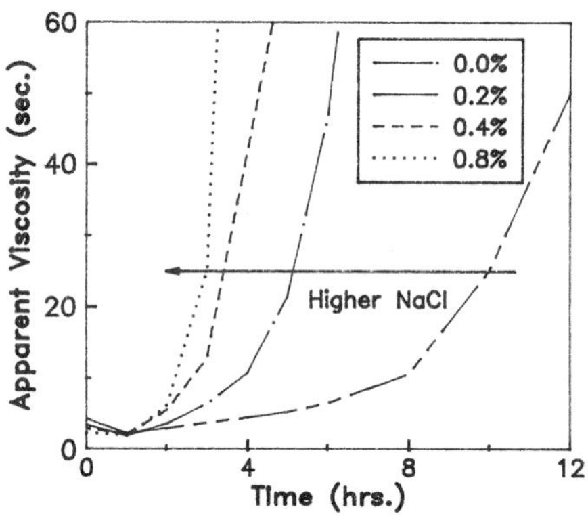

Figure 8 NaCl VARIATION
Gel Time Salinity Dependence

Increasing the salinity of the solution reduces the gel time (see Figure 8), and in some cases, decreases gel strength. Because of the polyelectrolytic nature of the hydrolyzed polyacrylamides, the impact of salinity on the gelation process is not surprising. These trends are consistent with the screening experiment predictions and dilute polymer solution behavior. The salt screening effect allows a close approach of individual polymer molecules resulting in a reduced gel time. While the intermolecular interactions speed gelation, the intramolecular forces are also reduced. The polymer assumes a coiled configuration limiting the intermolecular crosslinking sites. The resulting polymer network is weaker than a gel crosslinked in low salinity water with the polymer chains near full extension and a greater number of crosslinking sites exposed. In fact, it is possible that intramolecular crosslinking could occur in the salt screened polymer solutions causing further degradation in the gel strength. It should be noted that the same gel property dependence on salinity is not seen in nonionic polymer gelation [8].

Factorial Experiments. Statistical design experiments are used to define multiple variable effects and interactions. The majority of the factorial experiments that we use focus on three independent variables. The variable selection is determined by the results from the Plackett-Burman experiments and the extent to which those variables are controllable in field practice.

The factorial design method [9] spans a large variable space and is ideally suited for the use of regression techniques in the evaluation of the experimental responses. The accuracy of the regression models generated from the factorial design is often limited by the quality of the data and the selection of the model form. But the trends predicted by such correlations are valuable in developing a more fundamental understanding of the process.

Contour plots (see Figures 9 and 10) are generated from regression equations that show a given experimental response, like gel strength, as a function of two independent variables. The plots improve visualization of the trends that reaction variables produce on the gel properties. In addition, an overlay using the two plots (see Figure 11) allows for a simple optimization of the gel formulation and speeds the selection of conditions for core testing.

Core Tests

A large number of beaker tests can be rapidly conducted to screen polymer gel systems. For selected systems the evaluation is continued in core tests, since misleading conclusions could be drawn if the evaluation were to stop after the beaker evaluation phase. As in the beaker testing, three types of core evaluations are used. First, the base polymer is evaluated without the crosslinking agent in injectivity and retention testing. Then, the gel candidates are tested in the core screening panel. This preliminary testing identifies the gel systems that reduce permeability at reservoir conditions and attempts to relate beaker test results to core performance. The final core testing focuses on the placement and stability of the best gel systems.

The relationships that we have observed between the results from beaker test and core performance are discussed in detail in the results section. Examples of typical results from the other tests are included here.

Polymer Injectivity/Retention. Due to the increased viscosity, the rate of injection of a polymer solution into a porous matrix is less

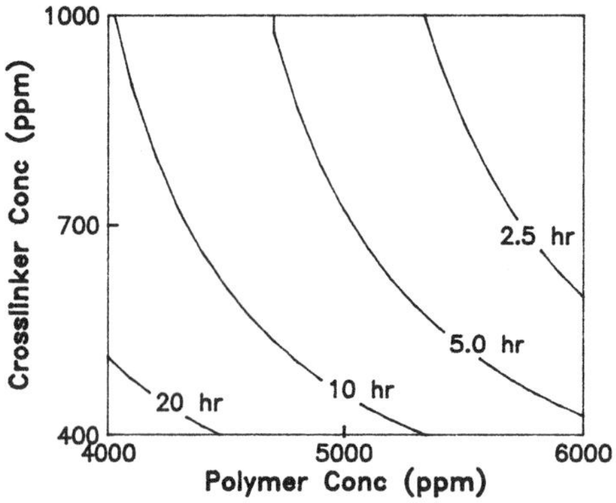

Figure 9 CONTOUR PLOT
Gel Time Response

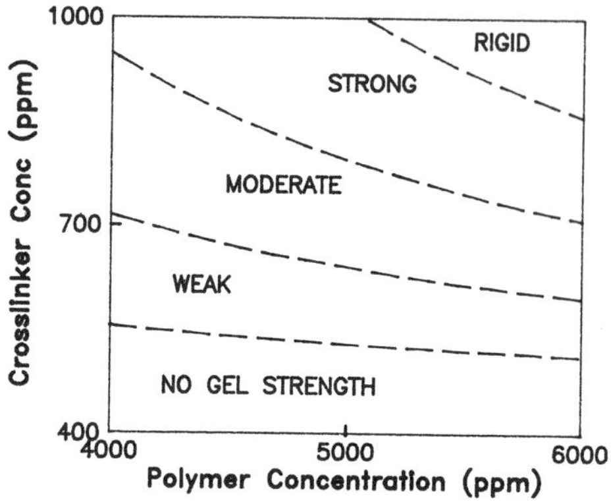

Figure 10 CONTOUR PLOT
Gel Strength Response

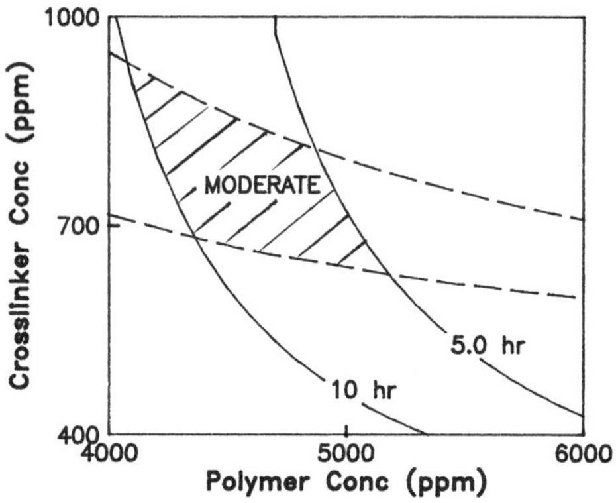

Figure 11 CONTOUR PLOT
Gel Time / Gel Strength Overlay

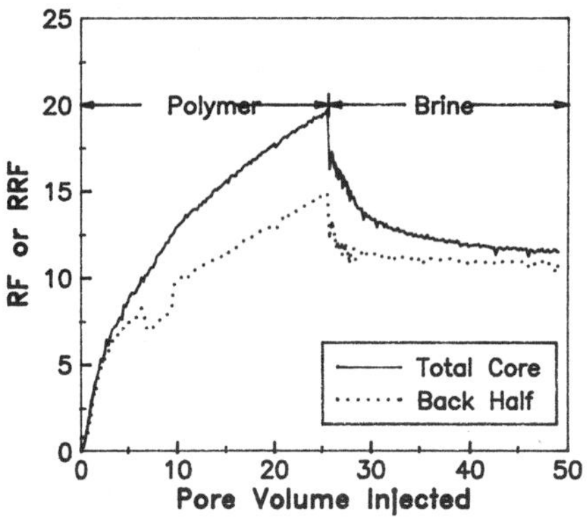

Figure 12 CORE INJECTIVITY TEST

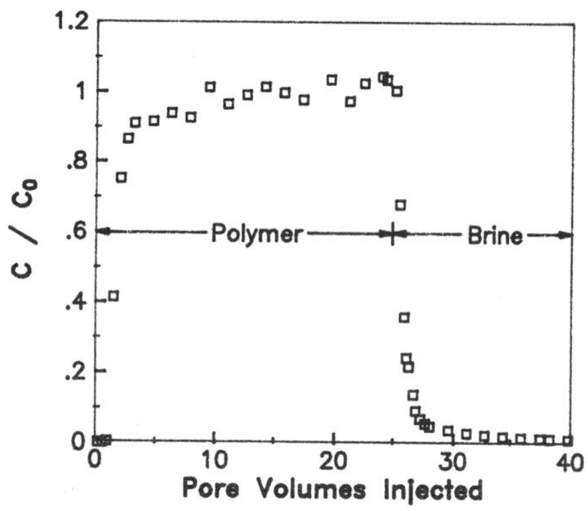

Figure 13 CORE INJECTIVITY TEST
Effluent Concentration Profile

than that of a brine at a given pressure drop. The magnitude of this reduction is critical in the design of a field treatment since it controls the amount of gel solution that can be injected prior to gelation.

The results from a typical injectivity test are given in Figure 12 where resistance factor (RF) and residual resistance factor (RRF) are plotted versus the number of pore volumes of polymer and brine injected. Resistance factor is the mobility of brine (k/μ) divided by the mobility of polymer and is a measure of the reduced injection rate the polymer produces in a given reservoir rock. Residual resistance factor is the mobility of brine before polymer injection divided by the mobility of brine after polymer injection. Thus, a permeability reduction of 99 percent corresponds to an RRF of 100.

Figure 12 shows that high RF values can be obtained during polymer injection. In addition, more polymer is retained in the front than in the back of the core since there is separation in the two RF curves for the total core length and the back of the core. Finally, there is a residual contribution from polymer remaining in the core since the RRF values during brine injection are around 10.

The effluent polymer concentration profile obtained during the polymer injectivity test just described is shown in Figure 13. Polymer was injected until the effluent C/C_o ratio reached a value of 1. The results from the effluent analysis would be used to estimate a value for polymer retention.

Permeability Reduction Screening. The intent of the preliminary gel screening in core is to determine if candidate polymer/crosslinker systems produce a lower permeability (higher RRF values) than the base polymer. Gels that meet the permeability reduction criteria for a specific field application are then evaluated fully in the final core test phase.

A comparison of two gel systems and the baseline polymers is shown in Figure 14. In both cases, the gels produce RRF values greater than the polymers alone. However, if in the field application a high permeability reduction were required, only the system with the highest RRF value would be tested in detail.

Gel Placement Testing. A schematic of a core sample that is used in this phase of the core testing is shown in Figure 15. Figures 16 and 17 show the results from two tests using core samples configured as in Figure 15. Figure 16 contains the RRF results from the test with polymer only while Figure 17 contains the RRF results from the test with polymer and crosslinker. The results from these two figures show that gelation occurred only in the front part of the core sample. If the RRF for the total core length were the only one measured, then the conclusion that gelation had occurred throughout the core sample would have been an erroneous one. The results also show a trend of increasing permeability with throughput that indicates a breakdown or instability of the gel with time.

RESULTS--RELATIONSHIP BETWEEN BEAKER TESTS AND CORE PERFORMANCE

Selected systems have been studied to compare the trends observed in beaker tests to core performance. If correlations can be developed between beaker test measurements and core test results, the screening of polymer/crosslinker systems can be expedited. These tests would ideally

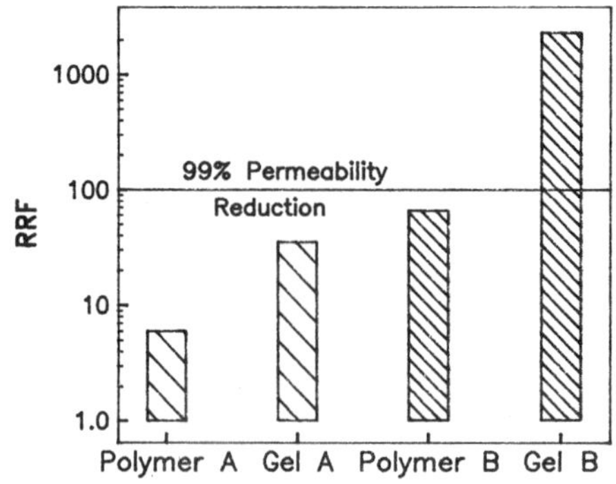

Figure 14 PERM REDUCTION SCREENING RRF Comparison

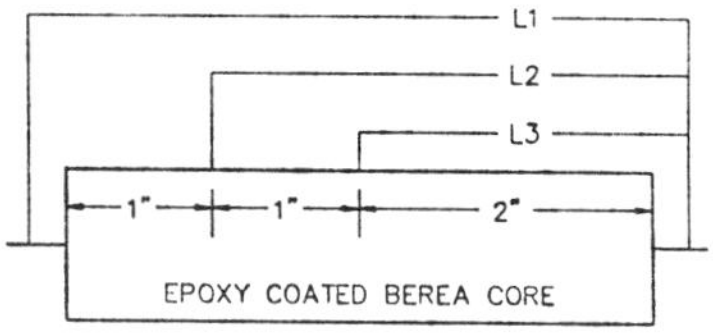

Figure 15 CORE SKETCH

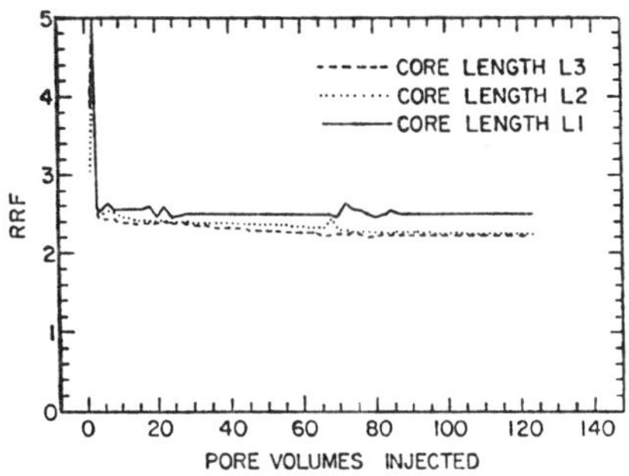

Figure 16 IN-DEPTH GELATION Polymer Only

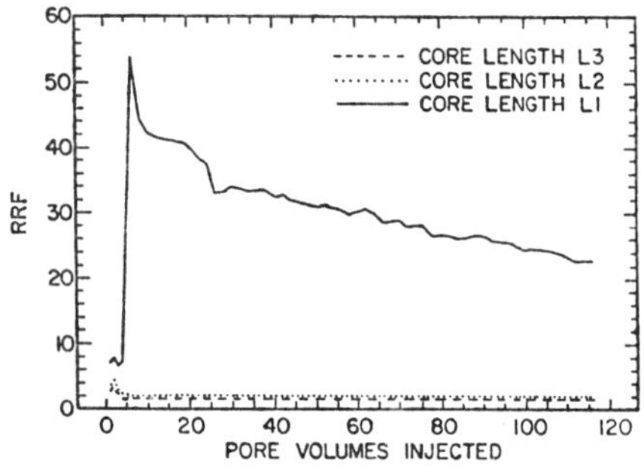

Figure 17 IN-DEPTH GELATION Polymer + Crosslinker

relate gel time to pump time, gel strength to initial permeability reduction, and gel stability to permeability reduction with time.

Gel Time

Pump time tests are designed to determine how long premixed solutions of polymer and crosslinker can be injected through a core plug before gelation occurs. The importance of conducting these tests is illustrated in Figure 18 where the pump time for a gel formulation is plotted along with the beaker gel time. The pump time with this system is less than the beaker gel time indicating that the volume of gel formulation which could be placed within the formation would be less than that expected based on the beaker tests. While the beaker gel times do not yield accurate pump time estimates for this system, Figure 18 also indicates that the trends identified in the gel time measurements are also followed in the pump time tests.

Gel Strength

The proper design of a profile modification treatment requires measurement of the permeability reduction caused by a given gel system. Core tests have shown that relative gel strength measurements with the capillary viscometer correlate with permeability reduction--increasing gel strength develops higher residual resistance factors.

A series of core tests was conducted with three gel formulations in the core screening panel. The RRF value from a core treated with polymer alone was compared with the RRF values of the three gel treatments in core (see Figure 19). While the baseline polymer developed an RRF of 3.6, the RRF values with gelled polymer ranged from 8.5 to 1000 as the relative gel strength of the formulations varied from a flocculated or precipitated polymer (no gel strength) to a stiff gel matrix.

Gel Stability

As a continuation of the tests discussed above, the permeability of the treated cores are measured periodically over the course of several weeks to several months. In order to develop a gel stability correlation, beaker samples can be also monitored over the same time frame to identify syneresis or gel strength degradation.

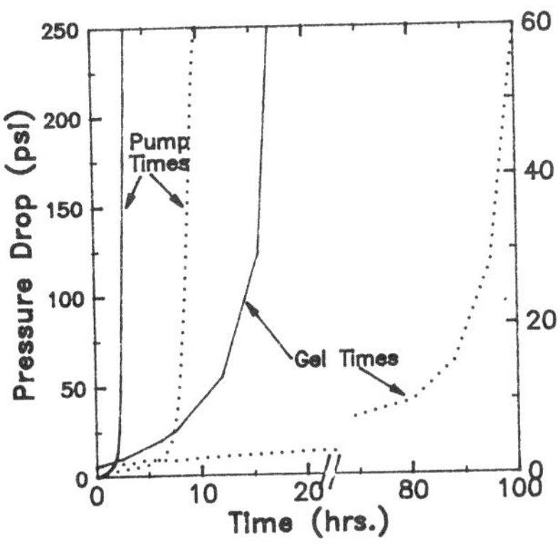

Figure 18 CORE TEST COMPARISON
Pump Time vs Gel Time

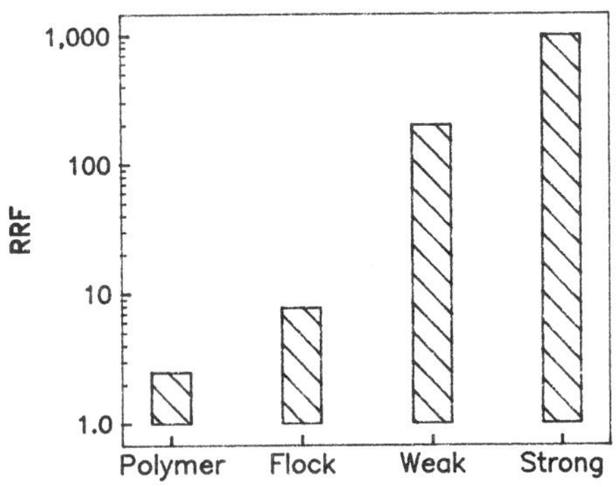

Figure 19 CORE TEST COMPARISON
Gel Strength vs. RRF

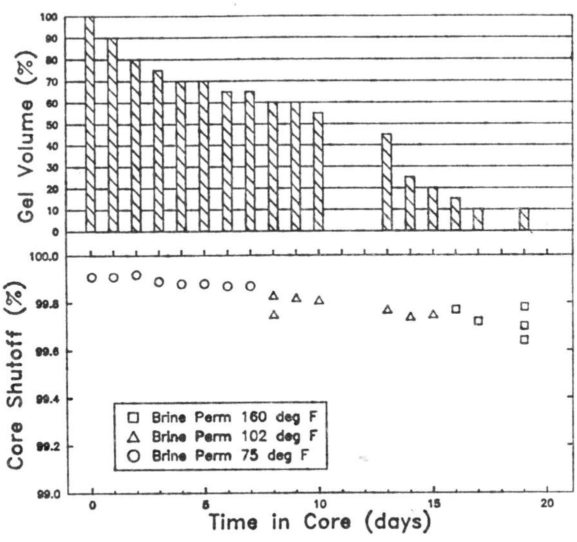

Figure 20 CORE TEST COMPARISON
Stability Trends

A core sample treated with gel and a beaker sample taken from the solution used to treat the core were aged for a period of 19 days. The temperature of both samples was increased from 75°F to 160°F during the test to accelerate the gel degradation process. It was observed that the gel shrank to less than 10 percent of the original gel volume in 17 days. While permeability measurements increased slightly, significant permeability reduction was maintained in the core despite the syneresis observed in the beaker sample (see Figure 20). As in the pump time example, there is a lack of a quantitative correlation between beaker test and core test stability measurements, but the trends are identified.

In general, the beaker test results are invaluable in the selection of candidates for a given application. However, the results presented above illustrate the necessity of core tests in the design and testing of gelled polymer formulations.

ACKNOWLEDGMENTS

The authors want to thank Conoco for its support of this work and allowing it to be published. The majority of the lab work was completed by Sarah Belitz and Lois Humble and their efforts are appreciated.

REFERENCES

1. Billheimer, J. S., and Parrette, R., Anal. Chem., 28, 272 (1956).

2. Marrs, W. M., and Wood, P. D., Photographic Gelation 2, Proc. Sym. (3rd), 101 (1974, Publ. 1976).

3. Bennett, K. E., Ph.D. Thesis, University of Minnesota, 1985.

4. Meister, J., SPE 13567, presented at the Symposium on Oilfield and Geothermal Chem., 1985.

5. DiGiacomo, P. M., and Schramm, C. M., SPE 11787 presented at the International Symposium on Oilfield and Geothermal Chem., 1983.

6. Plackett, R. L., and Burman, J. D., Biometrika, 33, 305 (1946).

7. Flory, P. J., "Principles of Polymer Chemistry," Cornell University Press, 356 (1953).

8. Prud'homme, R. K., and Uhl, J. T., SPE/DOE 12640 presented at Fourth Symposium on Enhanced Oil Recovery, 123, 1984.

9. Box, G.E.P., and Behnken, D. W., Technometrics 2, 455 (1960).

10. Van Wazer, Lyons, Kim and Colwell, "Viscosity and Flow Measurement," Interscience Publishers, 197 (1963).

APPENDIX A--DERIVATION OF GEL RANKING EQUATION

If the assumption is made that polymer gels behave somewhat like power law fluids, then for a capillary tube the volumetric flow rate [10] is:

$$Q = \pi(\Delta p/2k_1L)^{1/n}\ [n/(3n+1)]R^{(3n+1)/n} \tag{1}$$

where
- Q = volumetric flow rate
- Δp = pressure drop
- k = constant
- L = capillary length
- n = power law exponent, constant
- R = capillary radius

After taking the log of both sides of equation (1), it becomes

$$\log Q = \log A + 1/n \log (\Delta p/L) \tag{2}$$

where $A = \pi(1/2k_1)^{1/n}[n/(3n+1)]R^{(3n+1)/n}$

From the form of Equation (2), a plot of log Q versus log $\Delta p/L$ will give a straight line of slope 1/n.

For the capillary viscometer, L and R remain constant. Therefore, for any two measurements, Equation (2) becomes

$$\log \Delta P_2 - n \log Q_2 + n \log A = \log \Delta P_1 - n \log Q_1 + n \log A \tag{3}$$

Equation (3) can be simplified to

$$\log (Q_1/Q_2) = 1/n \log (\Delta P_1/\Delta P_2) \tag{4}$$

or

$$Q_1/Q_2 = (\Delta P_1/\Delta P_2)^{1/n} \tag{5}$$

But, for the capillary viscometer Q is proportional to 1/FT, and pressure drop is proportional to V where FT = flowing time in seconds and V is the applied vacuum in inches of mercury.

Therefore, Equation (5) becomes

$$FT_2/FT_1 = (V_1/V_2)^{1/n} \tag{6}$$

$$\text{or}\quad FT_2 = FT_1(V_1/V_2)^{1/n} \tag{7}$$

The value for n was estimated in the following fashion. FT data were collected on two polymer gel samples over a range of applied vacuum. The data were regressed by plotting log (1/FT) vs log V and the value for n was taken from the slope where n = 1/slope. One set of data gave a value of 0.7 for n (regression coefficient = 0.9897) while the other set of data gave a value of 0.75 for n (regression coefficient = 0.9711). The experimental data and FT predictions using (7) are given in Table A1.

TABLE A1
CAPILLARY VISCOMETER GEL STRENGTH COMPARISON
OF MEASUREMENTS AT DIFFERENT VACUUM LEVELS

Sample No.	Vacuum (in Hg)	Observed Flowing Time (Sec.)	Predicted[1] Flowing Time (Sec.)
1 (n = 0.7)	10	123.98	121.22
	15	57.80	67.92
	20	43.34	45.03
	25	32.74	32.74
2 (n = 0.75)	10	25.17	30.21
	15	17.60	16.93
	20	8.87	11.22
	25	8.16	8.16

[1]Predicted values are based on the measured values at V = 25 in Hg vacuum.

STUDY OF INTRA-MOLECULAR CROSSLINKING OF POLYACRYLAMIDE IN Cr(III)-POLYACRYLAMIDE GELATION BY SIZE-EXCLUSION CHROMATOGRAPHY, LOW-ANGLE LASER LIGHT SCATTERING, AND VISCOMETRY

T.S. Young, G.P. Willhite, and D.W. Green

Tertiary Oil Recovery Project, University of Kansas
4008 Learned Hall, Lawrence, Kansas 66045

INTRODUCTION

The application of crosslinked-polymer systems for permeability modification of petroleum reservoirs has received increasing attention in recent years. A crosslinked-polymer treatment, in general, involves an injection of a polymer solution into high-permeability zones or fractures which have been previously swept by the displacing fluid[1]. The polymer solution reacts, either before or after injection, to form a three-dimensional gel network which reduces the effective permeability of the invaded portions of the reservoir. Fluid subsequently injected is diverted to other, tighter regions of the formation, thereby improving overall volumetric sweep efficiency.

One method to apply a crosslinked-polymer system involves the cross-linking of polyacrylamide (PAAM) by Cr(III) in the formation. In this process, chromium is introduced in the (+6) state as sodium dichromate, and is subsequently converted to Cr(III) with a reducing agent such as thiourea. The Cr(III) then interacts with polyacrylamide in the formation to form a gel. By controlling the rate of reduction of Cr(VI) to Cr(III), gelation may be delayed until the gelling polymer solution has reached the desired portions of the formation. Gelation characteristics needed for the design of this process have been previously identified and studied[2-7].

It was observed in a previous study[7] that gelation would occur only when the polymer concentration exceeded a minimum value for a particular high molecular weight (MW) polyacrylamide. A preliminary study showed that below this minimum concentration, interaction between Cr(III) and PAAM led to a loss in fluid viscosity instead of gelation. Since viscosity is a measure of the hydrodynamic volume occupied by polymer in solution, this observation indicated a decrease of the hydrodynamic size of PAAM due to crosslinking. Because crosslinking of more than one polymer molecule (inter-molecular crosslinking) generally would lead to an eventual increase in average molecular size and hence solution viscosity, the observation of decreased viscosity suggests another type of crosslinking process. Little was known about the nature of the phenomenon and its effects on the performance of the gel treatment. This work was undertaken to investigate the properties of the non-gelling crosslinking solutions and to identify the factors that prohibit the PAAM gelation.

It has been known for years that high-MW, flexible polymers can also be crosslinked intra-molecularly[8-14]. The process of intra-molecular crosslinking (IMC) is similar to inter-molecular crosslinking except the chain segments bridged by the crosslink are of the same polymer molecule. Intra-molecular crosslinking is conceptually possible for a gel system consisting of a high-MW, flexible PAAM because the random-coil conformation of the polymer permits segments of the same molecule to approach one another through Brownian-type motion. Thus, both inter- and intra-molecular crosslinking may take place in PAAM gelation, competing with each other for available crosslinker (chromium in this case). Unlike inter-molecular crosslinking, IMC does not lead to increase in hydrodynamic size and contributes little to the development of elasticity of the gel network. Instead, IMC imposes entropic constraints upon polymer chains causing the polymer coils to shrink. Therefore, a crosslinking process which is dominated by IMC will show a decreased viscosity upon crosslinking and will not lead to gelation. This appears to be a logical explanation for the observed behavior of the dilute crosslinking PAAM solutions.

This paper reports experimental work that provided evidence for the occurrence of IMC in Cr(III)-PAAM gelation. The experiments were designed based on the concepts that (1) IMC should be more significant in relatively dilute crosslinking systems since the chance for inter-molecular chain collision would be relatively small, (2) IMC should lead to decreased hydrodynamic size and viscosity but not decreased PAAM molecular weight, and (3) Cr(III)-PAAM interaction must occur for the solution to lose viscosity by IMC. Crosslinking samples of varied PAAM concentration were characterized for viscosity, MW, hydrodynamic size, and degree of Cr(III)-PAAM interaction to test for the occurrence of IMC based on the above concepts. The MW and hydrodynamic size of the crosslinking PAAM were measured by size-exclusion chromatography (SEC) and low-angle laser light scattering (LALLS). The evidence for Cr(III)-PAAM interaction was obtained by equilibrium dialysis.

EXPERIMENTAL

Materials

Reagent grade polyacrylamide was purchased from Aldrich Chemical Co. in granular form and was used as received. The polyacrylamide sample was determined to be approximately 4% hydrolyzed by NMR. The polymer assay of the sample was determined by twice reprecipitating a 1% aq. polymer solution into acetone (solution to acetone volume ratio was 1:10). Sodium dichromate, sodium bisulfite, thiourea, and sodium chloride used for preparation of crosslinking samples were all of reagent grade and were obtained from Mallinckrodt, Inc.

Sample Preparation and Purification

Crosslinking of PAAM in solution was carried out by mixing a PAAM stock solution with stock solutions of dichromate and a reducing agent. Two series of crosslinking samples were prepared using either sodium bisulfite or thiourea as the reducing agent. These reducing agents have been previously used in either laboratory investigation,[2-5] or field applications[15]. Since sodium bisulfite is a stronger reducing agent than thiourea, the rate of Cr(III) generation and hence the rate of crosslinking of a bisulfite-reduced crosslinking system is several times faster than that of a thiourea-reduced system.

All the aq. stock solutions were prepared using a 2 wt.% NaCl synthetic brine. Concentrations of the redox agents were kept constant in each crosslinking solution while the polymer concentration was varied over a wide range. Compositions and characteristics of the solutions studied are listed in Table 1. Polymer solutions of the same composition as the crosslinking solution but without redox agents were used as controls. The crosslinking reactions in which bisulfite was used as the reducing agent were carried out at 30°C, whereas the ones using thiourea were carried out at 40°C to accelerate the reaction. Four replicate samples were made for a series of IMC samples (BIS-2 to -5 in Table 1) to examine the reproducibility of the crosslinking process.

All stock solutions except the PAAM solution were pressure filtered through a 0.2 μ Durapore membrane before use. The polymer solution was filtered by gravity through a 5 μ Nuclepore membrane five days after preparation. Selected IMC samples to be characterized for $\overline{M}_w$ and hydrodynamic size were purified by dialysis to remove low MW moieties that would interfere with the SEC-LALLS measurement. The dialysis was made with a Spectra/pore2 membrane tubing (Spectrum Medical Industries Inc.) with a MW cutoff of 12000-14000. The membrane tubings were pre-treated in hot water (ca. 90°C) three times to remove impurities such as sulfur compounds. The IMC samples were dialyzed with a large amount of distilled water (over 150 times sample volume) until no change in UV absorption of the dialysate was observed.

Detection of Cr(III)-PAAM Interaction by Equilibrium Dialysis

The amount of Cr(III) that reacted with PAAM in the IMC samples was determined by measuring the amount of unreacted chromium which was separable from the crosslinked polymer. The amount of unreacted chromium was determined by equilibrium dialysis[6,16,17] where the unreacted chromium was separated from the crosslinked polymer with a semi-permeable membrane. At dialysis equilibrium, the unreacted chromium which was free to move across the membrane should have the same activity on both sides of the membrane. The activity coefficients of free chromium ions on both sides of the membrane have been previously determined to be essentially

TABLE 1 - DESCRIPTION OF CROSSLINKING SAMPLES

Sample Number	Polymer Conc. (ppm)	Dichromate Conc. (ppm)	Mole* Ratio	Ionic* Strength
*THIO-1	250	500	23.5	0.354
2	500	500	23.5	0.354
3	1000	500	23.5	0.354
4	5000	500	23.5	0.354
5	10000	500	23.5	0.354
* BIS-1	200	500	4.8	0.362
2	250	500	4.8	0.362
3	500	500	4.8	0.362
4	1000	500	4.8	0.362
5	2000	500	4.8	0.362
6	4000	500	4.8	0.362

*Mole Ratio = [reducing agent]/[dichromate]; the ionic strength of control solutions was 0.349; THIO = thiourea; BIS = bisulfite.

identical[6]. Thus, a measurement of the chromium concentration of the dialysate allowed calculation of the total amount of unreacted chromium in the sample when the weights of the gel and the dialysate were known.

The equilibrium dialysis of IMC samples was made using pre-treated Spectra/pore2 membrane tubings. The samples were dialyzed at 25^0C for two weeks before being analyzed for chromium to insure equilibrium. The chromium analysis was made by atomic absorption at a wavelength of 359 nm using a Perkin-Elmer Model 460 Atomic Absorption Spectrophotometer.

The use of a cellulose-based membrane in chromium dialysis requires some precautions. The membrane could retain a substantial amount of Cr(III) when the sample pH is relatively high[18]. While only a small amount of Cr(III) is retained by the membrane at a pH below 4, Cr(III) loss can amount to 15% or higher when the pH is above 5. The cause for the Cr(III) retention is not clear, but is believed to relate to the hydrolysis[19] and olation[20] of Cr(III) compounds. In this study, the pH values of the IMC and dialysate samples were all below 4 because of the acidic nature of chromium solutions. Under this condition, the loss of chromium to the dialysis membrane is considered insignificant.

Viscosity Measurement

Viscosities of both the crosslinking and control solutions were measured at various times after the introduction of redox agents into the polymer solutions. The viscosities were measured with an Oswald-Fenske capillary viscometer at the same temperature as that of the crosslinking reaction. The temperature variation was controlled to within 0.1^0C, and the flow time measurement had a precision of better than 0.15%. The specific viscosity and the reduced viscosity, η_{sp}/c, were calculated from the flow time of the sample solution, t, and and that of the solvent, t_o ($\eta_{sp} = t/t_o - 1$).

Determination of $\bar{M}_w$ by SEC-LALLS

Weight-average molecular weights of both IMC and control polymer samples were determined by SEC-LALLS using a procedure previously reported[21]. In this procedure, the polymer $\bar{M}_w$ is determined by LALLS with the aid of a single SEC column. The use of the SEC column facilitates the LALLS measurement by minimizing dust interference which is a major problem with light scattering experiments. In a typical single-column SEC-LALLS experiment, the polymer sample is dispersed and separated from low MW impurities in a SEC column before being characterized for concentration and $\bar{M}_w$. The $\bar{M}_w$ of this dispersed but not fractionated sample is determined from the light scattering intensity detected at a small angle by[22]

$$\frac{K_\Theta c_i}{\bar{R}_{\Theta,i}} = \frac{1}{\bar{M}_w} + 2A_2c_i + \dots \tag{1}$$

where $\bar{R}_{\Theta,i}$ is the excess Rayleigh ratio measured for fraction i by LALLS, c_i is the polymer concentration of fraction i, K_Θ is an optical constant determined from solvent refractive index and specific refractive index increment of the polymer, and A_2 is the second virial coefficient for the polymer/solvent pair. The $\bar{M}_w$ is typically determined from a linear plot of $K_\Theta c_i/\bar{R}_{\Theta,i}$ versus c_i, where the $\bar{M}_w$ is identified with the intercept. In this study, the mobile phase for SEC was selected so that the second virial coefficient for the polymer-solvent pair was nearly zero. Thus, the second term on the right-hand side of Equation (1) can be neglected, and the $\bar{M}_w$ can be determined directly from a single pair of c_i and $\bar{R}_{\Theta,i}$ data.

The SEC-LALLS measurement was carried out with a Waters ALC/GPC system interfaced with a Chromatix CMX-100 LALLS photometer and an ISCO v[4] UV-VIS detector. The LALLS photometer, using a He-Ne laser operating at 633 nm, was connected to the column outlet to measure the Rayleigh ratio of the effluent (Figure 1). The UV-VIS detector was installed after the LALLS and measured the polymer concentration of the effluent by UV absorption at 214 nm. A Waters R401 refractive index detector was also connected to monitor the total retention volume (V_t) of each SEC run. Signals from both LALLS and UV-VIS detectors were recorded at real time by a chart recorder and a Data General Nova 3 Computer equipped with a 12-bit analog-to-digital converter. A typical single-column SEC-LALLS chromatogram is shown in Figure 2. The apparent $\overline{M}_w$ of the whole polymer sample was determined from the peak values of the LALLS and UV curves, h_1 and h_2, in Figure 2.

The SEC separation was made with a 75 cm long, 0.46 cm I.D. column packed with glyceryl-coated controlled-pore-glass (GCPG packing, Electro-Nucleonics, Inc.) which had a mean pore diameter of 75 Å and a particle diameter between 37μ and 74μ. The mobile phase was a 0.5 M sodium sulfate aq. solution and was pressure filtered through a 0.2 μ Durapore membrane before use. To minimize interference by dust particles with the LALLS measurement, the mobile phase was purified with a series of four inline filters (5μ, 2μ, 0.45μ and 0.1μ) before entering the injector. The flow rate of the mobile phase was 0.2 cm^3/min. The injected sample size was varied from 10 to 50 μl depending on the light scattering intensity of the sample.

The selection of the SEC column and the chromatographic conditions was based on several experimental concerns. The pore size of the SEC packing was chosen so that the column would totally exclude the polymer molecules while being totally permeable to low molecular weight moieties. The use of such a column packing optimized the separation of the polymer from low molecular weight impurities that could interfere with polymer concentration measurement by UV. The particle size of the porous glass packing was selected to avoid shear degradation of the polymer sample in the SEC column yet maintain an optimal separation efficiency. The combined use of the GCPG packing and the sodium sulfate mobile phase was designed to minimize adsorption of polymer in the column. To check for possible polymer loss by column adsorption, PAAM and crosslinked PAAM solutions of varied polymer concentration were injected into the column

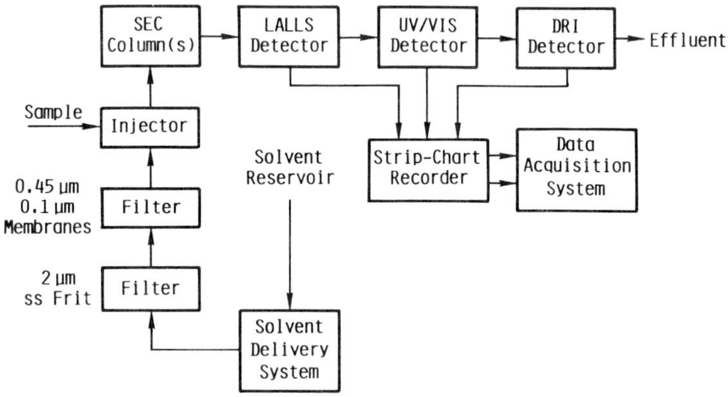

Figure 1. Flow Diagram of SEC-LALLS System.

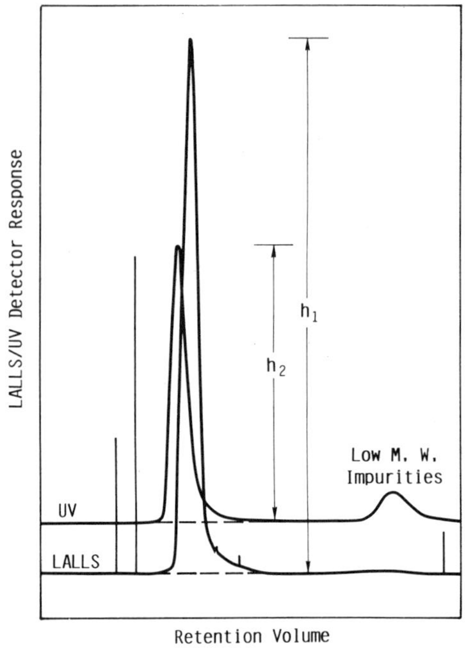

Figure 2. A Typical Single-Column SEC-LALLS Chromatogram

and the UV absorptions of their eluents monitored. The areas and heights of the eluent UV absorption peaks were plotted against sample concentration and are shown in Figures 3 and 4. If there was no polymer loss due to irreversible column retention, both the area and height plots should give a straight line which passes origin. As indicated by Figure 3, the

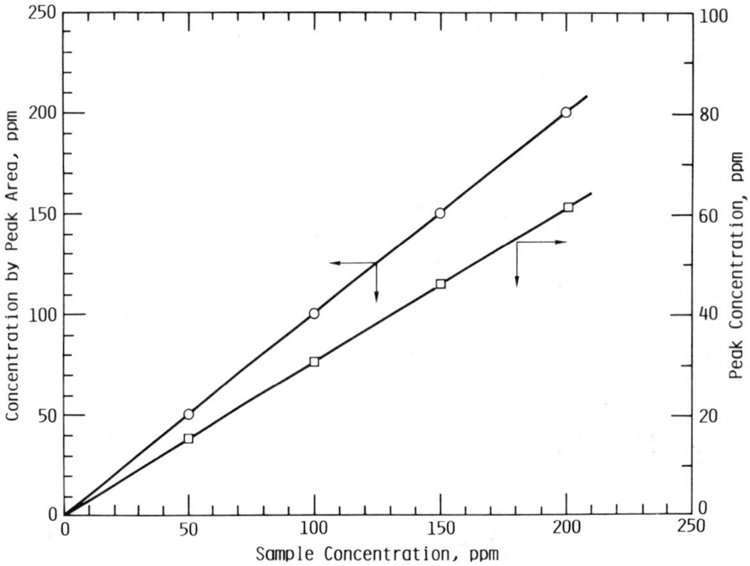

Figure 3. Polymer Concentrations of PAAM Samples Determined from Areas and Heights of UV Peaks.

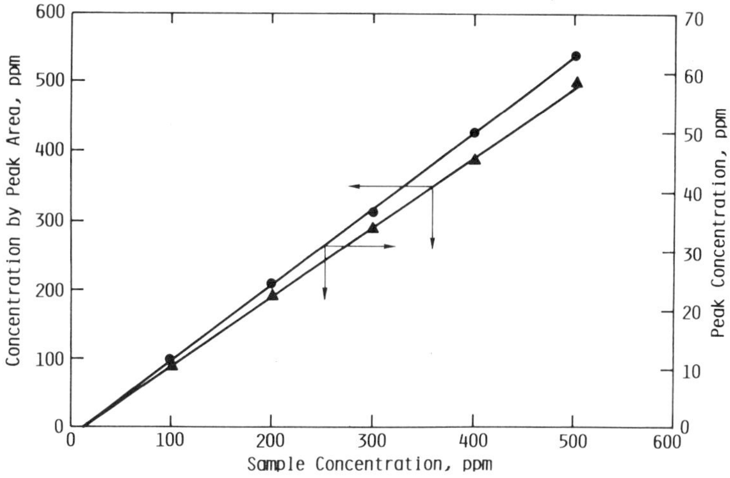

Figure 4. Polymer Concentrations of BIS-3 Solution Samples Determined from Areas and Heights of UV Peaks.

PAAM samples were not subject to significant polymer loss in the column. The crosslinked PAAM solutions that contained bound chromium showed a slight column retention according to Figure 4. The intercepts on the x-axis indicate a retention of approximately 12 ppm for the crosslinked sample (BIS-3) tested. This amount represented a 2.4% loss of the injected polymer (500 ppm for this particular sample). As we were primarily concerned with the relative $\bar{M}_w$ of the polymer samples, loss of this magnitude was considered insignificant.

Hydrodynamic Size Measurement by SEC-LALLS

Information on hydrodynamic sizes of IMC and PAAM samples was obtained by SEC-LALLS using a series of three GCPG columns (0.46 cm I.D., 75 Å x 75 cm + 1400 Å x 75 cm + 3000 Å x 100 cm). In the three-column SEC-LALLS experiment, the polymer sample was fractionated by size in the SEC columns, and the $\bar{M}_w$ of each fraction eluting the columns was determined by LALLS.

The same SEC flow system (Figure 1) and mobile phase as described in the previous section were used in the size measurement. The mobile phase flow rate was 0.6 cm^3/min and the sample size was 50 µl . The precision of the flow rate was better than 0.4% (at 95% C. I.). The three-column set had a total retention volume of 33.6 ± 0.1 cm^3. The amount of column dispersion of the column set was estimated from the peak contour of a potassium nitrate solution slug. Using the moment method described by Grushka[23], the band-broadening and peak asymmetry factors were determined to be 0.56 and 0.27 cm^3 respectively.

SEC-LALLS data were collected with the Data General data acquisition system where the detector signals were measured every second but were averaged over a 20-second period before being stored. The spikes in the LALLS signals were removed at real time before the signals were averaged and stored.

An internal calibration method based on the universal calibration concept[24] was used to establish a hydrodynamic-size calibration curve which gave the relationship between the retention volume (V_r) and the hard-sphere equivalent hydrodynamic diameter (D_η) of the polymer (Figure 5). The calibration curve was constructed based on the SEC-LALLS chromatogram of a 1000 ppm PAAM sample. In the calibration run, the $\bar{M}_w$ of a fraction i, $\bar{M}_{w,i}$ was converted into the hard-sphere equivalent hydrodynamic diameter by[25]

$$D_{\eta,i} = 2\left(\frac{3\ k}{10\ N\pi}\ \bar{M}_{w,1}^{a+1}\right)^{1/3} \qquad (2)$$

where N is the Avogadro number, and k and a are the Mark-Houwink constants for the polymer/solvent pair. As shown in Figure 5, linear size separation can be achieved with this column set over a D_η range of about 500 to 2,000 Å.

Theoretically, $\bar{M}_{w,i}$ used in Equation (2) should be the viscosity-average molecular weight ($\bar{M}_v$) of the fraction. Because we were concerned only with the relative magnitude of the hydrodynamic size and because $\bar{M}_{w,i}$ and $\bar{M}_{v,i}$ of a separated fraction probably would not greatly differ, $\bar{M}_{w,i}$ data obtained from the SEC-LALLS measurement were used directly in Equation (2) without further correction. The Mark-Houwink constants reported by Klein and Conrad[26] were used in the calculation of $D_{\eta,i}$. With the V_r-D_η calibration curve, we were able to compare the relative sizes of IMC and PAAM samples by means of the chromatogram.

RESULTS AND DISCUSSION

Viscometric Behavior of Crosslinking Solutions

Viscosity measurements were made to detect IMC through the viscosity change of polymer solution following introduction of crosslinking agents. The concentrations of the solutions were selected so that gelation would occur in solutions of high polymer concentrations but not in dilute solutions. The relatively concentrated solutions (10000 and 5000 ppm for THIO system) were found to gel within two days after showing an initial viscos-

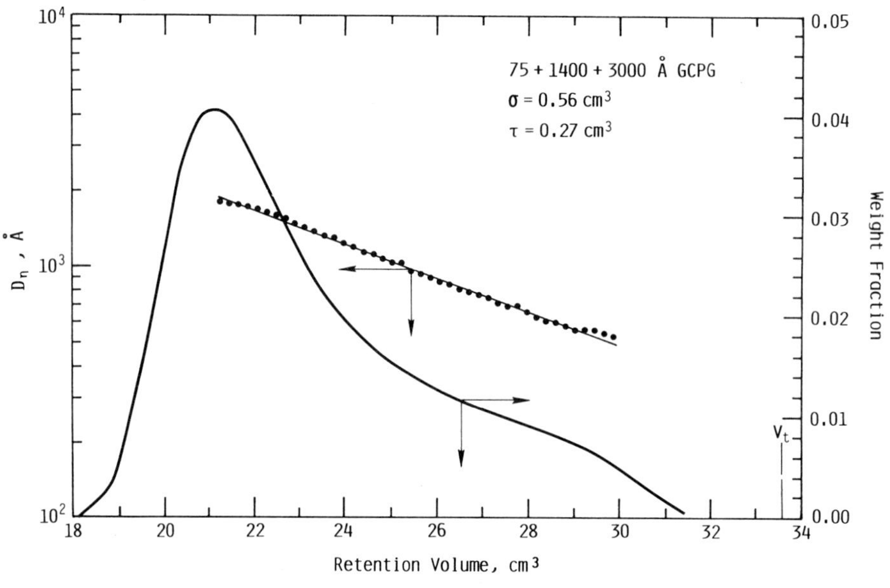

Figure 5. Hydrodynamic Size Calibration Curve Constructed Based on the Chromatogram of PAAM.

ity decrease. The less concentrated solutions never formed gels, and all showed decreasing viscosities for a period of over one week. The typical viscosity-time relationships for these solutions are shown in Figure 6. Viscosities of the crosslinking solutions were substantially lower than that of their corresponding controls, although both the control and crosslinking solutions showed viscosity decrease during the first few days. The initial drop in viscosity of the control solution is believed to be due to disentanglement of polymer chains, a physical process which is prevalent in concentrated solutions.

The percentage decreases in viscosity due to crosslinking, determined twelve days after mixing of dichromate with PAAM, are given in Table 2. The final viscosities of the crosslinking solutions were substantially lower than that of their corresponding controls. A solution of a lower polymer concentration showed a larger percentage decrease in viscosity, consistent with what would be expected from an IMC process. Since the solutions all contained 2% NaCl and the difference in ionic strength between the control and crosslinking solutions was insignificant (Table 1), the change in viscosity could not be attributed to the effect of charges on viscosity (polyelectrolyte effect). Thus, both crosslinking systems showed a clear trend of viscosity loss as a result of introducing the crosslinker, although the percentage loss of viscosity might have been exaggerated somewhat by the varying shear rate of the capillary viscosity measurement.

Observations from this viscosity study and previous studies with other gel systems suggest that there is a minimum polymer concentration for gelation for a particular polymer. Below this concentration, IMC may be overwhelming. For the Aldrich polyacrylamide, this minimum concentration appears to fall somewhere between 2000 and 4000 ppm. The minimum concentration is suspected to be related to the critical overlap concentration[27] commonly used to describe the transition between a dilute solution and a semi-dilute solution. The critical overlap concentration for the Aldrich PAAM was estimated from its intrinsic viscosity (5.0 g/dl)

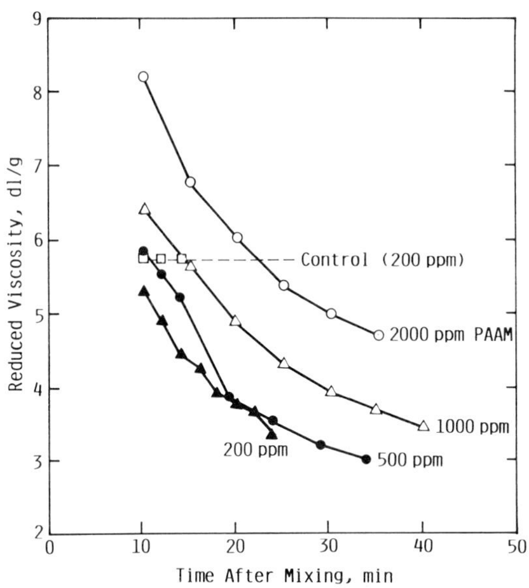

Figure 6. Viscosity-Time Dependence of Bisulfite-Reduced Crosslinking Solutions.

TABLE 2 - VISCOSITY CHANGE DUE TO THE CROSSLINKING

Sample Number	Polymer Conc. (ppm)	Final Specific Viscosity (Crosslinking)	(Control)	% Viscosity* Loss
THIO-1	250	0.022	0.228	90.4
2	500	0.043	0.275	84.4
3	1000	0.127	0.698	81.8
BIS-1	200	0.011	0.137	92.3
3	500	0.036	0.322	89.0
4	1000	0.095	0.715	86.8
5	2000	0.278	1.665	83.3

*Viscosity loss due to crosslinking measured 12 days after mixing.

using a model derived by Ewen, et al.[28] for polymers with flexible chains. The critical concentration was found to be approximately 2800 ppm, consistent with the minimum concentration range observed.

Evidence of Cr(III)-PAAM Interaction

Equilibrium dialysis experiments were carried out on a series of bisulfite-reduced crosslinking samples (BIS-2 to -5) to provide evidence for Cr(III)-PAAM interaction, which is necessary for IMC to occur. AA analysis of the dialysates showed that a significant amount of chromium had interacted with PAAM, the amount increasing with increased PAAM concentration. The fractional chromium conversions for the IMC samples are given in Table 3. This observation demonstrated that the viscosity loss of crosslinking solutions was indeed associated with a Cr(III)-PAAM interaction.

Another evidence came from the long-term dialysis experiment which was conducted to purify the IMC samples before SEC-LALLS analysis (see the Sample Preparation and Purification section). After all the dialyzable species were removed, the IMC samples still showed the green tint characteristic of Cr(III), indicating that chromium was bound to PAAM.

$\bar{M}_w$ of the Crosslinked Polyacrylamide

The BIS-series IMC samples were characterized for $\bar{M}_w$ to determine if the loss of fluid viscosity was due to degradation of PAAM in the presence of redox agents. The relative $\bar{M}_w$ values obtained from single-column SEC-LALLS, together with the relative UV absorption coefficients and specific refractive index increment (dn/dc) data used in the calculations are given in table 3.

All the IMC samples that had lower viscosity than the control PAAM solutions were found to have a substantially larger $\bar{M}_w$, indicating no degradation. The fact that the $\bar{M}_w$'s of the crosslinked samples were so much greater than the control PAAM indicates that in addition to IMC, inter-molecular crosslinking must have occurred to a significant degree. The increase of $\bar{M}_w$ due to the addition of the chromium alone could not account for the magnitude of increase observed. The apparent $\bar{M}_w$ ratios listed in Table 3 show a trend of increasing $\bar{M}_w$ with increasing PAAM concentration of the solution except for BIS-2 which had only 250 ppm of PAAM. It was noted that the BIS-2 sample was actually a fine dispersion, showing slight turbidity but no precipitation. The extremely large $\bar{M}_w$ of

TABLE 3 - PHYSICAL PROPERTIES OF PAAM AND IMC SAMPLES

	BIS-2	BIS-3	BIS-4	BIS-5	PAAM
Dichromate-to-PAAM Ratio*	500/250	500/500	500/1000	500/2000	--
Fractional Conversion of CR(III)	0.38	0.42	0.56	0.73	--
Reacted Cr(III)-to-PAAM Ratio* x 10^2	26.5	14.7	9.8	6.4	--
Apparent $\bar{M}_w$ Ratio	112.0	4.2	5.5	8.7	1
dn/dc	0.238	0.236	0.215	0.184	0.156
UV Absorption-to-DRI Ratio**	25.1	17.7	13.5	11.0	3.6

* Weight ratio; ** UV peak height/DRI peak height.

BIS-2 is considered to reflect the presence of those fine particles, probably formed by flocculation. Other than BIS-2, the greater increase in $\bar{M}_w$ of samples with higher PAAM concentrations appears to indicate an increasing effect of inter-molecular crosslinking.

It should be pointed out that the UV absorptivity of the crosslinked PAAM sample varied with the chromium content of the sample (Table 3). This variation made it difficult to assign the absolute concentration of polymer in the column eluent when the polymer was fractionated. This, however, did not constitute a major problem in the single-column SEC-LALLS measurement where the polymer was not fractionated. In this case, the absorptivity of each eluting fraction could be assumed to be constant and the absolute concentration could be determined for each eluting fraction by a mass balance on polymer.

Hydrodynamic Size of the Crosslinked Polyacrylamide

Hydrodynamic sizes of the BIS-series samples were also monitored by SEC-LALLS. Figure 7 shows the chromatograms of both IMC and PAAM samples that had been pre-purified by dialysis. Since the UV absorption coefficient varied with the dichromate-to-PAAM ratio (Table 3), it was impossible to assign an absolute concentration for each fraction of the eluted polymer. However, the relative sizes of the IMC samples may still be recognized from the normalized UV absorption elution curves.

As shown in the figure, some of the IMC samples seem to consist of two components. All the IMC samples had a large retention volume relative to the uncrosslinked PAAM, indicating a relatively small D_η for the crosslinked PAAM. The relative size of the IMC samples did not show a simple dependence on the dichromate-to-PAAM ratio. The overall size of BIS-5 was larger than BIS-4 as the relative population of smaller polymer molecules was higher for BIS-4. This appears to indicate a smaller degree of IMC for BIS-5 which had a higher PAAM concentration. This trend, however, seems to reverse with BIS-2 and BIS-3. The relatively large size of the dilute sample probably resulted from flocculation of PAAM, as discussed earlier.

Reproducibility of IMC

Viscosity measurement was made on replicate IMC samples to see if IMC is a reproducible process. Table 4 gives the deviation in relative

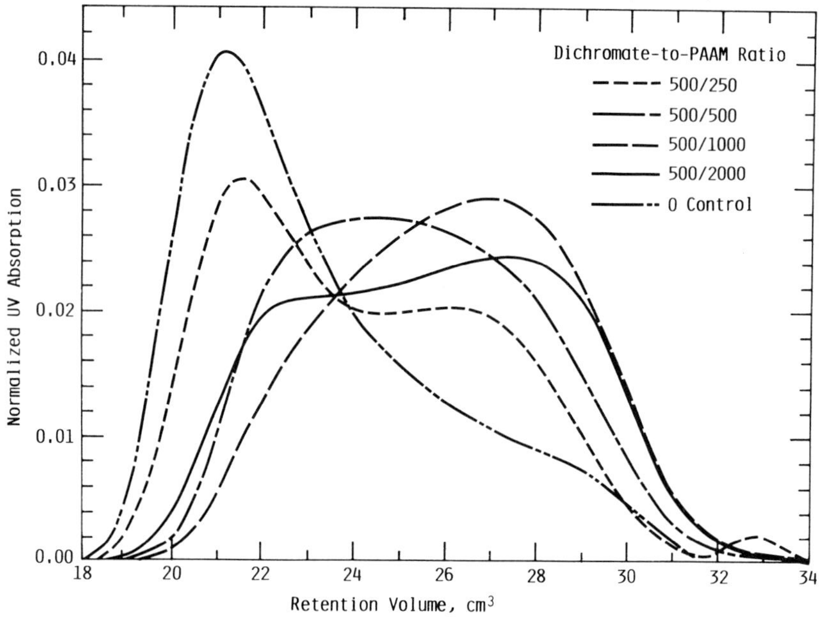

Figure 7. SEC Chromatograms of PAAM and IMC Samples of Various Dichromate-to-PAAM Ratios.

viscosity measured for a series of IMC solutions (BIS-2 to BIS-5). The estimated error was within 2.2% at a 95% C.I. for all the IMC samples tested.

The reproducibility of the IMC process was also examined by observing the sizes of four replicate IMC samples. The normalized SEC-LALLS chromatograms of the four samples were nearly identical (Figure 8). The percentage deviation for the peak position, apparent $\overline{M}_w$ measured at the peak position, and average apparent size (calculated assuming all fractions had the same UV absorption coefficient) were 0.5, 4.0 and 2.5, respectively at a 95% C.I.

Significance of IMC on the Gel Treatment

Based on the above observations, it was concluded that IMC can occur to a significant degree in a Cr(III)-PAAM gelation process. When this happens, IMC can greatly affect the gelation characteristics and the per-

TABLE 4 - REPRODUCIBILITY OF RELATIVE VISCOSITY OF IMC SAMPLES

Sample	*η_r,mean	Std. Deviation x 10^2	Estimated % Error
BIS-2	1.74	0.22	0.31
3	1.79	1.02	1.42
4	1.90	0.81	1.06
5	2.17	1.92	2.19

* From three replicate measurements.

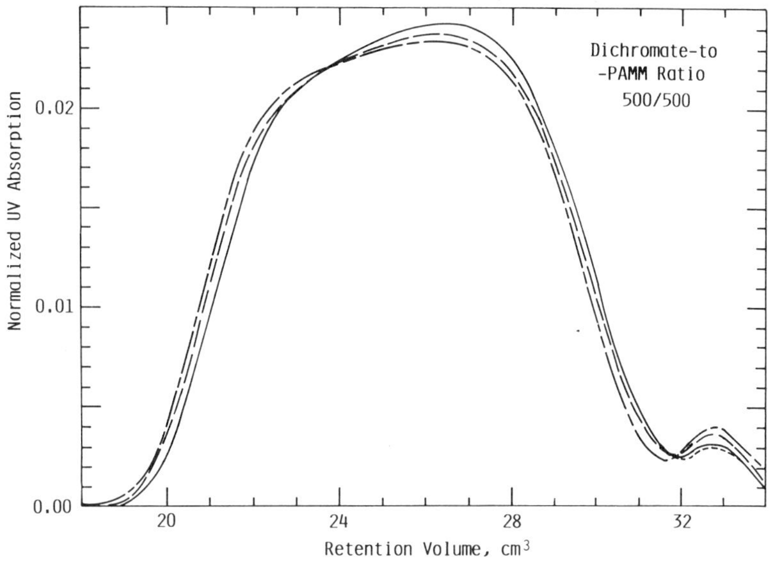

Figure 8. SEC Chromatograms of Four Replicate IMC Samples.

formance of a gel treatment. For instance, dispersion or diffusion of a gelling slug in the reservoir might lower the polymer concentration to a degree such that IMC would prevail and prevent gelation. This is not desirable for a gelling slug placed in the high-permeability zones, but it is desirable for the gelling solution that has invaded low-permeability zones. Since the amount of gelling solution in low-permeability zones is generally small relative to that placed in high-permeability zones or channels, dispersion or diffusion can occur to a much greater degree, increasing the chance of IMC, hence preventing gelation. Therefore, it might be possible to place a gel in the high-permeability zones without blocking low-permeability zones by using a gel system with a proper polymer concentration.

IMC also affects the timing of gelation and the properties of the developed gel. It consumes the chromium crosslinker that would otherwise be available for formation of intermolecular crosslinks, and thereby delays the development of the gel network. It is also suspected that IMC may be the primary factor in preventing a mechanically degraded Cr(III)-PAAM gel from reforming, as observed in our laboratory.

CONCLUSIONS

1. Crosslinking of PAAM in solution can lead to a decrease in hydrodynamic size and fluid viscosity when the polymer concentration is below a certain critical value.

2. Intra-molecular Crosslinking was determined to occur in the Cr(III)-PAAM gelation process. The effect of IMC is more pronounced in crosslinking systems with relatively low PAAM concentrations.

3. The timing of gelation, physical properties of developed gels, and gelation behavior in the reservoir can be significantly affected by IMC, when it occurs to an appreciable degree.

ACKNOWLEDGEMENTS

The authors would like to thank Bryan E. Adams and Maryam Charmchi for their assistance in sample preparation and viscosity measurement, and Mrs. R. Sleeper for preparation of this manuscript. This work was supported by the Tertiary Oil Recovery Project, University of Kansas, and grant DE-AC19-85BC10843 from the U.S. Department of Energy.

REFERENCES

1. R.L. Clampitt and J.E. Hessert, U.S. Patent 3,785,437 (January 15, 1974).
2. R.E. Terry, C.-G. Huang, D.W. Green, M.J. Michnick, and G.P. Willhite, Soc. Pet. Eng. J., **21**, 229 (1981).
3. D.S. Jordan, D.W. Green, R.E. Terry, and G.P. Willhite, Soc. Pet. Eng. J., **22**, 463 (1982).
4. C.-G. Huang, D.W. Green, and G.P. Willhite, SPE Reservoir Eng., **1**, 534 (1986).
5. S. Aslam, S. Vossoughi, and G.P. Willhite, Proceedings of the 4th SPE/DOE Symposium on EOR, **1**, 113 (1984).
6. T.S. Young, J.A. Hunt, D.W. Green, and G.P. Willhite, Proceedings of the 60th Annual SPE Meeting, SPE 14334 (September 1985).
7. U.S. DOE Report DOE/BC/10354-16, Chap. 2, National Technical Information Service, Springfield, Virginia (February 1986).
8. P.J. Flory, "Principles of Polymer Chemistry", Cornell University Press, Ithaca, NY (1953).
9. W.H. Stockmayer, in "Advancing Fronts in Chemistry", S.B. Twiss, ed., Van Nostrand Reinhold, New York, NY (1945).
10. W.M. Kulicke, R. Kniewske, and J. Klein, Prog. Polym. Sci., **8**, 373 (1982).
11. W. Kuhn and H. Majer, Makromol. Chem., **18/19**, 239 (1955).
12. W. Kuhn and G. Balmer, J. Polym. Sci., **57**, 311 (1962).
13. D. Braun, J. Polym. Sci., Symp., **50**, 149 (1975).
14. P. Molyneux and S. Vekavakayanondha, J. Chem. Soc. Faraday Trans., 1, **82**, 291 (1986).
15. J. Frank and L.G. Schoeling, Proceedings of the Sixth Tertiary Oil Recovery Conference, Wichita, Kansas, March 6-7, 113 (1985).
16. F. Karush and M. Sonnenberg, J. Amer. Chem. Soc., **71**, 1369 (1949).
17. F. Karush and S.S. Karush, in "Methods in Immunology and Immunochemistry", Vol. 3, C.A. Williams and M.W. Chase, eds., Academic Press, New York, NY (1971).
18. Annual Report for U.S. DOE Contract DE-AC19-85BC10843, in preparation.
19. M.J. Udy, "Chromium", Vol. 1, Van Nostrand Reinhold, New York, NY (1956).
20. C.L. Rollinson, in "Chemistry of Coordination Compounds", J.C. Bailer, ed., Van Nostrand Reinhold, New York, NY (1956).
21. J.A. Hunt, T.S. Young, D.W. Green, and G.P. Willhite, Proceedings of the 5th SPE/DOE Symposium on EOR, **2**, 325 (1986).
22. A.C. Ouano, J. Polym. Sci., Polymer Chem., **12**, 1151 (1974).
23. E. Grushka, Anal. Chem., **44**, 1733 (1972).
24. Z. Grubistic, R. Rempp, and H. Benoit, J. Polym. Sci., **B-5**, 753 (1967).
25. J.G. Kirkwood and J. Riseman, J. Chem. Phys., **16**, 565 (1948).
26. J. Klein and K.-D. Conrad, Makromol. Chem., **181**, 227 (1980).
27. P.-G. deGennes, "Scaling Concepts in Polymer Physics", Cornell University Press, Ithaca, NY (1979).
28. B. Ewen, D. Richter, J.B. Hayter and B. Lehner, J. Polym. Sci., Polym. Letter, **20**, 233 (1982).

DETECTION OF MICROGELS IN EOR POLYMERS USING MICROCAPILLARY FLOW

Jeffrey H. Sugarman and Robert K. Prud'homme

Department of Chemical Engineering
Princeton University
Princeton, NJ 08544

INTRODUCTION

The success of the use of polymers in enhanced oil recovery operations depends on the ability to inject large volumes of polymer solution into the oil bearing formation. Often, polymer solutions contain small, but significant amounts of insoluble material which can cause plugging of the reservoir [1]. Such damage to the reservoir can be irreversible. Thus, it is important to know the "residue" or "microgel" content of a polymer solution before the solution is injected.

Previous techniques of determining the microgel content of polymer samples include both gravimetric and volumetric assays [2-4]. These methods require the separation of insoluble material from the dissolved polymer. Because of this requirement, the techniques are not suitable for detecting very low microgel content, which, as mentioned above, can lead to formation damage. Another qualitative technique used to assess polymer quality is the filtration test. Here, the times for the filtration of polymer solutions (of equal concentration) through a membrane filter give an indication of the relative microgel contents, since the microgels plug the filter and slow the flow.

In this paper we describe a new chromatographic technique [5] for quantifying the absolute microgel content in polymer samples. The technique is based on the hydrodynamic separation of insoluble material from dissolved polymer during flow through a small capillary tube (25 microns i.d.). A fluorescence detector used in the chromatographic system is sensitive enough to detect individual microgel particles, resulting in an assay that is extremely sensitive.

MECHANISMS OF SEPARATION OF MICROGELS FROM DISSOLVED POLYMER

There is a wealth of theoretical and experimental evidence which demonstrates that particles convecting through capillary tubes move with a higher velocity than the suspending solvent. Two examples are blood cells moving through the microcirculatory system [6] and colloidal particles through chromatography tubing [7,8] (the latter example has been exploited as a technique for particle sizing). The same separation takes place in the

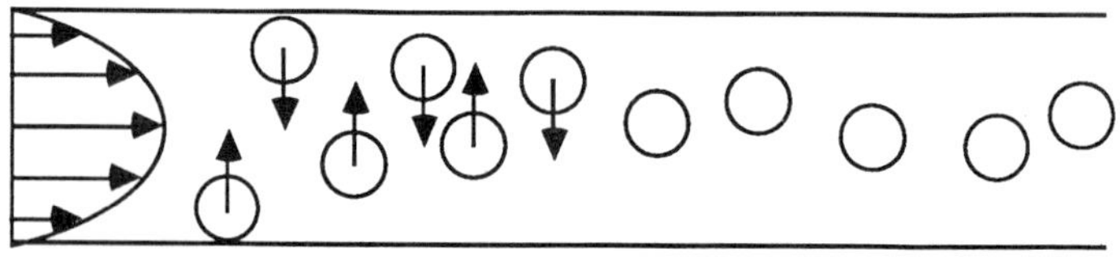

Fig. 1. Lateral migration. Inertial forces cause lateral migration of particles from slow moving regions near the capillary wall, resulting in particle velocities which are greater than the average solvent velocity.

microgel analysis technique-- insoluble microgel material moves with a higher velocity than the solvent and dissolved polymer, achieving the separation that enables the "visualization" of the amount of microgel.

Lateral Migration

Of the mechanisms which cause the enhanced velocity of particles, the most important is lateral migration or "tubular pinch" [9]. Here, because of inertial forces, the particles migrate away from the walls of capillary where the velocity is lowest (Figure 1). Deformable particles have been shown to concentrate near the tube center, where the velocity is highest [10]. The radial migration velocity is strongly dependent on the ratio of the particle and capillary sizes [11].

Wall Exclusion

Another mechanism that contributes to the enhanced velocity of particles is the exclusion of particles from the region near the capillary wall due to the finite size of the particles. Since it is in this excluded region that the velocity is lowest, large particles will, on average, move faster through the capillary than will small particles-- the hydrodynamic chromatography effect (Figure 2). This mechanism, operative in the pore spaces of a packed chromatography column as well as in capillaries, has been used to size latex particles and to determine molecular weight distributions of high molecular weight water-soluble polymers [12,13].

Transient Dispersion

The third mechanism that can result in the separation of large particles from dissolved polymer is transient dispersion. If, during flow through a capillary, a species with a low diffusion coefficient does not have sufficient time to diffuse radially, it will emerge from the capillary in a peak located well before that of a fast-diffusing species. The location and shape of the eluting peak are determined by the size (diffusion coefficient) of the eluting species. This mechanism has been used to determine the diffusion coefficients of monodisperse biopolymers [14,15].

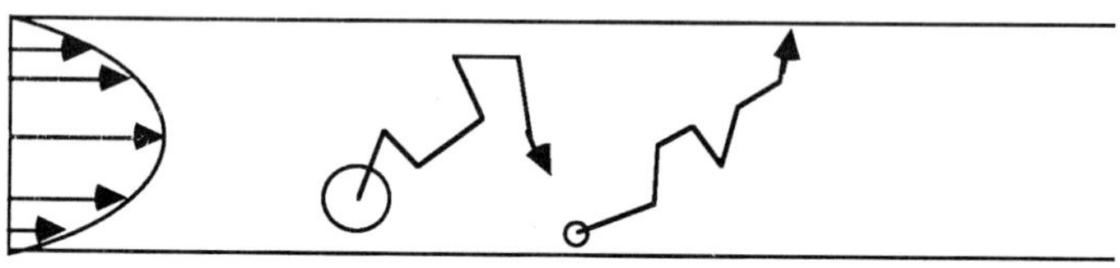

Fig. 2. Wall exclusion. Particles that are sterically excluded from the wall region move with a velocity greater than the average solvent velocity. The velocity increases with increasing particle size.

EXPERIMENTAL

The Microcapillary Flow System

The chromatographic system that is employed for the microgel analysis was originally designed for use in open-tubular liquid chromatography [16]. Shown schematically in Figure 3, the system is comprised of: a 60 cm length of vitreous silica capillary tubing (SGE Inc.) with an inner diameter of 25 microns; a solvent reservoir; and laser fluorescence detection optics. When the reservoir is pressurized to 5 p.s.i., a flow rate of 750 pl/sec (one drop every 18 hours) of aqueous solvent is produced. The solvent is 0.002 M sodium azide (added as a preservative) in filtered, de-ionized water. The capillary is connected to the system through a tee (Valco) via a vespel ferrule.

On-line Fluorescence Detection

For use with this microcapillary system, a laser-excited fluorescence detector which uses a small section of the capillary as the detector cell is employed [17], as shown in Figure 3. Such an on-line detector is needed because the entire capillary volume (300 nanoliters) is small compared to the detector cell volumes of conventional chromatography detectors (microliters). In addition, this configuration completely eliminates the band broadening associated with a capillary-detector connection. Fifty milliwatts of the blue line (488 nm wavelength) of a 2 watt argon ion laser (Model 85, Lexel) is used to excite the fluorescent tags on the polymer. The emitted fluorescent light is collected with a radius mirror and an aspheric condenser lens; it is measured with a photomultiplier tube (Model R763, Hamamatsu) operated at -1500 volts. The amplified output is collected on a strip chart recorder.

Due to fluorescence photobleaching, the detector signal is dependent on the flow rate. The flow rate dependence, however, is not a problem in these experiments since the analysis is performed at a constant flow rate. In fact, the photobleaching effect can be exploited to monitor the extremely low flow rates, which cannot be measured directly. A complete discussion of the detector's flow rate dependence is described elsewhere [18].

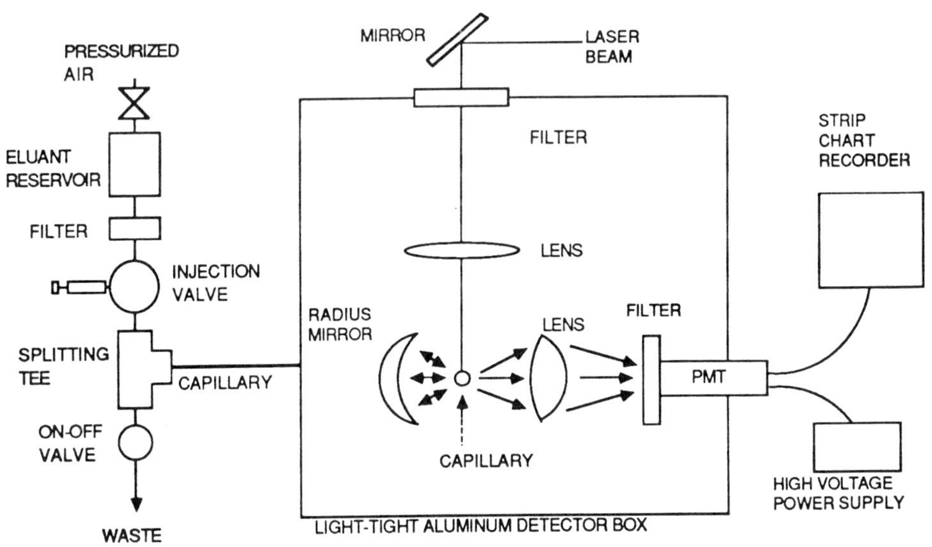

Fig. 3. Microcapillary flow system and laser fluorescence detector.

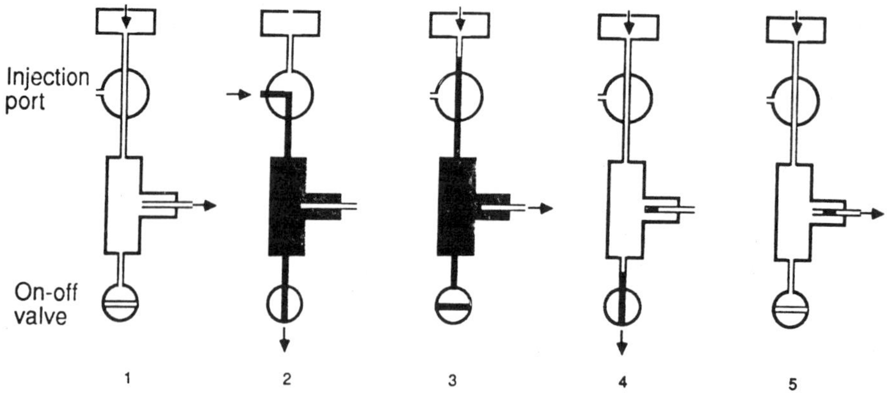

Fig. 4. Polymer injection sequence. (1) Normal flow; (2) filling the tee; (3) injecting into the capillary; (4) flushing the tee; (5) elution.

Injection of the Polymer

For the microgel analysis, a very small amount of polymer solution must be introduced into the flow system. To achieve this, a static splitting injection technique [16] is used. As shown in Figure 4, during normal flow the on-off valve is closed, and solvent flows directly from the pressurized reservoir to the capillary (1). To inject polymer, the on-off valve is opened, and polymer is introduced by syringe through the injection port, completely filling the splitting tee (2). During filling, little or no polymer will enter the capillary because of its large resistance relative to that of the valve. Next, the valve is closed, and polymer is forced into the capillary for several seconds (3). Afterwards, the valve is again opened to flush polymer from the injection tee and large diameter connecting tubing (4). Finally the valve is closed, and normal flow proceeds (5).

The microgel material begins to arrive at the fluorescence detector after approximately 3 minutes (at 5 p.s.i.g. driving pressure). Each microgel particle causes a spike in the detector output. The number of these spikes relative to the amount of polymer (as measured by the area of main chromatographic peak) indicates the microgel fraction in the sample. The analysis is complete after approximately 6 minutes.

Polymer Preparation

Without fluorescence detection, the sensitivity of the microgel analysis would not be possible. For this reason, the only requirement for a polymer to be suitable for analysis in the microcapillary system is that it be fluorescent or that it can be fluorescently tagged. Fortunately, it is possible to tag all water-soluble polymers which possess hydroxyl, carboxyl, or amide groups.

Polyacrylamides. Solutions of commercial partially hydrolyzed polyacrylamide (3,000 ppm) were fluorescently tagged using the following protocol [19]. First, 200 ml of the solution was adjusted to pH 6 using HCl. (Many commercial polyacryl- amides contain inorganic salts resulting in high pH solutions.) After adding 8 ml dimethyl sulfoxide in 12 ml water, the solution was cooled to below 20 degrees C. Next, 50 μl acetaldehyde (which vaporizes at 20.8 degrees C), 50 μl cyclohexyl isocyanide (Aldrich), and 3 ml of a solution of fluoresceinam- ine isomer I (Aldrich) in DMSO (1 mg/ml) were sequentially added with stir- ring. The mixture was shaken vigorously for 15 seconds to assure homogenei- ty. After 16 hours at room temperature (no stirring), 1 ml of the reaction mixture was precipitated in 20 ml acetone. The acetone was decanted, and the polymer precipitate dissolved in 20 ml aqueous sodium azide (0.002 M). Dis- solution was allowed to proceed without stirring over 24 hours. The tagged polymer was diluted to a final concentration of 65 ppm for the analysis. The 4-component condensation tagging reaction is shown in Figure 5.

Polysaccharides. Xanthan polysaccharide was tagged employing the reaction described above for polyacrylamide tagging. Hydroxypropyl guar and carboxymethyl hydroxypropyl guar were tagged using a fluorescent reagent, dichlorotriazinylaminofluorescein (DTAF), which links to hydroxyl groups [20]. The DTAF is prepared by reacting aminofluorescein with cyanuric chloride at 4 degrees C [21]. To tag the polymer, 30 ml of 3,000 ppm polysaccharide solution was added 200 ml DTAF. The mixture is maintained at pH 10 for 2 hours. Afterwards, the tagged polymer is recovered from solution in the manner described above for polyacrylamide.

RESULTS

Polyacrylamides

Figure 6 contains chromatograms of four polyacrylamides with different filterability characteristics. Sample S1 appears as a single smooth peak, indicating the absence of a detectable microgel fraction. The quality of this sample is confirmed by a filterability test: a 500 ppm polymer solution was filtered through a 25 mm diameter, 5 μm pore size Millipore filter under the pressure of 4 inches of solution. For sample S1, the measured flow rate was 10 ml in 4.7 minutes. S2 is a laboratory-prepared sample in which there is a significant amount of microgel, as evidenced by both the detector response preceeding the main polymer peak and the poor filterability (7.8 ml in 22 minutes).

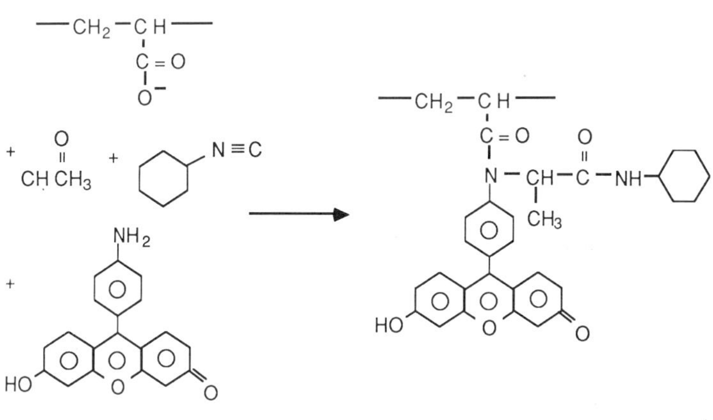

Fig. 5. Four-component isocyanide condensation empolyed to attach fluorescent tags to polyacrylamide and xanthan.

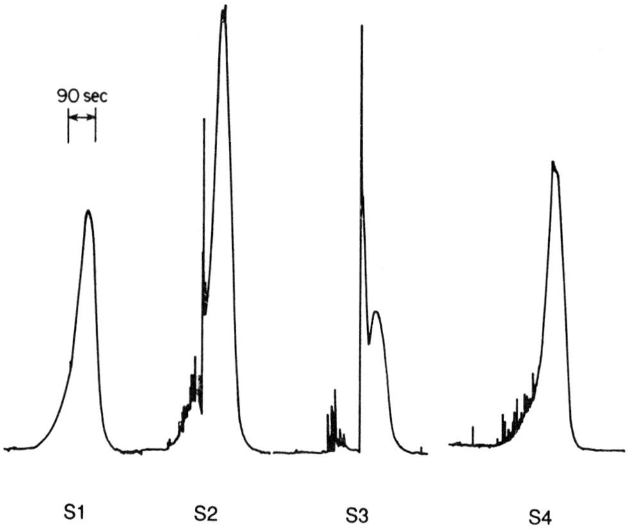

Fig. 6. Microgel analysis chromatograms of medium carboxyl content polyacrylamides. S1, a commercial polymer (Mw appx. $6x10^6$) free of microgels; S2 and S3, laboratory samples (Mw appx. $9x10^6$) showing significant microgel content; S3a, sample S3 after treatment with NaOH. The numbers refer to filtration test flow rates (through a 25-mm diameter 5μm pore size Millipore filter).

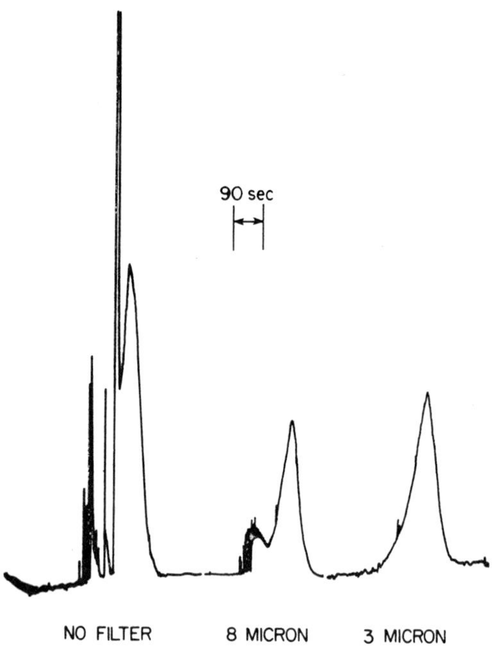

Fig. 7. Filtration study of sample S3. Sample S3 was filtered through Millipore filters before injection into the capillary.

Reduction of microgel fraction with base. S3, another laboratory polymer, prepared under severe conditions, is shown to have a significant amount of microgel-- more than any of the other samples tested. The filterability test on S3 yielded only 0.5 ml in 22 minutes. When this sample was treated with base (NaOH, pH 11, at room temperature for 16 hours), its filterability improved dramatically (7 ml in 22 minutes). By comparing the chromatograms of S3 and S3a (the treated sample), it is evident that the treated sample displays a decrease in the microgel fraction. It has been suggested that microgels in polyacrylamides are due to imide crosslinks formed during polymerization under severe conditions such as high temperature or acidity [22,23]:

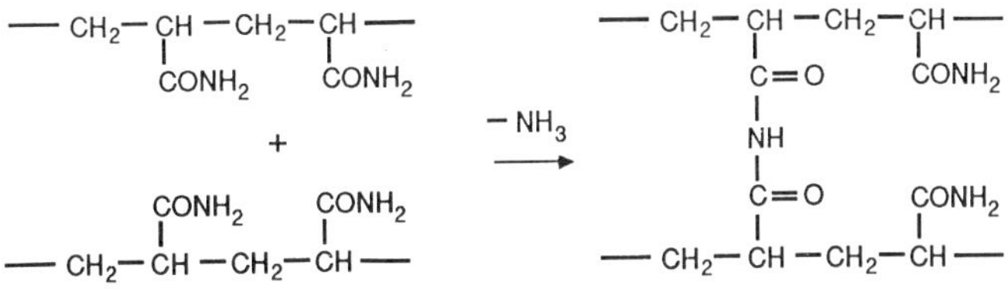

The demonstration of a decrease in microgels upon treatment with base for sample S3 supports the conjecture that imide crosslinks are a cause of microgels, since these bonds would be broken under basic conditions.

The size of microgel particles. In order to determine the approximate size of the microgels, the S3 sample was filtered through Millipore filters (8 and 3 microns pore size) before injection into the capillary. For these filtrations, the first drops of solution emerging from the filter cartridge were analyzed, avoiding the additional filtration that occurs as polymer covers the filter membrane's surface. Figure 7 shows that much of the microgel material is removed by the 8 micron filter, indicating that the majority of the microgel particles are in this size range. More accurate determinations of the particle sizes are not possible with Millipore filter membranes, which have a broad pore size distribution. The 3 micron filter removes essentially all of the detectable microgel material.

Polysaccharides

For comparison to the polyacrylamide samples, Figure 8 shows the microgel analysis for three polysaccharides: hydroxypropyl guar, carboxymethyl hydroxypropyl guar, and xanthan. Note that none of the samples tested were completely free of microgel, though the xanthan has considerably less insoluble material than the other polysaccharides.

SUMMARY

We have demonstrated the usefulness of a microcapillary chromatography system in the determination of the of the microgel content in polymers used in enhanced oil recovery operations. The microgel material, which causes poor sample filterability and can lead to formation damage, can be easily seen by the on-line laser fluorescence detector. Polyacrylamide microgels can be partially eliminated by treating a sample with base, which supports the hypothesis that the insoluble particles are caused by imide crosslinks between individual molecules. The microcapillary flow analysis, coupled with filtration through Millipore filters, provides information on the approximate size of the microgel particles.

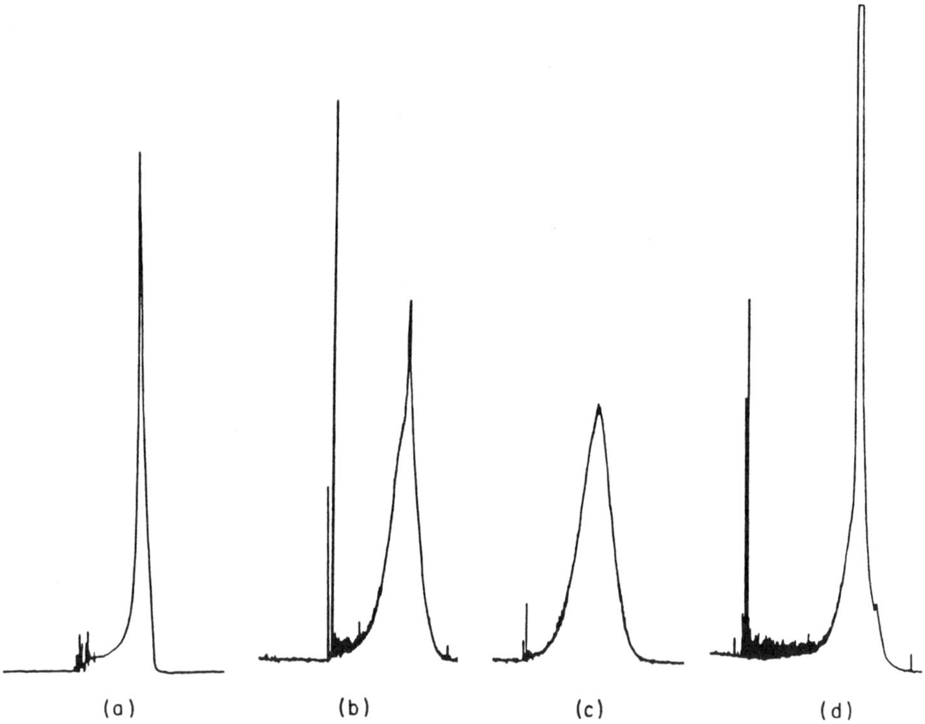

Fig. 8. Microgel analysis of polysaccharides. (a) hydroxypropyl guar, (b) carboxymethyl hydroxypropyl guar, (c) xanthan.

OTHER APPLICATIONS OF THE MICROCAPILLARY SYSTEM

Molecular Weight Separations

Additional experiments performed on the microcapillary system demonstrate the usefulness of the system in separating polymers by molecular weight and in determining the particle size and concentration of dilute latex suspensions. During flow through a 10 micron i.d. capillary, complete separation between a high and low molecular weight polyacrylamide is achieved. As shown in Figure 9, ultrasonic degradation of the polymer can be easily visualized. Molecular weight analysis of water-soluble polymers employing packed-bed hydrodynamic chromatography has been successfully achieved in this laboratory and elsewhere [13,24], but single capillary separations of molecules, which offer the advantages of low dispersion and the absence of polymer-deforming elongational stresses, have not previously been thoroughly investigated.

Latex Analysis

An example of latex analysis is shown in Figure 10a, where each of the spikes indicates the passage of a 0.57 micron tagged latex particle through the detector volume. Here, rather than measuring concentration, the detector counts individual events. For smaller particles, as shown in Figure 10b for 0.1 micron particles, the analysis produces a chromatographic peak because many particles can simultaneously reside in the detector volume, and the de-

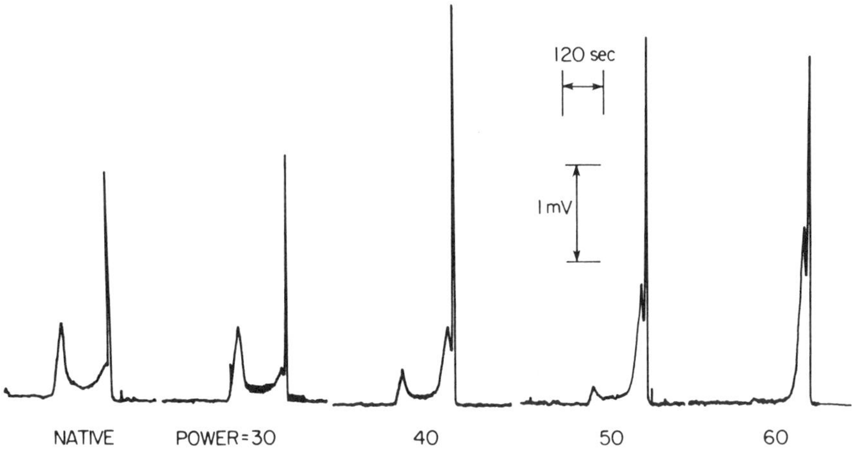

Fig. 9. The mechanism of the formation of imide crosslinks between polyacrylamide chains.

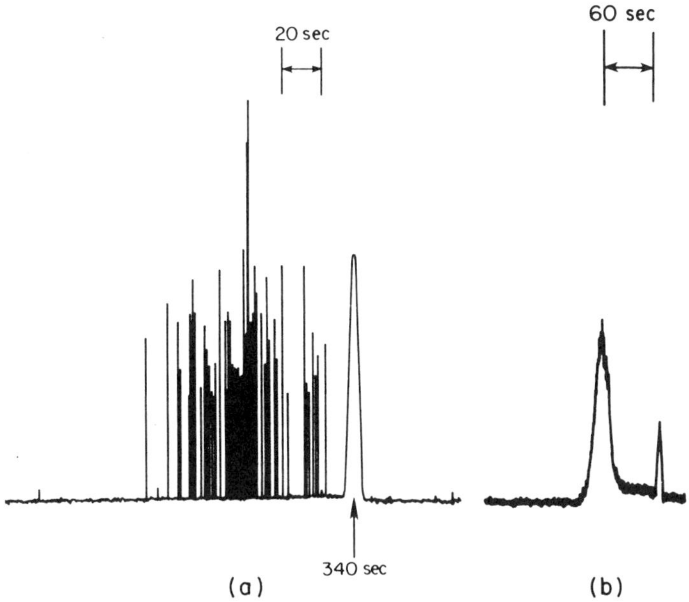

Fig. 10. Separation of molecules by molecular weight. Injections of a 50 ppm polyacrylamide solution into a 10 micron i.d. capillary results in a separation of high and low MW fractions. The sharp spike represents a low MW marker dye. Sonication of the sample for 30 seconds at increasing sonicator powers results in the reduction of high MW polymer.

tector measures concentration. In these analyses, the particles' average arrival time at the detector is directly related to the particles' size. The relation between particle size and velocity is determined with a calibration curve constructed with monodisperse latex standards.

In order to determine the particle concentration in a suspension, the suspension is injected continuously through the system (rather than as a pulse), and the particle flux is measured by the detector. The detector's output is connected to a Schmitt trigger circuit, which undergoes a discrete voltage transition each time a particle passes through the detector volume. The trigger circuit ouptut is sent to an IBM PC, which counts the transitions. By dividing the number of particles arriving at the detector in a given period of time by the volumetric flow rate (which can be calculated), the number concentration of particles is obtained. Concentrations up to 100,000/ml can be measured. More concentrated suspensions give erroneous results because the circuit cannot detect multiple particles simultaneously in the detector.

REFERENCES

1. G. Chauveteau and N. Kohler, SPE 9295, presented at the 55th Annual Fall Technical Conference and Exhibition of the Society of Petroleum Engineers of AIME, Dallas, TX, September 21-24, 1980.
2. J.R. Purdon, Jr. and R.D. Mate, *J. Polym. Sci. A-1*, 8:1306 (1970).
3. M.R. Ambler, R.D. Mate, and J.R. Purdon, Jr., *J. Polym. Sci. Polym. Chem.*, 12:1771 (1974).
4. S.W. Almond and W.E. Bland, SPE 12485, presented at the Formation Damage Control Symposium, Bakersfield, CA, February 1984.
5. J.H. Sugarman and R.K. Prud'homme, *J. Appl. Polym. Sci.*, in press.
6. H.W. Thomas, R.J. French, A.C. Groom, and S. Rowlands, in *Fourth Int. Congr. Rheol. Part 4*, Interscience (1965).
7. R.J. Noel, K.M. Gooding, F.E. Regnier, D.M. Ball, C. Orr, and M.E. Mullins, *J. Chromatogr.*, 166:373 (1978).
8. A.W.J. Brough, D.E. Hillman, And R.W. Perry, *J. Chromatogr*, 208:175 (1981).
9. G. Segre and A. Silberberg, *J. Fluid Mech.*, 14:136 (1962).
10. H.L. Goldsmith, *Fedn. Am. Socs. exp. Biol.*, 26:1813 (1971).
11. B.P. Ho and L.G Leal, *J. Fluid Mech.*, 65:365 (1974).
12. H. Small, *J. Coll. Sci.*, 48:147 (1974).
13. M.A. Langhorst, F.W. Stanley Jr., S.S. Cutie, J.H. Sugarman, L.R. Wilson, D.A. Hoagland, and R.K. Prud'homme, *Anal. Chem.*, 58:2242 (1986).
14. F.M. Kelleher and C.N. Trumbore, *Anal. Biochem.*, 137:20 (1984).
15 C.N. Trumbore, M. Grehlinger, M. Stowe, and F.M. Kelleher, *J. Chromatogr.*, 322:443 (1985).
16. J.W. Jorgenson and E.J. Guthrie, *J. Chromatogr.*, 255:335 (1983).
17. E.J. Guthrie, J.W. Jorgenson, and P.R. Dluzneski, *J. Chromatogr. Sci.*, 22:171 (1984).
18. J.H. Sugarman and R.K. Prud'homme, *Ind. Eng. Chem. Fund.*, submitted.
19. G. Holzwarth, *Carbohydr. Res.*, 66:173 (1978).
20. A.N. de Belder and K. Granath, *Carbohydr. Res.*, 30:375 (1973).
21. D. Blakeslee and M.G. Baine, *J. Immunol. Methods*, 13:305 (1976).
22. H.C. Haas and R.L. MacDonald, *J. Polym. Sci. A-1*, 9:3583 (1971).
23. R. Boyadjian, G. Seytre, P. Berticat, and G. Vallet, *Eur. Polym. J.*, 12:401 (1976).
24. D.A. Hoagland, Ph.D. Thesis, Princeton University (1985).

INDEX